Advances in Carbon Capture, Utilization and Storage (CCUS)

Advances in Carbon Capture, Utilization and Storage (CCUS)

Editors

Cheng Cao
Michael Zhengmeng Hou
Liehui Zhang
Yulong Zhao
Hejuan Liu

Basel • Beijing • Wuhan • Barcelona • Belgrade • Novi Sad • Cluj • Manchester

Editors
Cheng Cao
Southwest Petroleum
University
Chengdu
China

Michael Zhengmeng Hou
Clausthal University of
Technology
Clausthal-Zellerfeld
Germany

Liehui Zhang
Southwest Petroleum
University
Chengdu
China

Yulong Zhao
Southwest Petroleum
University
Chengdu
China

Hejuan Liu
Chinese Academy of Sciences
Wuhan
China

Editorial Office
MDPI AG
Grosspeteranlage 5
4052 Basel, Switzerland

This is a reprint of articles from the Special Issue published online in the open access journal *Energies* (ISSN 1996-1073) (available at: https://www.mdpi.com/journal/energies/special_issues/c_c_u_s).

For citation purposes, cite each article independently as indicated on the article page online and as indicated below:

Lastname, A.A.; Lastname, B.B. Article Title. *Journal Name* **Year**, *Volume Number*, Page Range.

ISBN 978-3-7258-2287-4 (Hbk)
ISBN 978-3-7258-2288-1 (PDF)
doi.org/10.3390/books978-3-7258-2288-1

Contents

Cheng Cao, Haonan Zhu and Zhengmeng Hou
Advances in Carbon Capture, Utilization and Storage (CCUS)
Reprinted from: *Energies* **2024**, *17*, 4784, doi:10.3390/en17194784 . . . . **1**

Zhengmeng Hou, Jiashun Luo, Yachen Xie, Lin Wu, Liangchao Huang and Ying Xiong
Carbon Circular Utilization and Partially Geological Sequestration: Potentialities, Challenges,
and Trends
Reprinted from: *Energies* **2023**, *16*, 324, doi:10.3390/en16010324 . . . . **4**

Yachen Xie, Jiaguo Qi, Rui Zhang, Xiaomiao Jiao, Gabriela Shirkey and Shihua Ren
Toward a Carbon-Neutral State: A Carbon–Energy–Water Nexus Perspective of China's Coal
Power Industry
Reprinted from: *Energies* **2022**, *15*, 4466, doi:10.3390/en15124466 . . . . **18**

Liangchao Huang, Zhengmeng Hou, Yanli Fang, Jianhua Liu and Tianle Shi
Evolution of CCUS Technologies Using LDA Topic Model and Derwent Patent Data
Reprinted from: *Energies* **2023**, *16*, 2556, doi:10.3390/en16062556 . . . . **42**

Maria Wetzel, Christopher Otto, Min Chen, Shakil Masum, Hywel Thomas,
Tomasz Urych, et al.
Hydromechanical Impacts of CO_2 Storage in Coal Seams of the Upper Silesian Coal Basin
(Poland)
Reprinted from: *Energies* **2023**, *16*, 3279, doi:10.3390/en16073279 . . . . **56**

Charli Sitinjak, Sitinjak Ebennezer and Józef Ober
Exploring Public Attitudes and Acceptance of CCUS Technologies in JABODETABEK:
A Cross-Sectional Study
Reprinted from: *Energies* **2023**, *16*, 4026, doi:10.3390/en16104026 . . . . **80**

Congxiu Guo, Ya Sun, Hongyan Ren, Bing Wang, Xili Tong, Xuhui Wang, et al.
Biomass Based N/O Codoped Porous Carbons with Abundant Ultramicropores for Highly
Selective CO_2 Adsorption
Reprinted from: *Energies* **2023**, *16*, 5222, doi:10.3390/en16135222 . . . . **95**

Jan Górecki, Maciej Berdychowski, Elżbieta Gawrońska and Krzysztof Wałęsa
Influence of PPD and Mass Scaling Parameter on the Goodness of Fit of Dry Ice Compaction
Curve Obtained in Numerical Simulations Utilizing Smoothed Particle Method (SPH) for
Improving the Energy Efficiency of Dry Ice Compaction Process
Reprinted from: *Energies* **2023**, *16*, 7194, doi:10.3390/en16207194 . . . . **107**

Felipe Cruz, Son Dang, Mark Curtis and Chandra Rai
Effect of Geochemical Reactivity on $ScCO_2$–Brine–Rock Capillary Displacement: Implications
for Carbon Geostorage
Reprinted from: *Energies* **2023**, *16*, 7333, doi:10.3390/en16217333 . . . . **119**

Dawei Feng, Wenchao Xu, Xinyu Gao, Yun Yang, Shirui Feng, Xiaohu Yang and Hailong Li
Carbon Emission Prediction and the Reduction Pathway in Industrial Parks: A Scenario
Analysis Based on the Integration of the LEAP Model with LMDI Decomposition
Reprinted from: *Energies* **2023**, *16*, 7356, doi:10.3390/en16217356 . . . . **136**

Stefania Moioli, Elvira Spatolisano and Laura A. Pellegrini
Techno-Economic Assessment for the Best Flexible Operation of the CO_2 Removal Section by
Potassium Taurate Solvent in a Coal-Fired Power Plant
Reprinted from: *Energies* **2024**, *17*, 1736, doi:10.3390/en17071736 . . . . **151**

Advances in Carbon Capture, Utilization and Storage (CCUS)

Cheng Cao [1], Haonan Zhu [1,*] and Zhengmeng Hou [2,*]

[1] State Key Laboratory of Oil and Gas Reservoir Geology and Exploitation, Southwest Petroleum University, Chengdu 610500, China; caochengcn@163.com

[2] Institute of Subsurface Energy Systems, Clausthal University of Technology, 38678 Clausthal-Zellerfeld, Germany

*   Correspondence: zhuhaonan86@163.com (H.Z.); hou@tu-clausthal.de (Z.H.)

Citation: Cao, C.; Zhu, H.; Hou, Z. Advances in Carbon Capture, Utilization and Storage (CCUS). *Energies* **2024**, *17*, 4784. https://doi.org/10.3390/en17194784

Received: 11 September 2024

Accepted: 24 September 2024

Published: 25 September 2024

Carbon Capture, Utilization, and Storage (CCUS) technology is essential for mitigating climate change as it captures CO_2 and either utilizes it for chemical applications or stores it in geological formations. Captured CO_2 can be injected into depleted oil and gas reservoirs to enhance oil recovery or can be stored directly in underground reservoirs, such as depleted oil and gas fields or deep saline aquifers. Despite their potential, most CCUS technologies are still in the early stages of development, with ongoing advancements and innovative solutions emerging rapidly. To help researchers stay abreast of the latest developments and promote the research, development, and field implementation of CCUS, we initiated a call for papers with the support of relevant academic journals. A total of ten papers related to CCUS technology and development were received and published in this Special Issue.

CO_2 capture is the first step in CCUS. Scholars are constantly searching for low-energy and efficient CO_2 separation and purification methods. Moioli et al. [1] proposed using a flexible potassium taurate process to treat flue gas from coal-fired power plants. Using potassium taurate solvent requires less energy than MEA solvent, and it is also less toxic and corrosive. Compared with the fixed mode, the degree of CO_2 separation in the flexible mode is controllable and can be flexibly adjusted according to changes in carbon tax prices and electricity prices to minimize energy and income losses. Therefore, using potassium taurate in flexible mode is economically feasible. Guo et al. [2] prepared N/O co-doped porous carbons (NOPCs) from corn silk accompanied by Na_2CO_3 activation. The optimized sample exhibited a large specific surface area, numerous ultramicropores, and a high pyrrolic nitrogen content, leading to enhanced CO_2 adsorption capacity and potentially offering a cost-effective alternative to CO_2 capture.

CO_2 utilization technologies, which are considered a cost-effective way to reduce carbon emissions, have drawn considerable attention from the academic community through experimental studies and numerical simulations. Cruz et al. [3] conducted an experiment to quantify the impact of chemical reactions caused by the dissolution of supercritical CO_2 ($SCCO_2$) in formation water on capillary phenomena. Thin sections were prepared from various reservoir rocks and allowed to statically react with brine rich in s_CCO_2 for 21 days. Post-reaction samples were then observed using instruments to assess the geochemical reactions. The results indicated that the geochemical reactions preferentially dissolved calcite, while other minerals were preserved. Short-term geochemical reactions did not significantly affect the physical properties of the samples or the capillary displacement mechanisms. Górecki et al. [4] demonstrated the applicability of the Smoothed Particle Method (SPH) for the numerical simulation of loose dry ice compaction processes. This study aimed to optimize the energy efficiency of this process by systematically exploring the effects of particle packing density (PPD) and mass scaling (MS) parameters on the consistency of the simulated and experimental outputs. This study will help reduce the energy consumption of these production processes. Wetzel et al. [5] assessed the hydromechanical impacts of CO_2 storage in coal seams through numerical simulations, evaluating the risks associated with CO_2 injection in Poland's Upper Silesian Coal Basin. Their findings indicate that while CO_2 injection can cause vertical displacements, the risk of fault reactivation is low under the simulated conditions, providing insights for safe and sustainable CO_2 storage operations.

Accurate estimation of carbon emissions is critical for formulating targeted emission reduction plans. Feng et al. [6] developed a hybrid analysis framework that integrates the Logarithmic Mean Divisia Index (LMDI) decomposition method with the Long-range Energy Alternatives Planning (LEAP) system. Applying various data processing methods and Tapio decoupling theory, they assessed and forecasted CO_2 emissions in the Yancheng Economic Development Zone from 2020 to 2035 and identified optimal emission reduction strategies for the region. Xie et al. [7] explored the carbon–energy–water (CEW) nexus in China's coal power industry within the context of carbon neutrality. By developing the ATPCC assessment tool, the research evaluates trade-offs between carbon emissions, energy consumption, water usage, and financial profits within the industry. The findings suggest that while CCUS can reduce emissions, it may also increase energy and water consumption, necessitating a comprehensive understanding of the CEW nexus.

Raising public awareness and advancing the development of new CCUS technologies are essential to facilitate their implementation. Sitinjak et al. [8] investigated public perception towards CCUS in the JABODETABEK region of Indonesia. This study reveals a generally favorable attitude towards carbon capture and utilization (CCU) but resistance to carbon capture and storage (CCS) due to lack of knowledge and fear. The research underscores the importance of public education to enhance understanding and acceptance of CCUS initiatives. Huang et al. [9] employed the Latent Dirichlet Allocation (LDA) topic model and Derwent patent data to examine the evolution of CCUS technology topics. The study found a shift in application levels, with emerging fields like computer science gaining attention. This forecasts the maturity and trends in key CCUS technologies, highlighting the growing role of universities and research institutes in China's R&D landscape. Hou et al. [10] proposed six strategic priorities for China to achieve its dual carbon goals. This paper discusses the potential and challenges of conventional and advanced CCUS technologies. It introduces the concept of Carbon Capture, Circular Utilization, and Storage (CCCUS), emphasizing the role of biomethanation for renewable energy storage and the carbon circular economy, with significant potential for CO_2 storage and energy storage.

The papers published in this Special Issue cover cutting-edge research on advanced CCUS technologies, technology trends, and risk assessment. We believe that these studies will contribute to the development of CCUS-related theoretical, experimental, and numerical simulation methods and provide guidance for the engineering application of CCUS technology. To promote the development of this technology, it is necessary to strengthen exploration in these research directions, including improving the economy and efficiency of CCUS technologies, developing new technologies and methods, optimizing the integration and scalability of these technologies, etc.

Funding: This research received no external funding.

Acknowledgments: The Guest Editors would like to thank the reviewers for their work, which helped the authors improve their manuscripts.

Conflicts of Interest: The authors declare no conflicts of interest.

References

1. Moioli, S.; Spatolisano, E.; Pellegrini, L.A. Techno-Economic Assessment for the Best Flexible Operation of the CO2 Removal Section by Potassium Taurate Solvent in a Coal-Fired Power Plant. *Energies* **2024**, *17*, 1736. [CrossRef]
2. Guo, C.; Sun, Y.; Ren, H.; Wang, B.; Tong, X.; Wang, X.; Niu, Y.; Wu, J. Biomass Based N/O Codoped Porous Carbons with Abundant Ultramicropores for Highly Selective CO2 Adsorption. *Energies* **2023**, *16*, 5222. [CrossRef]
3. Cruz, F.; Dang, S.; Curtis, M.; Rai, C. Effect of Geochemical Reactivity on ScCO2–Brine–Rock Capillary Displacement: Implications for Carbon Geostorage. *Energies* **2023**, *16*, 7333. [CrossRef]

4. Górecki, J.; Berdychowski, M.; Gawrońska, E.; Wałęsa, K. Influence of PPD and Mass Scaling Parameter on the Goodness of Fit of Dry Ice Compaction Curve Obtained in Numerical Simulations Utilizing Smoothed Particle Method (SPH) for Improving the Energy Efficiency of Dry Ice Compaction Process. *Energies* **2023**, *16*, 7194. [CrossRef]

5. Wetzel, M.; Otto, C.; Chen, M.; Masum, S.; Thomas, H.; Urych, T.; Bezak, B.; Kempka, T. Hydromechanical Impacts of CO_2 Storage in Coal Seams of the Upper Silesian Coal Basin (Poland). *Energies* **2023**, *16*, 3279. [CrossRef]

6. Feng, D.; Xu, W.; Gao, X.; Yang, Y.; Feng, S.; Yang, X.; Li, H. Carbon Emission Prediction and the Reduction Pathway in Industrial Parks: A Scenario Analysis Based on the Integration of the LEAP Model with LMDI Decomposition. *Energies* **2023**, *16*, 7356. [CrossRef]

7. Xie, Y.; Qi, J.; Zhang, R.; Jiao, X.; Shirkey, G.; Ren, S. Toward a Carbon-Neutral State: A Carbon–Energy–Water Nexus Perspective of China's Coal Power Industry. *Energies* **2022**, *15*, 4466. [CrossRef]

8. Sitinjak, C.; Ebennezer, S.; Ober, J. Exploring Public Attitudes and Acceptance of CCUS Technologies in JABODETABEK: A Cross-Sectional Study. *Energies* **2023**, *16*, 4026. [CrossRef]

9. Huang, L.; Hou, Z.; Fang, Y.; Liu, J.; Shi, T. Evolution of CCUS Technologies Using LDA Topic Model and Derwent Patent Data. *Energies* **2023**, *16*, 2556. [CrossRef]

10. Hou, Z.; Luo, J.; Xie, Y.; Wu, L.; Huang, L.; Xiong, Y. Carbon Circular Utilization and Partially Geological Sequestration: Potentialities, Challenges, and Trends. *Energies* **2022**, *16*, 324. [CrossRef]

Perspective

Carbon Circular Utilization and Partially Geological Sequestration: Potentialities, Challenges, and Trends

Zhengmeng Hou [1,2], Jiashun Luo [1,2,3,*], Yachen Xie [1,2,4,*], Lin Wu [1,2,3], Liangchao Huang [1,2,5] and Ying Xiong [3,6]

[1] Institute of Subsurface Energy Systems, Clausthal University of Technology, 38678 Clausthal Zellerfeld, Germany
[2] Research Centre of Energy Storage Technologies, Clausthal University of Technology, 38640 Goslar, Germany
[3] State Key Laboratory of Oil and Gas Reservoir Geology and Exploitation, Southwest Petroleum University, Chengdu 610500, China
[4] State Key Laboratory of Geomechanics and Geotechnical Engineering, Institute of Rock and Soil Mechanics, Chinese Academy of Sciences, Wuhan 430071, China
[5] Sino-German Research Institute of Carbon Neutralization and Green Development, Zhengzhou University, Zhengzhou 450001, China
[6] School of Earth & Space Sciences, Peking University, Beijing 100871, China
[*] Correspondence: jiashun.luo@tu-clausthal.de (J.L.); yx91@tu-clausthal.de (Y.X.); Tel.: +49-5323-72-2680 (J.L.)

Abstract: Enhancing carbon emission mitigation and carbon utilization have become necessary for the world to respond to climate change caused by the increase of greenhouse gas concentrations. As a result, carbon capture, utilization, and storage (CCUS) technologies have attracted considerable attention worldwide, especially in China, which plans to achieve a carbon peak before 2030 and carbon neutrality before 2060. This paper proposed six priorities for China, the current world's largest carbon emitter, to achieve its dual carbon strategy in the green energy transition process. We analyzed and summarized the challenges and potentialities of conventional carbon utilization (CU), carbon capture utilization (CCU), and CCUS. Based on the current development trend, carbon dioxide capture, circular utilization, and storage (CCCUS) technology that integrates carbon *circular* utilization and partial sequestration, with large-scale underground energy storage were proposed, namely biomethanation. Technically and economically, biomethanation was believed to have an essential contribution to China's renewable energy utilization and storage, as well as the carbon circular economy. The preliminary investigation reveals significant potential, with a corresponding carbon storage capacity of 5.94×10^8 t~7.98×10^8 t and energy storage of 3.29×10^{12} kWh~4.42×10^{12} kWh. Therefore, we believe that in addition to vigorously developing classical CCUS technology, technical research and pilot projects of CCCUS technology that combined large-scale underground energy storage also need to be carried out to complete the technical reserve and the dual-carbon target.

Keywords: carbon neutrality; carbon circular utilization; partially geological sequestration; renewable energy; underground biomethanation

Citation: Hou, Z.; Luo, J.; Xie, Y.; Wu, L.; Huang, L.; Xiong, Y. Carbon Circular Utilization and Partially Geological Sequestration: Potentialities, Challenges, and Trends. *Energies* **2023**, *16*, 324. https://doi.org/10.3390/en16010324

Academic Editor: Nikolaos Koukouzas

Received: 29 November 2022
Revised: 21 December 2022
Accepted: 23 December 2022
Published: 28 December 2022

1. Introduction

Greenhouse gases (GHG) are gases that trap heat in the Earth's atmosphere, of which carbon dioxide (CO_2) is a major component and is emitted through human activities such as burning fossil fuels (coal, natural gas, and oil), biological respiration, and specific industrial reactions (e.g., manufacture of cement) [1]. In 2020, CO_2 accounted for about 76% of all greenhouse gas emissions from human activities [2]. Moreover, the concentration of atmospheric CO_2 has significantly increased from a preindustrial baseline of 280 ppm, reaching 415 ppm in 2022 [3,4]. Particularly, it has maintained an annual growth of 2.4 ppm over the past decade, which has aggravated the greenhouse effect and led to an increase in global temperatures. To limit these impacts, the Paris Agreement, which aimed to constrain warming to 1.5–2 °C, was proposed in 2015 [5]. Apparently, this goal related to the whole global effort requires drastic reductions in GHG emissions. According to the IPCC's report, limiting temperature rise to 1.5 °C by the end of the century, global annual emissions need to be reduced from 50 GtCO$_2$e (billion metric tons of CO$_2$ equivalent) to 25–30 GtCO$_2$e by 2030 [6,7]. Evidently, various efforts have been undertaken globally, to date, more than 130 countries worldwide have committed to carbon neutrality, with the developed world's universal label proposing to achieve carbon neutrality by 2050 and Germany by 2045. As the world's largest emitter of CO$_2$ with a population of 1.3 billion, China has also put forward the target of carbon peaking and carbon neutrality before 2030 and 2060, respectively. Regarding China's external dependence on oil and gas has exceeded 70% and 40% [8], respectively, and coal remains China's main fossil energy source [9], achieving carbon neutrality is of critical strategic importance to ensuring China's energy security, improving ecological management, and enhancing national competitiveness. Furthermore, among the dual carbon strategies, significantly improving energy efficiency, increasing the proportion of renewable energy, promoting a green and low-carbon energy transition, and building a new type of power system mainly based on renewable energy are necessary measures and essential approaches.

Based on the analysis of the practical experience overseas, as well as the consideration of China's status quo, "Six Priorities" are proposed by the authors as follows: (1) Energy saving and energy efficiency prior to renewable energy; (2) Local renewable energy prior to remote energy; (3) Renewable energy prior to a carbon sink; (4) Natural carbon sink prior to anthropogenic carbon sink like CCUS; (5) Carbon reduction prior to a carbon sink; (6) Electrification prior to hydrogenation. However, China's land natural carbon sink capacity is restricted, with an estimated around 1.1 billion tons, and it is expected to grow steadily to nearly 1.5 billion tons in the future by expanding forestry, planting trees, etc. [10]. Comparatively, in 2019, China's CO$_2$ emissions from fossil fuel combustion reached 9.8×10^9 t, and the total carbon emissions exceeded 1×10^{10} t in 2018 [11]. Therefore, achieving a balance of carbon emissions and carbon sink also requires anthropogenic carbon uptake, like CCUS, which is an important part of China's technology portfolio for achieving its carbon neutrality target. According to the current technologies development, in 2050 and 2060, the emission reductions that need to be realized through CCUS technologies are $0.6{\sim}1.4 \times 10^9$ t and $1.0{\sim}1.8 \times 10^9$ t of CO$_2$, respectively [12]. In addition, considering the compatibility of sources and sinks in China, the emission reduction potential offered by CCUS technology can basically fulfill the requirements for realizing the carbon neutrality goal [13]. Meanwhile, the proportion of wind and solar power in total electricity production is expected to reach 85% by 2060 from 11.5% in 2021. At that time, wind and solar power will generate nearly 2.6×10^{13} kWh resulting in high demand for large-scale energy storage due to the fluctuation and instability. Comparatively, the underground energy storage technologies (e.g., underground hydrogen storage, methane storage, etc.) are more favorable due to their large capacity (more than 10^{12} kWh) and long-term (a few months) storage characteristics. Consequently, CCUS technologies combined with underground energy storage will experience unprecedented development opportunities, especially those that incorporate underground energy storage capacity, and will play a key role in carbon neutrality processes.

This paper briefly reviews and analyses the technologies of CU, CCU, and CCUS first. Moreover, based on CO$_2$ circular utilization, geological storage, and underground large-scale energy storage, the concept of carbon dioxide capture, circular utilization, and storage (CCCUS) is proposed. As a carbon circular utilization technology for realizing underground artificial natural gas production, hydrogen, and methane storage, biomethanation's technical mechanism, principles, and site selection criteria are described in detail, and the potential and development challenges are analyzed as well. Finally, development suggestions are made combined with China's actual situation and strategy in the hope of providing a reference for the development of China's CCUS industry and carbon neutrality targets.

2. CU/CCU

For several decades, CO_2 has been utilized directly in various activities with/without chemical or biological conversion. These include, but are not limited to, carbonated beverages, inerting agents, and extractants in chemical industries (e.g., for the decaffeination of coffee and drinking water abstraction). Additionally, CO_2 also can be used for refrigeration, fire suppression, and plastics production. It is also used as a supplement injected into metal castings, respiratory stimulant (added to medical O_2), aerosol can propellant, etc. Moreover, through chemical or biological conversion, fuels, biomass, or bioproducts such as aquaculture feed can be achieved. The above-mentioned applications are already relatively mature technologies with complete supply, production, and sales chains. However, the above often have a small scope. CO_2 utilization is not expected to deliver emissions reductions on the same scale as carbon capture and storage (CCS) but is still a part of an "all technologies" approach. Particularly, carbon-based energy sources and chemicals will play an important role in the future defossilized industrial society. Based on CO_2 capture (industrial or direct air capture) and power-to-X technologies, collected CO_2 and electrolytically generated hydrogen can be used to produce possible products and materials. Theoretically, these technologies are available where carbon carriers are required for material use, especially in transportation, basic chemicals, and industries. Among them, are organic base materials such as methanol and high-value chemicals ethylene, propylene, butane, and butadiene as well as the aromatics benzene, toluene, and xylene. In addition, synthetic methane can be produced both surface and subsurface, as detailed in Chapter 4. The future relevance of producing synthetic fuels from carbon dioxide and hydrogen, namely power-to-fuels, is primarily in aviation and maritime transport, and to a lesser extent heavy-duty transport.

Currently, nearly 2.3×10^8 t of CO_2 is used per year globally contains CO_2 enhanced oil recovery (CO_2-EOR) and is still increasing rapidly, with preliminary estimates of over 2.7×10^8 t by 2025 [14]. The primary customer is fertilizer production, which annually consumes around 1.3×10^8 t CO_2 for urea synthesis. Subsequently, the food and beverage industry each consumed 3% of the total, corresponding to 6.9×10^6 t. In addition, Metal fabrication and other uses accounted for 2% and 4%, corresponding to 4.6×10^6 t and 9.2×10^6 t, respectively [14]. Notably, China's CO_2 demand has grown rapidly in recent years, with its market share already increasing from 21% in 2018 to 28% in 2020, overtaking the US as the world's largest CO2-consuming market. According to estimation China's CO_2 consumption in the future will reach 6.2×10^8 t~8.7×10^8 t in 2060 [12].

3. Geological CCUS

Undoubtedly, the amount of sequestration that can be achieved by geological storage of CO_2 is enormous compared to fixed products, such as building materials, food, beverage, etc. Estimations have shown that China's geological storage potential of CO_2 is roughly 1.21×10^{12}~4.13×10^{12} t [12]. Obviously, CO_2 can be widely used in underground energy development and storage. Among them, using carbon dioxide to flood oil or gas is one of the methods of CO_2 utilization (Figure 1a,b), which has been practiced worldwide in the oil and gas industry [15]. Injection of CO_2 into hydrocarbon formations is able to provide significant underground storage for CO_2 while enhancing hydrocarbon recovery (CO_2-EOR, CO_2-EGR) which cuts down the expenses. Moreover, as emerging technologies with significant potential and application interest, CO_2 fracturing and CO_2 application in geothermal energy development have also attracted increasing attention from both academics and industries (Figure 1c,d). Generally, considering a portion of carbon dioxide is sequestrated permanently below the surface in geological utilization, CCUS technologies can accomplish negative carbon emissions.

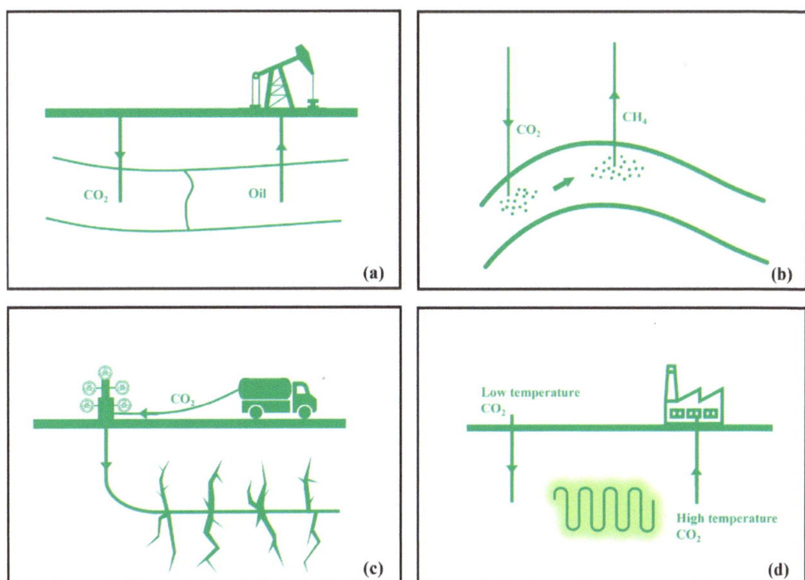

Figure 1. The diagrammatic sketch of geological CCUS technologies. (**a**) CO_2-enhanced oil recovery; (**b**) CO_2-enhanced gas recovery; (**c**) CO_2 fracturing; (**d**) CO_2-enhanced geothermal system (modified from [16–19]).

3.1. CO_2-EOR

Using CO_2 as an injectant to enhance oil recovery has gained increasing attention in the oil and gas industry. It is not only an environmentally friendly method of greenhouse gas treatment, but also an important technology for the development of conventional and unconventional oil reservoirs [20,21]. The primary displacement mechanisms of CO_2-EOR are: (a) reduced oil viscosity and density; (b) reduced oil swelling due to CO_2 solution; (c) vaporization and extraction of light hydrocarbon component; (d) dissolved gas drive; (e) reduced interfacial tensions; and (f) improved permeability due to acidification effect [22–24]. In terms of reservoir fluid properties and reservoir pressure and temperature conditions, the injection scheme of a typical CO_2-EOR process can be classified as miscible or immiscible based on CO_2 and oil miscibility. Basically, the main mechanism depends on the CO_2 solubility in oil in the rock matrix and fractures, which can reduce oil density and viscosity, increase oil mobility, and result in improved oil recovery [25]. Moreover, diffusion is a crucial influencing factor for CO2-enhanced oil recovery in tight or shale oil reservoirs with well-developed natural and artificial fractures generated by hydraulic fracturing at the initiation of development. Hence, the characterization of the flow behavior of CO_2 and oil in low permeability and pressure conditions, as well as the diffusion mechanism are encouraged to be emphasized in future studies. Since the 1960s, USA and China have been pioneers in CO_2-EOR [26]. Recently, it has been gaining more attention worldwide, including in Brazil, Norway, and Trinidad, which have recognized CO_2-EOR storage as a vital step toward reducing carbon emissions from industrial and energy sources via large-scale carbon capture projects [27]. Wei et al. [28] found that the technical potential for CO_2-EOR in China is approximately 1.0 Gt, with the attendant capability of storing 2.2 Gt of CO_2 in the process. Furthermore, according to a recent report by the Ministry of Environmental Protection of China, CO_2-EOR can sequester approximately 5.1 Gt of CO_2 in the subsurface [12]. China has focused heavily on developing CO_2-EOR projects to utilize and eventually store CO_2 to meet the carbon neutralization target.

However, when compared to reservoir characteristics and properties in North America, the geological conditions of oil reservoirs in China are relatively poor, particularly for tight reservoirs, which are characterized by ultra-low permeability, low brittleness of rock, and poor mobility. Furthermore, gas displacement requires a large-scale and stable supply of CO_2, so the match between carbon sources and carbon demand is also crucial. Usually, CO_2 is separated from nearby coal-fired power plants and piped to the fields. Thereby, this also places a demand on the corrosion resistance of equipment such as pipes and pumping equipment. It should be noted that economic issues are vital concerns in the practice process of CO_2-EOR. As current carbon capture and storage technologies for enhanced oil recovery are not cost-competitive with low oil prices, thus many planned carbon capture, utilization, and storage projects have been canceled over the past few years. Technically, CO_2-EOR also has some requirements, including reservoir characteristics (reservoir thickness, porosity, permeability, heterogeneity, geochemical and geomechanical parameters, etc.), well pattern and spacing, and CO_2 leakage control, etc. Generally, the main challenges of the CO_2-EOR process including:

(1) Control the cost of the overall industrial chain covering carbon capture and carbon transportation to improve economics.
(2) Reservoir screening and evaluation (seepage and storage conditions, mechanical and chemical properties).
(3) Stable carbon source and large-scale CO_2 transport technology.
(4) Underground carbon dioxide monitoring technologies and anti-corrosion, anti-blocking pipes, operating strategies, and service life extension.
(5) Infrastructure investments, such as pipelines and surface separation and recycling facilities for CO_2, should be improved.

3.2. CO_2-EGR

CO_2-EGR is another CCUS technology that improves gas recovery by injecting CO_2. The gas displacement effect and re-pressurization in depleting or depleted gas reservoirs are the primary mechanisms. On the one hand, CO_2 injection can lower the dew point pressure of reservoir fluids in wet gas reservoirs. The separated CO_2 from the produced gas, on the other hand, can be injected back into the reservoir to improve gas recovery [29]. After natural gas production and pressure reduction, gas reservoirs have sufficient pore space to store injected CO_2 molecules. Moreover, they enable the long-term sequestration of CO_2 within a reservoir sealed by impermeable cap rocks. Notably, depleted gas reservoirs showed a greater potential to store CO_2 compared to oil reservoirs due to the high primary recovery in gas reservoirs (>60%), which is nearly double that of oil recovery [30]. However, the process is complicated because of the gas adsorption on the surface of reservoir rocks, the mixture of CO_2 and natural gas, and CO_2 breakthrough production wells. As is described in many studies, an incremental recovery of up to 11% can be achieved by CO_2 injection [31]. Furthermore, several experimental and numerical simulation studies have been conducted to assess the feasibility of CO_2-EGR [32–35]. According to the observations, the CO_2 breakthrough is the most significant concern in EGR. Early CO_2 breakthroughs actually resulted in unfavorable gas production. As a result, geological formations, particularly microstructures, are thought to have a remark impact on CO_2 flow behavior and sweep the region. Furthermore, irreducible water saturation influences the mixing of CO_2 and CH_4. In addition, engineering parameters also play an important role in determining CO_2-EGR performance. The findings showed that injecting CO_2 with a horizontal well into the lower reservoir while extracting CH_4 from the upper reservoir would reduce CO_2 breakthrough into the production well. Likewise, CO_2 injection during the early decline phase of natural gas production is advantageous for achieving a high CH_4 recovery because it ensures supercritical displacement and reduces CO_2 and CH_4 mixing. To ensure the supercritical phase in the displacement process, it is suggested that CO_2 be injected at relatively high pressure. In terms of injection rate, numerous studies have shown that a high injection rate is beneficial for gas recovery. Moreover, it is proposed that the CO_2 injection rate be lower than the CH_4 production rate to prevent early CO_2 breakthrough. Furthermore, parameters such as diffusion coefficient, viscosity, and permeability are important in gas displacement as well.

Several CO_2-EGR projects have been implemented in North America, Europe, and Asia. China's gas reservoirs are primarily distributed in the Ordos Basin, Sichuan Basin, Bohai Bay Basin, and Tarim Basin, and CO2-enhanced natural gas extraction technology can sequester approximately 9 billion tons of CO_2 [12].

As it is still a developing technology, the main technological challenges of CO_2-EGR cover the following four parts:

(1)	Key technologies (e.g., CO_2 breakthrough) and engineering parameters should be overcome and optimized to improve technical maturity.
(2)	Reservoir storage capacity assessment and evaluation of site selection criteria.
(3)	Similarly, the amount of CO_2 consumed by EGR poses specific demands on the carbon source, as well as sufficient and cost-effective CO_2 is the basis for the commercialization of CO_2-EGR.
(4)	Monitoring technology of downhole CO_2 and anti-corrosion and anti-blocking technologies of equipment and pipelines.

3.3. CO_2 Fracturing

Carbon dioxide fracturing refers to a new type of fracturing technology that uses carbon dioxide as the fracturing fluid. According to the phase state of CO_2 after reaching the reservoir, carbon dioxide fracturing can be divided into liquid CO_2 fracturing (carbon dioxide dry fracturing) and supercritical CO_2 fracturing, and the latter generally has a wellhead heating device, as shown in Figure 1c. During the fracturing process, after carbon dioxide is filtered into the matrix, part of the carbon dioxide reacts with water to form H_2CO_3, which is further dissociated to form HCO_3^- and CO_3^{2-}, so carbon dioxide will be dissolved and stored in reservoir solution in the form of $CO_2(aq)$, H_2CO_3, HCO_3^-, CO_3^{2-} [36,37]. On the other hand, the injection of CO_2 will form a large number of acidic zones, which will lead to the dissolution of feldspar, clay minerals, and carbonate minerals, resulting in a large number of metal cations such as Ca^{2+} and Fe^{2+}, which are combined with CO_3^{2-} and other anions to produce carbon-containing minerals such as calcite, dolomite, etc., and realize the mineral sequestration of carbon dioxide [38]. When CO_2 fracturing technology is applied to unconventional oil and gas reservoirs, carbon dioxide can compete with methane adsorbed on the rock surface, thereby realizing the adsorption and storage of CO_2 [39]. In addition, compared with hydraulic fracturing, carbon dioxide fracturing also has the advantages of reducing rock breakdown pressure, forming complex fracture networks, and reducing reservoir damage and water resources dependence [40,41]. Therefore, the use of carbon dioxide for fracturing can also realize the geological sequestration of CO_2 during the efficient development of oil and gas reservoirs.

As for potential, current field data show that one fracturing well consumes around 1000 t of liquid carbon dioxide [42]. Furthermore, according to the US Energy Agency (EIA), the number of horizontal wells fractured in North America in 2020 exceeded 15.3×10^4 [43]. With China's continued investment in unconventional oil and gas development, it is roughly estimated that the number of fractured wells in China will reach between 1×10^5 and 1.5×10^5. Consequently, the potential of CO_2 utilized in fracturing could be 1×10^8 t~1.5×10^8 t.

However, the low viscosity and high filtration of carbon dioxide fracturing fluid lead to narrow fracture width, poor sand carrying effect, and easy sand plugging [36], which limits the large-scale popularization and application in the field and causes several challenges:

(1)	The fracturing mechanisms still need to be further investigated, including supercritical carbon dioxide fracturing initiation and fracture expansion mechanism.
(2)	The improvement of sand-carrying ability and the development of a thickening agent, resistance-reducing agent, and new lightweight proppant.
(3)	Precision phase control and strict requirements of equipment sealing and corrosion resistance.

(4)   Cost control and economics for large-scale industrial applications.

3.4. CO_2-EGS

Earth is regarded as a tremendous thermal energy resource due to its hot core and the decay of radioactive minerals [44,45]. Furthermore, geothermal energy has a higher utilization rate than solar and wind energy (5.2 times and 3.5 times, respectively) [46]. Consequently, the exploitation of geothermal energy sources has attracted significant attention because of their exceptional properties of being stable, sustainable, environmentally friendly, and weather-independent. Despite sufficient geothermal energy resources, commercial geothermal energy extraction is constrained to shallow geothermal energy and hydrothermal resources that enable fluid circulation through natural high-permeable formations currently. To exploit geothermal energy in deep and low permeability formations, a number of studies and pilot projects are being carried out, namely the development of an enhanced geothermal system (EGS) based on hot dry rock (HDR) through hydraulic fracturing. Conventionally, water has been used as fracturing and heat extraction fluid due to its high density and specific heat levels in EGS. Generally, a large amount of water will be consumed for continuous operations during fracturing and heat extraction, with more than 20% water loss [47,48]. Employing supercritical CO_2 to replace water for EGS development has been proposed due to its significant advantages, such as reduced pumping power requirements, lower fluid density, less risk of scaling, and improved thermodynamic efficiency [49]. Evidence also suggests that supercritical CO_2 has more compressibility and expansivity in HDR, making it a mobile fluid of interest. According to estimates, heat flow rates utilizing CO_2 as the working fluid might be up to five times higher than those using formation brine [50].

China's geothermal resources are significant, accounting for 7.9% of the world's reserves, with an annual availability of 3.06×10^{18} kWh [51]. Specifically, the annual recoverable resources of shallow geothermal energy and hydrothermal resources are equivalent to 700 million tons and 1.865 billion tons of standard coal, respectively. In addition, the resources of HDR within 5500 m are approximately 106 trillion tons of standard coal, which is also the emphasis for future geothermal energy exploration and development [52]. Only considering the EGS development of HDR geothermal energy, referring to the Soultz project in France and the Hijiori project in Japan, the water consumption of a single EGS well is between 30,000 and 60,000 m^3. Therefore, it can be calculated that the potential for applying CO_2 in developing geothermal energy from hot dry rock in China will be 3.3×10^7 t~6.6×10^7 t (density of liquid CO_2: 1101 kg/m^3).

Despite the considerable advantages of CO_2-EGS and numerous numerical efforts, it is still at the conceptual stage around the world to the best of our knowledge, and the following challenges are still needed to be addressed:

(1)   Reasonable and acceptable CO_2 capture cost is a necessary prerequisite for CO_2 application in geology.
(2)   The potential threats of CO_2 leakage during storage in deep formations have been the major constraints in promoting CO_2 geological storage techniques.
(3)   Complex phase transition mechanisms and migration processes of CO_2 in geothermal reservoirs.
(4)   Generation of large-area hydraulic fracture network and heat exchange surface.

Nevertheless, combining geothermal heat extraction and CO_2 geological storage in deep EGS is a promising alternative to improve the economic feasibility of CCUS.

4. CCCUS

Aiming to promote carbon cycle economy, underground energy storage, and enhance the scale of CO_2 utilization, the authors propose a new concept of CO_2 capture, circular utilization, and storage, namely underground biomethanation (Figure 2). H_2 and CO_2 are mixed or injected sequentially into the depleted or depleting natural gas reservoirs in the CCCUS process. A very small percentage of injected carbon dioxide is sequestrated underground through mineralization or dissolution, and most of the injected gas is bio-converted to water and renewable methane under the catalysis of methanogens during the long-time well shut-in [53]. The reaction equation of methanation is as follows:

$$CO_2 + 4H_2 \rightarrow CH_4 + 2H_2O; \ \Delta G = -136 \ kJ/mol \tag{1}$$

Natural gas composed of renewable methane and other impurities can be produced for industrial or domestic usage. CO_2 in the industrial exhaust gas or emitted into the atmosphere is captured and then used for the next stage of biomethanation, to realize the circular utilization and continuous underground storage of carbon dioxide. In the CCCUS process, H_2 can be produced by electrolysis of water from renewable energy (e.g., wind and solar), or by coal-based gasification, an industrial by-product, and so on. As a result, this technology can also realize the storage and partially circular utilization of CO_2. China has become the world's largest H_2 producer, and its annual output is about 33 million tons. Almost 80% and 20% of the output comes from H_2 produced by fossil energy and industrial by-product H_2, respectively [54].

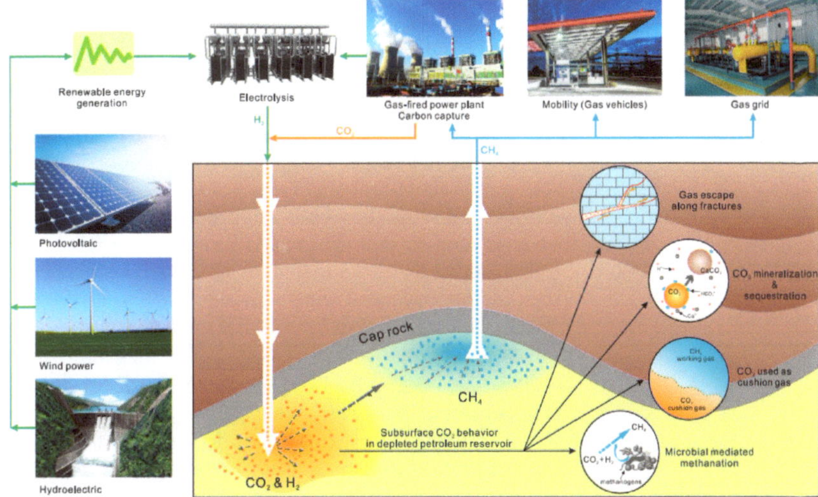

Figure 2. Simplified concept of biomethanation [55].

To realize the ideal biomethanation of injected H_2 and CO_2 underground, two conditions need to be met: (i) the injected gas will not leak underground; (ii) the methanogens have high biological activity. As for the first aspect, trapping (e.g., anticline and fault) and impermeable cap rock are necessary to prevent the migration and leakage of the injected gas. As a large amount of oil and gas was stored in depleted reservoirs before production, the sealing performance of depleted reservoirs is more reliable than that of aquifers. In terms of the methanogens' activity, it is affected by temperature, porosity, water saturation, pH, salinity, and so on (Figure 3): (1) Methanogen growth also requires pore space and water. Strobel et al. [56] suggested that the minimum reservoir porosity and permeability for biomethanation are 10% and 10mD, respectively, and pores should be connected for easier injection. Moreover, the minimum water saturation is 10%; (2) Higher and lower temperatures affect the activity of enzymes in methanogens, and even lead to the death of methanogens. Most methanogens can survive in the temperature range of 15 to 90 °C, and the favorable temperature range of 30 to 70 °C [57]; (3) The brine pH affects the methanogen's growth by directly affecting metabolism or indirectly through redox reactions. Most methanogens will die if the pH range is outside the range of 4 to 9.5, and the favorable pH range is 6.5 to 7.5; (4) The salinity affects the osmotic pressure of methanogen cells, and higher salinity causes methanogen cells to lose too much water and die. The upper limit of salt content for the growth of methanogen is 150 g/L. The environmental requirements for underground biomethanantion are summarized in Table 1.

According to the requirements of formation tightness and a suitable environment for the biomethanation process (Table 1) mentioned above, the depleted reservoirs or aquifers suitable for biomethanation can be screened out. If there are no natural methanogens in the formation water, the methanogens cultivated on the ground can be artificially injected into the underground bio-reactors. Moreover, when the temperature is so high that the activity of methanogens is low, pre-injection or circulating injection of cold fluid can be used to reduce the temperature.

Table 1. Environmental requirements for underground biomethanation.

Items	Description
Underground structure	Trapping (e.g., anticline and fault) and impermeable cap rock
Porosity	Minimum: 10%
Permeability	Minimum: 10 mD
Water saturation	Minimum: 10%
Temperature	Survival range: 15–98 °C; favorable range: 30–70 °C
pH	Survival range: 4–9.5; favorable range: 6.5–7.5
Salinity	Maximum: 150 g/L

The CCCUS technology stores and utilizes among others green H_2, which means that this method enables large-scale underground storage of renewable hydrogen while consuming a large amount of green power. Moreover, this method effectively couples zero (negative) emission economic utilization of impure H_2, CO_2 recycling, and sequestration, underground natural gas synthesis and storage, and geothermal parallel development, thus boosting the development of a low-carbon circular economy. In terms of operation management, safety requirements, etc., it is also necessary to develop unified norms and standards. Underground biochemical mechanization technology is currently at the stage of mechanism research and small-scale field trials, even in Europe and USA. Before large-scale industrial applications, the following challenges are still needed to be addressed:

(1) Screening of specific strains and large-scale efficient and low-cost nurture.
(2) Dynamic tracking and monitoring system for biochemical catalytic reaction processes.
(3) Unknown biochemical processes and influencing factors in real reservoirs.
(4) Activation and control of microbial populations in the reservoir environment and inhibition of competing organisms.
(5) Energy conversion efficiency and site selection and evaluation criteria.
(6) Uncertainty of economic benefits, operating mechanisms, and business models. It is needed for quantitative financial model assessment on field tests (the Sichuan Province of China is suitable for such tests) to find the appropriate business mechanism.
(7) The full supply chain of hydrogen, including production (cost-effective, method, etc.), transportation (related materials, mode, etc.), and storage (corrosion, leakage, etc.) are also key issues of CCCUS.

Regarding H_2 being extremely flammable and easy to form an explosive gas mixture with air, safety management measures need to be implemented during transportation and injection, mainly from the following four aspects: (1) Hydrogenation facilities should be placed, installed, and running in designated dedicated locations. Lightning protection facilities, safety fences, anti-collision columns, and grounding devices should be set up in the operation area; (2) Implement uninterrupted monitoring and periodic maintenance system to improve equipment reliability; (3) Vehicles transporting hydrogen should comply with relevant regulations, use special motor vehicles, and the bottom of the vehicle should be equipped with static-conducting mopping tapes that meet the regulations; (4) Personnel engaged in the safety management of hydrogenation facilities should undergo professional training in pressure vessel safety management to understand the characteristics of pressurized hydrogenation equipment, the use characteristics of packaging containers, and emergency measures.

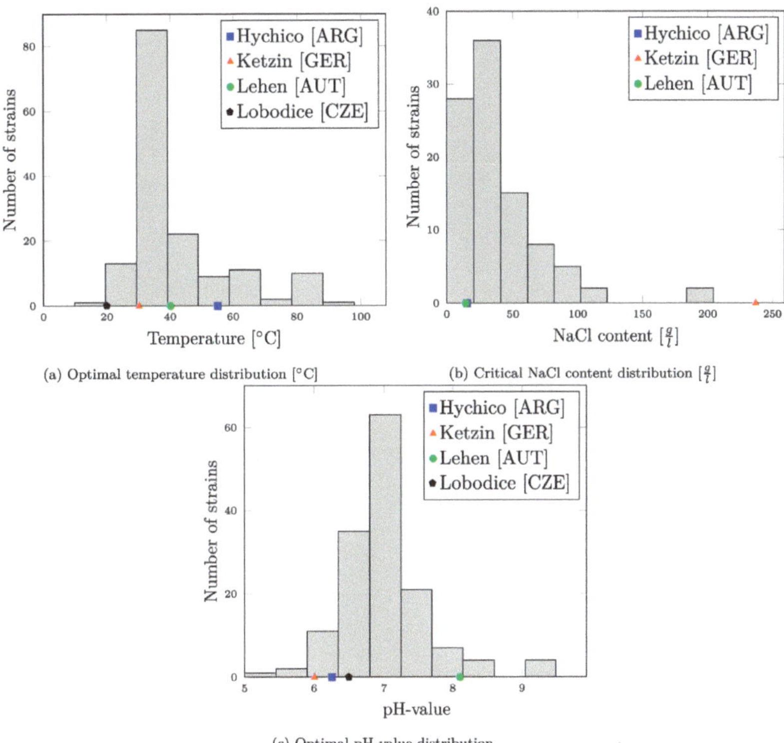

(a) Optimal temperature distribution [°C]

(b) Critical NaCl content distribution [$\frac{g}{l}$]

(c) Optimal pH-value distribution

Figure 3. Effects of (**a**) temperature, (**b**) salinity, and (**c**) pH on the activity of methanogens [56].

As a key carbon-negative technology, CCCUS will have an unprecedented development opportunity and significant potential. According to preliminary estimates, depleted gas reservoirs can sequester approximately 1.53×10^{10} tons of CO_2 in China, which shows that China has sufficient underground gas reservoir space for CCCUS. In addition, in terms of energy storage, the demand for reserved power in China would be 5×10^{11} kWh~1×10^{12} kWh in 2030, and 6×10^{12} kWh~7×10^{12} kWh in 2060 based on our studies. Among them, China's large-scale underground energy storage will reach 4.8×10^{12} kWh~5.6×10^{12} kWh (80% of the total power storage demand) in 2060. In the case of full capacity utilization to produce hydrogen and the industrial by-product hydrogen, underground biochemical synthesis of methane with CO_2, energy storage of 3.29×10^{12} kWh~4.42×10^{12} kWh, and consumption of 5.94×10^8 t~7.98×10^8 t of carbon dioxide can be achieved (energy efficiency: 68.6~79.2%).

5. Outlook

As a crucial artificial carbon sink and underground large-scale energy storage method in the future, CCCUS technology represented by underground biomethanation has a promising development prospect, particularly in China, where renewable energy and hydrogen are booming significantly. According to conservative estimations, underground biomethanation has significant potential, with a corresponding energy storage capacity of over 4×10^{12} kWh and a carbon dioxide uptake capacity close to 8×10^8 t. Under the limited growth of natural carbon sinks, vigorously promoting CCUS development and proposing new CCCUS modes will play a key role, particularly in China, where the planned period from carbon peak to carbon neutrality is strictly restricted. Comparatively, underground biomethanation incorporates renewable energy consumption, reduces wind and solar abandonment, and simultaneously enables energy storage resulting in an efficient carbon circular economy, hence making the investigation highly valuable as a potential carbon neutrality technology and solution.

According to the analysis of the potentialities and challenges in each method (see Table 2), as well as the evaluation of its development trend, combined with the consideration of China's status quo, the following suggestions have been made: deepening field studies of depleted oil and gas reservoirs as natural biochemical reactors to synthesize methane is necessary. Moreover, the primary control parameters and evaluation indicators to prepare for field testing can be investigated. Combined with the author's investigation and research, it is recommended that the pilot work of CCCUS should be implemented in all localities appropriately. For instance, the Sichuan Basin, where gas reservoirs are widely distributed, also has a large number of depleted oil and gas reservoirs, the pilot study of depleted oil and gas reservoirs as a natural biochemical reactor to synthesize methane can be organized.

Table 2. Characteristics and potentialities of various technologies in China.

Technical Type		Main Characteristics	Technology Maturity	Potential
CU/CCU	Direct or conversion utilization	CO_2 fixation is simple and convenient	Mature	CO_2 consumption: 6.2×10^7 t~8.7×10^7 t
Geological CCUS	EOR	Large-scale, partially permanent sequestration of CO_2, more economical than pure CO_2 sequestration	Almost mature	CO_2 storage: 5.1×10^9 t
	EGR		Almost mature	CO_2 storage: 9.0×10^9 t
	Fracturing		Pilot stage	CO_2 consumption: 1×10^8 t~1.5×10^8 t
	Geothermal		Theoretical stage	CO_2 consumption: 3.3×10^7 t~6.6×10^7 t
CCCUS	Underground Biomethanation	Designed carbon sink, large scale, low cost of underground methanation, storage of renewable methane, carbon circular economy	Theoretical stage	CO_2 consumption: 5.94×10^8 t~7.98×10^8 t Energy storage: 3.29×10^{12} kWh~ 4.42×10^{12} kWh

Note: These summaries of characteristics and maturity is based on references [12,14,16–19,55,58,59].

Nevertheless, there are also various development opportunities for geological CCUS considering the significant CO_2 uptake capacity, particularly, CO_2 fracturing and CO_2-EGS, although they are not as widespread and commercialized as CO_2-EOR and CO_2-EGR. Establishing a complete set of site selection and evaluation criteria, and promoting demonstration projects of geological CCUS are necessary. Likewise in CO_2 fracturing and CO_2-EGS development, relevant research needs to be implemented urgently. Furthermore, investigations on key technologies related to carbon dioxide applications, such as CO_2 transportation, phase control, and related equipment can be performed.

Author Contributions: Conceptualization, Z.H.; Investigation, J.L. and Y.X. (Yachen Xie); Methodology, Y.X. (Yachen Xie) and L.H.; Supervision, Z.H.; Writing—original draft, J.L. and L.W.; Writing—review and editing, Y.X. (Yachen Xie) and Y.X. (Ying Xiong). All authors have read and agreed to the published version of the manuscript.

Funding: Henan Institute for Chinese Development Strategy of Engineering & Technology (Grant No. 2022HENZDA02), Science & Technology Department of Sichuan Province (Grant No. 260 2021YFH0010) and China Scholarship Council (CSC File No. 201808510186).

Data Availability Statement: Data available in a publicly accessible repository.

Conflicts of Interest: The authors declare no conflict of interest.

References

1. Kweku, D.W.; Bismark, O.; Maxwell, A.; Desmond, K.A.; Danso, K.B.; Oti-Mensah, E.A.; Quachie, A.T.; Adormaa, B.B. Greenhouse effect: Greenhouse gases and their impact on global warming. *J. Sci. Res. Rep.* **2018**, *17*, 1–9. [CrossRef]
2. Kazemifar, F. A review of technologies for carbon capture, sequestration, and utilization: Cost, capacity, and technology readiness. *Greenh. Gases Sci. Technol.* **2021**, *12*, 200–230. [CrossRef]
3. National Oceanic and Atmospheric Administration. Carbon Dioxide Now More than 50% Higher than Pre-Industrial Levels. Available online: https://www.noaa.gov/news-release/carbon-dioxide-now-more-than-50-higher-than-pre-industrial-levels (accessed on 6 June 2022).
4. Lowe, R.; Drummond, P. Solar, wind and logistic substitution in global energy supply to 2050–Barriers and implications. *Renew. Sustain. Energy Rev.* **2022**, *153*, 111720. [CrossRef]
5. Schleussner, C.-F.; Rogelj, J.; Schaeffer, M.; Lissner, T.; Licker, R.; Fischer, E.M.; Knutti, R.; Levermann, A.; Frieler, K.; Hare, W. Science and policy characteristics of the Paris Agreement temperature goal. *Nat. Clim. Chang.* **2016**, *6*, 827–835. [CrossRef]
6. Change, P.C. *Global Warming of 1.5 °C*; World Meteorological Organization: Geneva, Switzerland, 2018.
7. Roelfsema, M.; Fekete, H.; Höhne, N.; den Elzen, M.; Forsell, N.; Kuramochi, T.; de Coninck, H.; van Vuuren, D.P. Reducing global GHG emissions by replicating successful sector examples: The 'good practice policies' scenario. *Clim. Policy* **2018**, *18*, 1103–1113. [CrossRef]
8. Yu, L.; Zhao, L. Research on China's Energy External Dependence and Energy Strategy. *E3S Web Conf.* **2021**, *245*, 01047.
9. Xie, Y.; Qi, J.; Zhang, R.; Jiao, X.; Shirkey, G.; Ren, S. Toward a Carbon-Neutral State: A Carbon–Energy–Water Nexus Perspective of China's Coal Power Industry. *Energies* **2022**, *15*, 4466. [CrossRef]
10. Wang, J.; Feng, L.; Palmer, P.I.; Liu, Y.; Fang, S.; Bosch, H.; O'Dell, C.W.; Tang, X.; Yang, D.; Liu, L.; et al. Large Chinese land carbon sink estimated from atmospheric carbon dioxide data. *Nature* **2020**, *586*, 720–723. [CrossRef]
11. Our World in Data. *China: CO_2 Country Profile*; Our World in Data: Oxford, UK, 2020.
12. BoFeng, C.; Li, Q.; Zhang, Q. *China Carbon Dioxide Capture Utilization and Storage (CCUS) Annual Report (2021)—China CCUS Pathway Study*; Ministry of Ecology and Environment, Chinese Academy of Sciences, The Administrative Center for China's Agenda 21: Beijing, China, 2021. (In Chinese)
13. Xie, Y.; Hou, Z.; Liu, H.; Cao, C.; Qi, J. The sustainability assessment of CO_2 capture, utilization and storage (CCUS) and the conversion of cropland to forestland program (CCFP) in the Water–Energy–Food (WEF) framework towards China's carbon neutrality by 2060. *Environ. Earth Sci.* **2021**, *80*, 468. [CrossRef]
14. Agency, I.E. *Putting CO_2 to Use-Creating Value from Emissions*; International Energy Agency (IEA): Paris, France, 2019.
15. Luo, J.; Hou, Z.; Feng, G.; Liao, J.; Haris, M.; Xiong, Y. Effect of Reservoir Heterogeneity on CO_2 Flooding in Tight Oil Reservoirs. *Energies* **2022**, *15*, 3015. [CrossRef]
16. Kuuskraa, V.A.; Godec, M.L.; Dipietro, P. CO_2 utilization from "next generation" CO_2 enhanced oil recovery technology. *Energy Procedia* **2013**, *37*, 6854–6866. [CrossRef]
17. Hamza, A.; Hussein, I.A.; Al-Marri, M.J.; Mahmoud, M.; Shawabkeh, R.; Aparicio, S. CO_2 enhanced gas recovery and sequestration in depleted gas reservoirs: A review. *J. Pet. Sci. Eng.* **2021**, *196*, 107685. [CrossRef]
18. Nianyin, L.; Jiajie, Y.; Chao, W.; Suiwang, Z.; Xiangke, L.; Jia, K.; Yuan, W.; Yinhong, D. Fracturing technology with carbon dioxide: A review. *J. Pet. Sci. Eng.* **2021**, *205*, 108793. [CrossRef]
19. Zhang, L.; Li, X.; Zhang, Y.; Cui, G.; Tan, C.; Ren, S. CO_2 injection for geothermal development associated with EGR and geological storage in depleted high-temperature gas reservoirs. *Energy* **2017**, *123*, 139–148. [CrossRef]
20. Ferguson, R.C.; Nichols, C.; Van Leeuwen, T.; Kuuskraa, V.A. Storing CO_2 with enhanced oil recovery. *Energy Procedia* **2009**, *1*, 1989–1996. [CrossRef]
21. Cooney, G.; Littlefield, J.; Marriott, J.; Skone, T.J. Evaluating the climate benefits of CO_2-enhanced oil recovery using life cycle analysis. *Environ. Sci. Technol.* **2015**, *49*, 7491–7500. [CrossRef]
22. Ghedan, S.G. Global laboratory experience of CO_2-EOR flooding. In Proceedings of the SPE/EAGE Reservoir Characterization and Simulation Conference, Abu Dhabi, United Arab Emirates, 19–21 October 2009; OnePetro: Richardson, TX, USA, 2009.
23. Andrei, M.; De Simoni, M.; Delbianco, A.; Cazzani, P.; Zanibelli, L. Enhanced oil recovery with CO_2 capture and sequestration. In Proceedings of the Conference: 21st World Energy Congress (WEC) 2010, Montreal, QC, Canada, 12–16 September 2010.
24. Alfarge, D.; Wei, M.; Bai, B. CO_2-EOR mechanisms in huff-n-puff operations in shale oil reservoirs based on history matching results. *Fuel* **2018**, *226*, 112–120. [CrossRef]
25. Jia, B.; Tsau, J.-S.; Barati, R. A review of the current progress of CO_2 injection EOR and carbon storage in shale oil reservoirs. *Fuel* **2019**, *236*, 404–427. [CrossRef]
26. Hill, B.; Hovorka, S.; Melzer, S. Geologic carbon storage through enhanced oil recovery. *Energy Procedia* **2013**, *37*, 6808–6830. [CrossRef]
27. Liu, Z.-x.; Liang, Y.; Wang, Q.; Guo, Y.-j.; Gao, M.; Wang, Z.-b.; Liu, W.-l. Status and progress of worldwide EOR field applications. *J. Pet. Sci. Eng.* **2020**, *193*, 107449. [CrossRef]
28. Wei, N.; Li, X.; Dahowski, R.T.; Davidson, C.L.; Liu, S.; Zha, Y. Economic evaluation on CO_2-EOR of onshore oil fields in China. *Int. J. Greenh. Gas Control* **2015**, *37*, 170–181. [CrossRef]

29. Al-Hasami, A.; Ren, S.; Tohidi, B. CO_2 injection for enhanced gas recovery and geo-storage: Reservoir simulation and economics. In Proceedings of the SPE Europec/EAGE Annual Conference, Madrid, Spain, 13–16 June 2005; OnePetro: Richardson, TX, USA, 2005.

30. Kühn, M.; Förster, A.; Großmann, J.; Lillie, J.; Pilz, P.; Reinicke, K.; Schäfer, D.; Tesmer, M.; Partners, C. The Altmark natural gas field is prepared for the enhanced gas recovery pilot test with CO_2. *Energy Procedia* **2013**, *37*, 6777–6785. [CrossRef]

31. Odi, U. Analysis and potential of CO_2 Huff-n-Puff for near wellbore condensate removal and enhanced gas recovery. In Proceedings of the SPE Annual Technical Conference and Exhibition, San Antonio, TX, USA, 8–10 October 2012; OnePetro: Richardson, TX, USA, 2012.

32. Clemens, T.; Secklehner, S.; Mantatzis, K.; Jacobs, B. Enhanced gas recovery, challenges shown at the example of three gas fields. In Proceedings of the SPE Europec/Eage Annual Conference and Exhibition, Barcelona, Spain, 14–17 June 2010; OnePetro: Richardson, TX, USA, 2010.

33. Khan, C.; Amin, R.; Madden, G. Carbon dioxide injection for enhanced gas recovery and storage (reservoir simulation). *Egypt. J. Pet.* **2013**, *22*, 225–240. [CrossRef]

34. Klimkowski, Ł.; Nagy, S.; Papiernik, B.; Orlic, B.; Kempka, T. Numerical simulations of enhanced gas recovery at the Załęcze gas field in Poland confirm high CO_2 storage capacity and mechanical integrity. *Oil Gas Sci. Technol. Rev. d'IFP Energ. Nouv.* **2015**, *70*, 655–680. [CrossRef]

35. Eliebid, M.; Mahmoud, M.; Shawabkeh, R.; Elkatatny, S.; Hussein, I.A. Effect of CO_2 adsorption on enhanced natural gas recovery and sequestration in carbonate reservoirs. *J. Nat. Gas Sci. Eng.* **2018**, *55*, 575–584. [CrossRef]

36. Zhang, R.; Yin, X.; Winterfeld, P.H.; Wu, Y.-S. A fully coupled thermal-hydrological-mechanical-chemical model for CO_2 geological sequestration. *J. Nat. Gas Sci. Eng.* **2016**, *28*, 280–304. [CrossRef]

37. Luo, Z.; Wu, L.; Zhao, L.; Zhang, N.; Chen, W.; Liang, C. Numerical study on filtration law of supercritical carbon dioxide fracturing in shale gas reservoirs. *Greenh. Gases Sci. Technol.* **2021**, *11*, 871–886. [CrossRef]

38. Gundogan, O.; Mackay, E.; Todd, A. Comparison of numerical codes for geochemical modelling of CO_2 storage in target sandstone reservoirs. *Chem. Eng. Res. Des.* **2011**, *89*, 1805–1816. [CrossRef]

39. Fan, C.; Elsworth, D.; Li, S.; Zhou, L.; Yang, Z.; Song, Y. Thermo-hydro-mechanical-chemical couplings controlling CH4 production and CO_2 sequestration in enhanced coalbed methane recovery. *Energy* **2019**, *173*, 1054–1077. [CrossRef]

40. Wang, M.; Huang, K.; Xie, W.; Dai, X. Current research into the use of supercritical CO_2 technology in shale gas exploitation. *Int. J. Min. Sci. Technol.* **2019**, *29*, 739–744. [CrossRef]

41. Wu, L.; Hou, Z.; Luo, Z.; Xiong, Y.; Zhang, N.; Luo, J.; Fang, Y.; Chen, Q.; Wu, X. Numerical simulations of supercritical carbon dioxide fracturing: A review. *J. Rock Mech. Geotech. Eng.* **2022**. [CrossRef]

42. Wang, F.; Wang, Y.; Zhu, Y.; Duan, Y.; Chen, S.; Wang, C.; Zhao, W. Application of liquid CO2 fracturing in tight oil reservoir. In Proceedings of the SPE Asia Pacific Oil & Gas Conference and Exhibition, Perth, Australia, 25–27 October 2016; OnePetro: Richardson, TX, USA, 2016.

43. US Energy Information Administration (EIA). *Drilling Productivity Report*; EIA: Washington, DC, USA, 2022.

44. Olasolo, P.; Juárez, M.; Morales, M.; Liarte, I. Enhanced geothermal systems (EGS): A review. *Renew. Sustain. Energy Rev.* **2016**, *56*, 133–144. [CrossRef]

45. Haris, M.; Hou, M.Z.; Feng, W.; Luo, J.; Zahoor, M.K.; Liao, J. Investigative coupled thermo-hydro-mechanical modelling approach for geothermal heat extraction through multistage hydraulic fracturing from hot geothermal sedimentary systems. *Energies* **2020**, *13*, 3504. [CrossRef]

46. Li, D.; Wang, Y. Major issues of research and development of hot dry rock geothermal energy. *Earth Sci* **2015**, *40*, 1858–1869.

47. Harto, C.; Schroeder, J.; Horner, R.; Patton, T.; Durham, L.; Murphy, D.; Clark, C. *Water Use in Enhanced Geothermal Systems (EGS): Geology of US Stimulation Projects, Water Costs, and Alternative Water Source Policies*; Argonne National Lab. (ANL): Argonne, IL, USA, 2014.

48. Wang, X.-x.; Wu, N.-y.; Su, Z.; Zeng, Y.-c. Progress of the enhanced geothermal systems (EGS) development technology. *Prog. Geophys.* **2012**, *27*, 355–362.

49. Brown, D.W. A hot dry rock geothermal energy concept utilizing supercritical CO_2 instead of water. In Proceedings of the Twenty-Fifth Workshop on Geothermal Reservoir Engineering, Stanford University, Stanford, CA, USA, 24–26 January 2000; pp. 233–238.

50. Randolph, J.B.; Saar, M.O. Combining geothermal energy capture with geologic carbon dioxide sequestration. *Geophys. Res. Lett.* **2011**, *38*. [CrossRef]

51. Hou, J.; Cao, M.; Liu, P. Development and utilization of geothermal energy in China: Current practices and future strategies. *Renew. Energy* **2018**, *125*, 401–412. [CrossRef]

52. China Geological Survey. *China Geothermal Energy Development Report*; China Petrochemical Press: Beijing, China, 2018.

53. Amigáň, P.; Greksak, M.; Kozánková, J.; Buzek, F.; Onderka, V.; Wolf, I. Methanogenic bacteria as a key factor involved in changes of town gas stored in an underground reservoir. *FEMS Microbiol. Ecol.* **1990**, *6*, 221–224.

54. Gmw. With an Annual Output of about 33 Million Tons, China Has Become the World's Largest Hydrogen Producer. 2021. Available online: https://ptx-hub.org/factsheet-on-china-the-worlds-largest-hydrogen-producer-and-consumer/ (accessed on 21 December 2022).

55. Xiong, Y.; Hou, Z.; Xie, H.; Zhao, J.; Tan, X.; Luo, J. Microbial-mediated CO_2 methanation and renewable natural gas storage in depleted petroleum reservoirs: A review of biogeochemical mechanism and perspective. *Gondwana Res.* **2022**. [CrossRef]
56. Strobel, G.; Hagemann, B.; Huppertz, T.M.; Ganzer, L. Underground bio-methanation: Concept and potential. *Renew. Sustain. Energy Rev.* **2020**, *123*, 109747. [CrossRef]
57. Thaysen, E.M.; McMahon, S.; Strobel, G.J.; Butler, I.B.; Ngwenya, B.T.; Heinemann, N.; Wilkinson, M.; Hassanpouryouzband, A.; McDermott, C.I.; Edlmann, K. Estimating microbial growth and hydrogen consumption in hydrogen storage in porous media. *Renew. Sustain. Energy Rev.* **2021**, *151*, 111481. [CrossRef]
58. Middleton, R.S.; Carey, J.W.; Currier, R.P.; Hyman, J.D.; Kang, Q.; Karra, S.; Jiménez-Martínez, J.; Porter, M.L.; Viswanathan, H.S. Shale gas and non-aqueous fracturing fluids: Opportunities and challenges for supercritical CO_2. *Appl. Energy* **2015**, *147*, 500–509. [CrossRef]
59. Haris, M.; Hou, M.Z.; Feng, W.; Mehmood, F.; bin Saleem, A. A regenerative Enhanced Geothermal System for heat and electricity production as well as energy storage. *Renew. Energy* **2022**, *197*, 342–358. [CrossRef]

Article

Toward a Carbon-Neutral State: A Carbon–Energy–Water Nexus Perspective of China's Coal Power Industry

Yachen Xie [1,2], Jiaguo Qi [1,2], Rui Zhang [1], Xiaomiao Jiao [3], Gabriela Shirkey [1,2] and Shihua Ren [3,4,*]

[1] Department of Geography, Environment and Spatial Sciences, Michigan State University, East Lansing, MI 48824, USA; xieyache@msu.edu (Y.X.); qi@msu.edu (J.Q.); zhangr50@msu.edu (R.Z.); shirkeyg@msu.edu (G.S.)
[2] Center for Global Change and Earth Observations, Michigan State University, East Lansing, MI 48823, USA
[3] Technology Support Center, China Coal Research Institute, Beijing 100013, China; jiaoxm@pku.edu.cn
[4] School of Management, China University of Mining & Technology (Beijing), Beijing 100083, China
* Correspondence: ren@cct.org.cn

Abstract: Carbon neutrality is one of the most important goals for the Chinese government to mitigate climate change. Coal has long been China's dominant energy source and accounts for more than 70–80% of its carbon emissions. Reducing the share of coal power supply and increasing carbon capture, utilization, and storage (CCUS) in coal power plants are the two primary efforts to reduce carbon emissions in China. However, even as energy and water consumed in CCUS are offset by reduced energy consumption from green energy transitions, there may be tradeoffs from the carbon–energy–water (CEW) nexus perspective. This paper developed a metric and tool known as the "Assessment Tool for Portfolios of Coal power production under Carbon neutral goals" (ATPCC) to evaluate the tradeoffs in China's coal power industry from both the CEW nexus and financial profits perspectives. While most CEW nexus frameworks and practical tools focus on the CEW nexus perturbation from either an external factor or one sector from CEW, ATPCC considers the coupling effect from C(Carbon) and E(Energy) in the CEW nexus when integrating two main carbon mitigation policies. ATPCC also provides an essential systematic life cycle CEW nexus assessment tool for China's coal power industry under carbon-neutral constraints. By applying ATPCC across different Chinese coal industry development portfolios, we illustrated potential strategies to reach a zero-emission electricity industry fueled by coal. When considering the sustainability of China's coal industry in the future, we further demonstrate that reduced water and energy consumption results from the energy transition are not enough to offset the extra water and energy consumption in the rapid adoption of CCUS efforts. However, we acknowledge that the increased energy and water consumption is not a direct correlation to CCUS application growth nor a direct negative correlation to carbon emissions. The dual effort to implement CCUS and reduce electricity generation from coal needs a thorough understanding and concise strategy. We found that economic loss resulting from coal reduction can be compensated by the carbon market. Carbon trading has the potential to be the dominant profit-making source for China's coal power industry. Additionally, the financial profits in China's coal power industry are not negatively correlated to carbon emissions. Balance between the carbon market and the coal industry would lead to more economic revenues. The scenario with the most rapid reduction in coal power production combined with CCUS would be more sustainable from the CEW nexus perspective. However, when economic revenues are considered, the scenario with a moderately paced energy transition and CCUS effort would be more sustainable. Nevertheless, the ATPCC allows one to customize coal production scenarios according to the desired electricity production and emission reduction, thus making it appropriate not only for use in China but also in other coal-powered regions that face high-energy demands and carbon neutrality goals.

Keywords: carbon–energy–water nexus; CCUS; energy transition; life cycle coal power production

Citation: Xie, Y.; Qi, J.; Zhang, R.; Jiao, X.; Shirkey, G.; Ren, S. Toward a Carbon-Neutral State: A Carbon–Energy–Water Nexus Perspective of China's Coal Power Industry. *Energies* **2022**, *15*, 4466. https://doi.org/10.3390/en15124466

Academic Editor: Adam Smoliński

Received: 25 April 2022
Accepted: 17 June 2022
Published: 19 June 2022

Publisher's Note: MDPI stays neutral with regard to jurisdictional claims in published maps and institutional affiliations.

1. Introduction

1.1. Background on Carbon Neutrality and China's Coal Industry

Greenhouse gases absorb infrared radiation and retain heat, warming Earth's surface and driving global warming. Anthropogenic emissions have contributed to most atmospheric CO_2 over the past 150 years [1]. As global warming imposes significant impacts on extreme weather, rising sea levels, environmental stress on terrestrial and aquatic ecosystems, and food insecurity, there is a need to constrain global warming to 1.5 °C through carbon neutrality by 2050 [1].

Carbon neutrality means that emitted CO_2 is equal to the eliminated/sequestered CO_2 in the atmosphere within the same period. To mitigate climate change and global warming triggered by carbon emissions, reaching carbon neutrality is one of the world's most urgent missions [2]. Present efforts largely include sustainable economic growth and energy consumption goals [3]. 29 countries or regions have already declared carbon-neutral climate goals [4].

Globally, electricity and heat-related energy production accounts for around 31% of carbon emissions and is the most significant contributor by sector. As the most populous country with the most rapidly developing economy globally, China has the largest energy consumption demand, which is still increasing. With 3000 million tons of oil equivalent (Mt) energy consumed each year, China accounts for about 28% (9.8 Gt/year) of all greenhouse gas (GHG) emissions. Additionally, the type of primary energy used to generate electricity is a significant contributor to China's GHG emissions. For electricity generation alone, traditional thermal sources account for around 70% of the total share, compared to a 30% share of green energy sources, such as wind, hydro, and nuclear. Coal contributes to about 92% of all thermal energy sources, making it the dominant energy source in China for electricity generation [5]. In 2019, coal accounted for greater than 60% of the total energy consumption in the country.

Coal has long been one of the biggest shares in the mix of energy sources due to its low cost and high accessibility in China [6,7]. However, the economic benefit comes hand in hand with negative impacts on the environment [8–10]. Coal combustion has long been recognized to account for an enormous carbon emission [11]. Approximately 80% of CO_2 emissions in China between 2000 and 2013 were from coal combustion alone [12]. Far beyond that, carbon emission from the process of coal mining is also significant. Mining activities release a large amount of methane (CH_4), the second most important greenhouse gas after CO_2, as well as CO_2 and other gases from coal and surrounding rock strata [13]. Additionally, emissions during the process of mining and washing, as well as transportation, significantly contribute to the total carbon emissions of generating coal power [14,15]. Therefore, it is essential to consider the entire coal power generation process in calculating total carbon emission and the environmental cost of utilizing coal power.

In the 2020 United Nations General Assembly, President Xi Jinping declared that China would aim to cut peak emissions before 2030 and pledged to achieve carbon neutrality before 2060 [16]. To accomplish carbon neutrality, the energy sector, especially coal power generation, should be the first emission source that needs urgent action. Generally, two pathways of action can reduce carbon emissions in power generation in countries where coal is the primary power source. The first is to reduce the share of coal power in the country's total power supply, and the second is to reduce carbon emissions from coal power generation by applying the technology for carbon capture, utilization, and storage (CCUS).

The best and only approach to, while ensuring sufficient power, reduce the emissions is energy transition, defined here as shifting the energy sector from fossil-based production and consumption systems to renewable energy sources [17,18]. Currently, China is enthusiastically promoting zero-emission green energy, such as wind, solar, and hydropower, to replace coal power to reduce emissions. Over the past 30 years, China has reduced the share of coal power by ~20% and vastly increased the percentage of green energy by distributing subsidies to the industry.

In addition to reducing the coal energy, investing in and applying the technology of CCUS is another effective component of the national strategy to reduce carbon emissions in China [19,20]. The CCUS is a processing chain that aims to capture and compress carbon emissions at the source plant and then transport the emissions for another utilization cycle or geological sequestration [21–23]. This technique is among the most cost-effective approaches to reducing carbon emissions and has the potential to reduce 20% of total emissions across the industry sector, which is projected to be above 28 Gt CO_2 by the year 2060. Currently, China is currently trying to develop and utilize CCUS in coal power plants. By the end of 2017, around 26 sites of CCUS had been put into service throughout the country [24,25].

When considering a sustainable coal power system, the water sector must also be considered as it provides a significant contribution to the life cycle of coal electricity production. Water is not only consumed in the mining, washing, and refining process, but also the combustion at power plants [26]. Most coal mines are in the arid region in western China, an area vulnerable to water stress. The energy transition process would significantly reduce water stress since most alternative power sources, except nuclear power, consume less water. However, extra water consumption is required for CCUS application in coal power plants. Studies show that the consumption of water in power plants increased by 50–90% after equipping CCUS [27–29]. Therefore, while water stress caused by coal-based power generation could be alleviated in future energy transitions, it will be exacerbated by integrating CCUS into coal-based electricity generation.

1.2. Literature Review on CEW Nexus Approach

As CCUS applications would significantly increase energy and water consumption at electricity generation facilities, it is crucial to understand the tradeoffs on water, energy, and carbon. This perspective is afforded by coupling two approaches to carbon reduction. The carbon–energy–water (CEW) nexus approach could offer a sustainability assessment perspective as it explores the effects of interactions between factors and the functionality of the entire system. Dynamics within the CEW nexus have been widely discussed, focusing on different driving forces. Some studies explored how external factors affected the CEW dynamics. For example, Yu et al. [30] assessed the effects of agricultural activities on the CEW nexus, and Li et al. [31] and Liang et al. [32] both considered socioeconomic cost as a significant external driving factor when investigating the CEW nexus. Internal driving forces are also widely discussed. Lim et al. [33] performed an energy-centric study that assessed how each factor in the nexus affects energy generation and the ultimate achievement of long-term energy plans in the United Arab Emirates. Water-centric [34–37], and carbon-centric [38,39] studies on the internal dynamics within the CEW nexus were also conducted. However, most of these studies primarily investigated only one individual sector's fluctuation in the nexus as an intrinsic driving factor for the nexus.

Focusing on one single centric driver of the CEW nexus can increase the understanding of the change in one sector and identify the relationship between the centric sector and other sectors in the system but may also lose the information on interactions between two or more sectors and their coupled effects on the dynamics of the nexus. The coupling effect needs to be considered, including both carbon and energy sectors as intrinsic driving factors to the CEW dynamics. In this research study, we explore the relationship between the two sectors and investigate how they are coupled to impact the CEW system.

CEW studies also set implementation goals in various industries or sectors, which implies the versatility and significance of CEW research. Wang et al. [40] made an effort to assist China's iron and steel industry achieve water and energy cost-effectiveness goals while reducing carbon emissions. Similar applications of the CEW were also explored in food and beverage products [41] and ceramic tile production [42]. Scott et al. [43], Gu et al. [44], and Trubetskaya et al. [37] attempted to put forward policy recommendations on water management and wastewater treatment. Emissions by sectors (e.g., agriculture, urban household, energy generation, and industry) were broadly estimated and discussed in the CEW framework [40,45–48]. Coal-based power generation sectors must be critically investigated as China's most crucial energy sector supplier and emitter. However, CEW studies on the coal sector are still lacking. To narrow this knowledge gap, the CEW nexus of China's coal power sector is first explored here in this study.

1.3. Literature Review on CEW Model Selection

Due to the complex interconnections between sectors in the CEW system, a comprehensive calculation of the perturbation in each component and the impact on the nexus is highly dependent on counting multiple sectors in multiple steps. Therefore, process-based models are widely adopted in CEW nexus studies. One of the most applied modeling techniques is the environmental input–output (EIO) model and life cycle assessment. The EIO model evolved from the economic input–output model that represented the interdependencies between different sectors of the economy [49]. EIO is suitable for CEW nexus analysis because it can adequately address direct and indirect contributions from each sector to the system's dynamics. It was widely applied in studies that calculated the direct and indirect effects of separate sectors to the CEW nexus [31,49,50] as well as the individual and combined contributions from different regions to the entire study area [51,52].

On the other hand, life cycle assessment is to evaluate processes within all stages of a product's life cycle and is commonly used to assess environmental impacts from the entire process [14]. Life cycle assessment is one of the most suitable approaches for those CEW nexus studies that need to comprehensively count procedural impacts on the CEW system from a sector [36,51,53] or industry [41,42]. However, previous studies mostly focused on a specific case study using EIO or life cycle assessment. In our opinion, it would be instrumental in generalizing these methods into a flexible tool to allow users in different research and geographic areas to fit in their cases and obtain their desired outcomes. Therefore, we applied life cycle assessment in the coal power generation process and proposed this method as a tool for customized uses. In addition to this generalization, we also improved the tool by adding an external factor, CCUS implementation, which integrates a more complex feedback loop to account for the interactions among CEW components. Adding CCUS as a factor in the CEW nexus is a significant improvement to the tool as it provides practical and valuable impacts on the coal electricity industry in China.

Since CEW nexus studies often aim at providing policy recommendations or environmental management solutions, scenario analysis is also widely adoptedto evaluate and compare the effectiveness of potential environmental acts [33,39,54–56]. The Assessment Tool for Portfolios of Coal power production under Carbon neutral goals (ATPCC) (Figure 1) is a scenario-based tool that can be used to inform future energy policy, especially for the policymakers that concerningthe coal electricity industry in China. The ATPCC offers scenarios to sustainably develop portfolios for the coal-based electricity industry to achieve the carbon neutrality mission in China.

1.4. Research Gaps and Goals

There are three major research gaps in the CEW literature. Theoretically, most approaches studied the external drivers or focused on one sector within the CEW system, rarely studying the coupled effects of two or more sectors. Second, there is a lack of a comprehensive and systematic life cycle assessment to address tradeoffs in the CEW nexus and financial benefits to the coal power industry. Third, the tradeoff analysis is rarely studied for the coupled effect of energy transition and CCUS application for China's coal electricity system at the national level.

In this study, we first developed a state-of-the-art life cycle assessment tool that includes the following features:

- A general framework to analyze the perturbation of the CEW nexus driven by coupled effects of carbon–energy sectors;

- A mechanism to account for the coupled effects of energy transition and CCUS applications in China with carbon mitigation synergy and sustainability tradeoffs for the CEW nexus and financial profits;
- A life cycle analysis of China's coal power industry; and
- Luxuriant and diverse empirical data for China's coal power industry from experts.

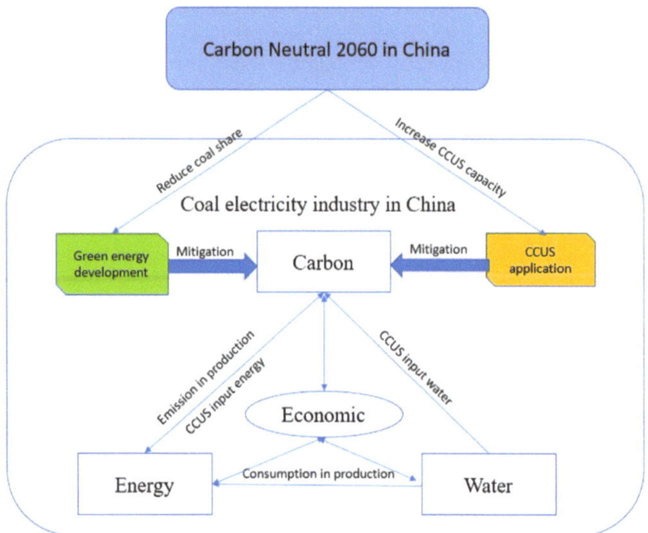

Figure 1. Conceptual framework of ATPCC design.

We then analyzed three coal power development portfolios that represent different pathways for the future of China's coal-based electricity generation based on an extensive literature review. We finally applied these three portfolios in the ATPCC assessment tool to address the following questions: (1) Is there a potential for China's coal power industry to achieve zero-emission under the present carbon neutrality goals; and (2) what are the tradeoffs when coupling CCUS and energy transitions in the CEW nexus and economic profit and do they vary by adoption scales?

2. Methodology

2.1. Conceptual Framework of ATPCC Design

The production portfolio for coal-based electricity under the carbon neutrality goal is crucial for policymakers in China, who seek to build a sustainable, profitable future. The feasibility of a given scenario relies on the relationship between the policies for energy transition and CCUS contribution to the CEW security and economy. This is not only an issue for the central government but also crucial to policymakers in China, especially for sectors where there is a risk of natural resources such as water or environmental costs or financial profits. There is a need to find a way toward a sustainable and profitable coal system under a carbon-neutral perspective.

The life cycle of coal electricity production is directly linked to energy, water, carbon emission, and financial profits. It includes coal mining and washing, coal transportation, and coal electricity production in power plants. Each process has a carbon–energy–water footprint and is associated with economic measurement. This is also true for the implementation of CCUS in coal industries with the extra energy, water footprint, and the financial profit regarding carbon trade. Therefore, the preparation of energy portfolio development and policies should integrate the assessment tradeoffs and synergies of energy, carbon, water, and economic profit (Figure 1).

By applying scenarios that represent different levels (intensities) of coal power share reduction and CCUS implementation, we produce quantitative water/energy consumption outputs, which is crucial for natural resources management. Considering the total amount of water/energy consumption, including reduced consumption in life cycle coal power production and the increased amount of water/energy used through CCUS application, is necessary from a resource management perspective. The total carbon emission from the coal power system is critical to achieving the expected carbon-neutrality goal. The economic profits from China's coal power system are also important for the economy.

Therefore, we provide a comprehensive assessment tool to help better understand the tradeoffs in China's coal power industry between environmental and economic outcomes.

Given these prerequisites, we constructed the ATPCC (Figure 1) and subsequently applied it to address the following specific tradeoffs:

- Energy consumption tradeoffs between the levels of energy transition and CCUS applications: Reducing more share of coal electricity would lead to less energy consumption. More carbon captured from CCUS would meanmore energy consumption.
- Water consumption tradeoffs between the levels of energy transition and CCUS adoptions: Reducing coal-generated electricity would lead to less water consumption. More carbon captured from CCUS would lead to more water consumption.
- Economic revenue tradeoffs between the levels of energy transition and CCUS applications: Reducing coal-based electricity would lead to less economic profits from electricity sales. More CCUS adoptions would increase economic profit from carbon trade.

2.2. The ATPCC

To study the tradeoffs and synergies of two main carbon-neutral policy impacts on the CEW nexus in the China's coal power system with complex interconnections, we propose a framework, "The Assessment Tool for Portfolios of Coal power production under goals or ATPCC". This tool is the first to integrate all components in China's coal production processes, including carbon, energy, water, and profit, with a CEW nexus approach. The scenario enabled ATPCC enables policymakers to create coal electricity portfolios based on carbon-neutral policies. Policymakers could assess portfolio scenarios by evaluating China's coal industry's CEW and economic sustainability.

The detailed structure of the proposed ATPCC, illustrated in Figure 2 provides specific factors in the life cycle of coal power production in four processes: coal mining and washing/refining, coal transportation to power plants, coal-based electricity production in power plants, and the CCUS adoption in coal power plants. The quantitative parameters, factors, and predicted future trends from 2020 to 2060 are sourced from the literature and in consultation with experts. The energy consumption includes nine energy sources: raw coal, coke, crude oil, gasoline, kerosene, diesel oil, fuel oil, natural gas, and electricity. They are all estimated as standard coal transformation coefficients [57].

In the coal mining and washing/refining process, there are around 4700 coal mine pits in China, and approximately 7% of the coal was imported from other countries. Based on the national statistics on coal processing, we used the average parameter values for all the domestic and imported coal, where most of the imported coal is raw material and still goes through the washing/refining process. Parameters include energy consumption, carbon emission, water consumption, and financial cost. The energy consumption aggregates the whole energy inputs of all mechanical equipment used in the entire process, including shearers, road headers, washing equipment, transportation equipment in the well, and power boilers for workers. The carbon emission is the sum of energy-related emissions, emissions equivalent to underground mine gas emissions, and post-mining emissions. Water consumption is the sum of water usage in mining and cooling without considering the grey water footprint. The financial cost is the price of the sum of the energies used.

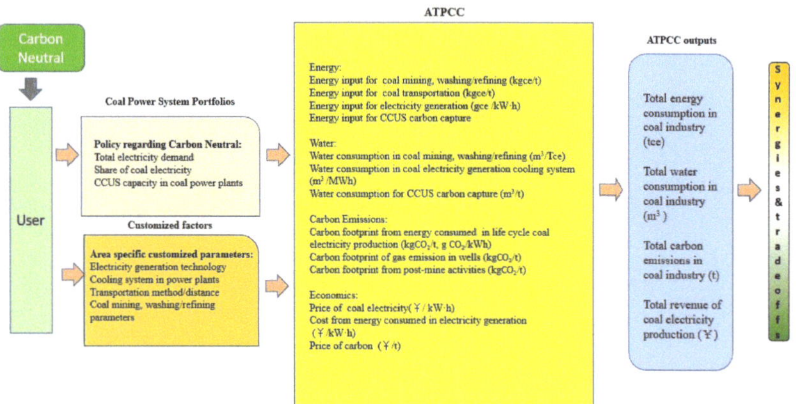

Figure 2. The Assessment Tool for Portfolios of Coal power production under Carbon neutral goals (ATPCC) structure.

The ATPCC provides the average distance of coal transportation to power plants on highways and railways in the transportation process. The energy consumption factors come from the sum of diesel and electricity usage. Carbon emission parameters are calculated based on energy consumption. The energy consumption also calculates financial costs. The ATPCC provides two options for input scenarios in coal-based electricity generation. By changing the parameters of selected types of electricity generation sets, the portfolio can provide a unique factor value for each scenario. The ATPCC provides factors and sets the parameters for average energy consumption, carbon emission related to energy, water consumption, and the cost of energy used in all processes.

CCUS is also considered in the ATPCC. Total extra energy and water consumption are the product of the carbon captured by CCUS facilities in coal power plants. The amount of economic costs for energy consumed from CCUS is also considered. With the given carbon price under global carbon-neutral missions, the financial profits gained from carbon sales can be calculated.

The ATPCC is a policy-driven, carbon–energy coupling effective tool, allowing users to apply it to any governance level or geographic area with the coal industry and under different carbon-neutral requirements. The user input portfolios comprise two major sub-scenarios: the total electricity generated from the coal industry and the total carbon captured by CCUS in the coal industry. There is also a customized option for users to provide specific parameters for electricity generation and cooling in the power plant, or users can directly replace parameters and their trends in ATPCC based on their status quo.

Given the scenario inputs and the user-defined parameter values, ATPCC is used to quantify the following outputs of the life cycle coal industry for electricity production in China: total water consumption, energy consumption, total carbon emissions, and total financial profits. Thus, a tradeoff analysis can be drawn based on ATPCC outputs and help policymakers find a sustainable pathway for coal power development. The customization feature of ATPCC can help users generate area-specific scenarios and the related tradeoff analysis.

3. ATPCC Model Parameters

3.1. Energy Consumption in Life Cycle Coal Electricity Production

The energy consumption in the life cycle of coal-based electricity production (Equation (1)) can be presented as a function that aggregates the energy input from three processes:

mining and washing/refining, transport, and coal combustion in power plants. Nine types of energy are accounted for, including raw coal and electricity.

$$E_{\text{Life Cycle}} = E_{\text{mine and wash/refine}} + E_{\text{transportation}} + E_{\text{electricity production}} \tag{1}$$

3.1.1. Energy Consumption in Coal Mining and Washing/Refinement

The energy used in the mining, washing, and refining can be expressed in Equations (2) and (3):

$$E_{\text{mine and wash/refine}} = M * e_{\text{mine and wash/refine}} \tag{2}$$

$$e_{\text{mine and wash/refine}} = \sum_{i}^{9} \alpha_i \times T_i \tag{3}$$

where M is the total raw coal consumed in a power plant. $e_{\text{mine and wash/refine}}$ is the energy consumption factor representing the average energy consumption of coal supply. α_i represents the conversion coefficient for each energy source to standard coal (Ce). i = 1–9 indicates for raw coal, coke, crude oil, gasoline, kerosene, diesel oil, fuel oil, natural gas, and electricity, respectively. T_i is the mean energy input for unit coal production. Parameters α_i, T_i and average $e_{\text{mine and wash/refine}}$ are shown in Table 1 [58].

Table 1. The conversion coefficient of standard coal with different energy sources and unit energy consumption in coal mining, washing, and refining process and transportation.

Energy Category	Comprehensive	Coal	Coke	Oil	Gasoline	Kerosene	Diesel	Fuel Oil	Natural Gas	Power
Convert coefficient (kgce/kg) or (kgce/m^3) or (kgce/kW·h)		0.7143	0.9714	1.4286	1.4714	1.4714	1.4571	1.4286	1.330	0.1720
Energy consumption in coal mining and washing/refinement by sources		27.3	0.0	0.0	0.0	0.0	0.2	0.0	2.4	9.3
Energy consumption of coal mining and washing converted for standard coal (kgce/t)	24.6	19.5	0.0	0.0	0.0	0.0	0.3	0.0	3.2	1.6
Energy consumption in coal transportation (kgce/t)	4.2				0.5		2.5			1.2

The current value of $e_{\text{mine and wash/refine}}$ is 24.6 kgce/t, and there is a continuous improvement in mechanization for coal mining and washing in China that has reduced energy consumption [59]. The average energy consumption in coal mining and washing decreased rapidly from 30.6 kgce/t in 2010 to 24.6 kgce/t in 2020, with an average annual declining rate of 2.1% [60]. Energy consumption in China's coal-mining/washing/refining process could decline more rapidly in the future through continuous optimization of the coal production system under the carbon-neutral goal. However, the energy consumption level in most of China's coal mining and washing facilities is close to reaching an advanced level worldwide [61]. Therefore, we assumed that the comprehensive energy consumption of coal mining and washing would continue to decline, but with a shrinking magnitude. It would decrease at an average annual rate of 4.0% before 2030, 3% from 2030 to 2040, 2% from 2040 to 2050, and 1% from 2050 to 2060; the predicted factors are shown in Table 2.

Table 2. Prediction of the future trend in comprehensive energy consumption.

Year	2020	2025	2030	2035	2040	2045	2050	2055	2060
Comprehensive coal consumption of coal mining and washing (gce/t)	24.6	20.1	16.4	14.0	12.1	10.9	9.9	9.4	8.9
Comprehensive coal consumption in power generation and power supply (gce/kWh)	351.2	342.5	334.1	325.8	317.8	309.8	302.1	294.7	287.4
Coal consumption of coal power generation and power supply (gce/kWh)	305.5	297.9	290.6	283.4	276.4	269.5	262.8	256.3	250.0

3.1.2. Energy Consumption in Coal Power Transportation

Considering only domestic transportation, the average coal transport distance is 651 km by railway and 162.7 km on the highway. The railway coal transport is around 80% and 20% for highways [58]. The average energy consumed on the railway has 55% of diesel and 45% electricity. The mean fuel mix of highway transport is 68% diesel and 32% gasoline. The energy level (intensity) for coal transportation on railways and highways is 5.06 gce/t·km and 46.53 gce/t·km, respectively [62]. Therefore, the energy consumption of coal transportation is:

$$E_{transportation} = M \times e_{transportation} \tag{4}$$

$$e_{transportation} = \sum_{i}^{2} \theta_i \times D_i \times \ni_i \tag{5}$$

where θ_i represents the percentage of coal transported by transportation types, $\ni_i$ represents the average energy consumption by distance, D_i represents the average distance for transportation types, $i = 1$ or 2 representing railway or highway.

The calculated average $e_{transportation}$ value is 4.15 Kgce/t and is applied in this study (Table 1). Waterway coal transportation also exists in China, but most studies and the statistic yearbook suggest only considering railway and highway transportation of coal. This may lead to an underestimation of energy consumption and the related carbon emission by a magnitude of $\times$ 32–38 [63]. However, we still use this number due to the lack of historical data and references. We also assume the energy consumption factor in coal transportation would remain the same in the future.

3.1.3. Energy Consumption in Coal Power Generation

There are many kinds of coal electricity generator sets in the Chinese coal-fueled power industry, separated by the type (i.e., domestic, subcritical, supercritical, ultra-supercritical), capacity and the cooling method (air/water cooling). We specified nine typical sets and provided the unit consumption parameters for each energy source in Table 3. Thus, the total energy consumption in coal-based electricity can be calculated by:

$$E_{electricity\ production} = \sum_{i,j}(Q \times \phi_{ij})\, e_{ij} \tag{6}$$

where Q is the total production of coal electricity, ϕ_{ij} is the share of different typical sets used, e_{ij} is the unit energy consumption for each typical set. The converted standard coal is based on the coefficient in Table 1, with results shown in Table 3 [64].

The current energy consumption factor in electricity generation $e_{electricity\ production}$ is 351.2 gce/kW·h (Table 4). According to the China Energy Big Data Report (2021) [5], the average coal consumption of Chinese coal-fueled power generation also had a declining trend in the last decade. Consumption reduced from 385.4 gce/kW·h in 2010 to 362.1 gce/kW·h in 2015, with an average annual reduction of 1.2%. It slowly decreased to 351.2 gce/kW·h in 2020, with a slight yearly average decrease of 0.6%. A continuous declining trend is expected in the future. China's average energy consumption of coal-fueled units in 2060 could reach the currently most advanced state, the world's first 1350 MW ultra-supercritical secondary reheat coal-fueled generating unit with a comprehensive energy consumption level of 289 gce /kW·h [65]. Therefore, we assume the annual reduction rates in coal power energy consumption will remain at 0.5% till 2060. The specific predicted energy and coal consumption are shown in Table 2, assuming that the proportion of energy consumption in coal power production will not change.

Table 3. Energy consumption of power generation under different capacities of typical units.

Type	Capacity (MW)	Comprehensive (gce/kW·h)	Coal (gce/kW·h)	Coke (mgce/kW·h)	Oil (mgce/kW·h)	Gasoline (mgce/kW·h)	Kerosene (mgce/kW·h)	Diesel (mgce/kW·h)	Fuel Oil (mgce/kW·h)	Natural Gas (gce/kW·h)	Power (gce/kW·h)
Domestic	100	417.9	363.5	114.6	0.45	85.1	8.1	195.1	13.3	15.96	38.01
Domestic	125	342.7	298.1	94.0	0.37	69.8	6.7	160.0	94.0	13.1	31.2
Subcritical	300 Water-cooling	326.9	284.3	89.6	0.35	66.6	6.3	152.6	89.6	12.5	29.7
Supercritical	660 Water-cooling	314.2	273.3	86.2	0.34	64.0	6.1	146.7	86.2	12.0	28.6
Subcritical	600 Water-cooling	321.9	280.0	88.3	0.34	65.6	6.2	150.3	88.3	12.3	29.3
Ultra-supercritical	660 Water-cooling	294.2	255.9	80.7	0.31	59.9	5.7	137.4	80.7	11.2	26.8
Ultra-supercritical	600 Air-cooling	341.1	296.7	93.5	0.36	69.5	6.6	159.3	93.5	13.0	31.0
Subcritical	600 Air-cooling	337.7	293.7	92.6	0.36	68.8	6.6	157.7	92.6	12.9	30.7
Ultra-supercritical	1000 Water-cooling	303.4	263.9	83.2	0.32	61.8	5.9	141.7	83.2	11.6	27.6

Table 4. Unit energy consumption in coal power plants and the proportion of energy consumption.

Energy Category	Integrated	Coal	Coke	Oil	Gasoline	Kerosene	Diesel	Fuel Oil	Natural Gas	Power
Energy consumption in coal power generation (gce/kW·h)	351.2	305.5	0.1	0.0	0.1	0.0	0.2	0.0	13.4	32
Proportion of energy consumption (%)	100.00	86.98	0.03	0.00	0.02	0.00	0.05	0.00	3.82	9.10

3.2. Carbon Emission in Life Cycle Coal Electricity Production

The carbon emissions in the life cycle coal-based electricity production process can be divided into three categories: carbon emissions from energy consumption, gas emissions in mining (carbon emissions equivalent) and post-mine coal emissions [66]. Energy-related emissions are the emissions through the consumption of input energy. Gas emissions in mining mainly represent the CH_4 escaping from wells and open-pit mining before and during mining (converted to carbon emission). Carbon emissions from the post-mine activities are from open-pit mining, abandoned mines, and fugitive emissions during transportation, washing, refining, and storing raw coal. The function of total carbon emission is:

$$C_{\text{Life cycle}} = C_{\text{energy consumption}} + C_{\text{gas}} + C_{\text{postmine}} \tag{7}$$

The carbon emission related to energy consumption can be calculated by the total energy consumption times the carbon intensity coefficient widely used in the Chinese industry, $\delta = 2.66 \text{ kg CO}_2/\text{kgce}$ [67].

Therefore, we can obtain the carbon emission from energy input in the following equation:

$$C_{\text{energy consumption}} = \delta \times E_{\text{life cycle}} \tag{8}$$

The future carbon emission intensity trends follows the energy intensity trend (Table 5).

The carbon equivalent coal mine gas is one of the most important sources of carbon emissions in coal production [68]. The carbon emission in coal mining and washing mainly lies in the direct emptying after gas extraction [69]. In recent years, with the improvement in the utilization rate of gas extraction, the carbon emission intensity of gas per ton of coal has shown a trend of gradually declining, from 123.7 kgCO_2/t in 2010 to 67.6 kgCO_2/t in

2020, with an average annual reduction rate of 5.8% [60]. With carbon neutrality efforts, coal mine gas extraction will be strengthened, the gas utilization rate will be improved, and the carbon emission of the coal mine will be reduced [70,71]. Therefore, we assume the carbon emission intensity of coal gas emissions per ton of gas will continue to decrease at an average annual rate of 5.8%, as shown in Table 5.

Table 5. Prediction of the future trend in carbon emission intensity life cycle coal electricity production.

Year	2020	2025	2030	2035	2040	2045	2050	2055	2060
Carbon emission intensity of coal mining and washing/refinement (kg CO_2/t)	65.4	53.5	43.6	37.2	32.2	29.0	26.3	25.0	23.7
Carbon emission intensity of coal transportation (kg CO_2/t)	11.46	11.46	11.46	11.46	11.46	11.46	11.46	11.46	11.46
Carbon emission intensity of coal-fired power supply (g CO_2/kWh)	934.3	911.0	888.7	866.7	845.3	824.2	803.7	783.8	764.5
Gas carbon emission intensity per ton of coal (kgCO_2/t)	67.6	50.1	37.2	27.6	20.5	15.2	11.3	8.4	6.2
Carbon emission intensity of post-mine activities (kgCO_2/t)	18.0	16.5	15.2	13.9	12.8	11.7	10.8	9.9	9.1

Carbon emissions equivalent from post-mine activities refer to the amount of gas discharged during storage, transportation, and stacking of coal after it leaves the mine. In recent years, with the widespread application of mine gas drainage prevention and control technologies, the carbon emission intensity of coal-mining activities has also shown a gradual decrease trend from 21.5 kgCO_2/t in 2010 to 18.0 kgCO_2/t in 2020, with an average annual reduction rate of 1.7% [60]. Under carbon neutrality, it is increasingly urgent to strengthen the supervision and control of gas emissions from post-mine activities and reduce the problem of unorganized gas emissions from coal mines. Therefore, we assume that it will continue to decrease at an average annual rate of 1.7%, as shown in Table 5.

3.3. Water Consumption in the Life Cycle of Coal Electricity Production

The function below (Equation (9)) can express the water consumption in the life cycle of coal electricity production for water consumption in coal mining and washing/refining and water consumption in the cooling system of coal electricity production. We ignored the water consumption in coal transport since it accounts for less than 1% of the total water consumption [72].

$$W_{\text{life cycle}} = W_{\text{mine and wash/refine}} + W_{\text{electricity production}} \tag{9}$$

3.3.1. Water Consumption in Coal Mining and Washing/Refinement

The water consumption in coal mining and washing/refining is the summation of mining, washing, processing, and dressing. Water consumption varies greatly in different regions of China, which is mainly determined by the water resources, economic conditions, and mineral conditions in different regions, ranging from 0.34 to 3.5 m^3/t [73]. The average coal mining and washing/refining consumption is estimated as 3.1 m^3/t in 2020 [74]. A continuously declining water consumption trend in coal mining, washing and refining is predicted. It is assumed that after 2030, the water consumption will reach the level of water areas in western China [75]. The predicted water consumption is shown in Table 6.

Table 6. Prediction of the future trend in water consumption factors.

Year	2020	2025	2030	2035	2040	2045	2050	2055	2060
Water consumption in coal mining and washing	3.1	2.55	2	1.35	0.85	0.68	0.47	0.33	0.14
Water consumption in coal power generation (m^3/MWh)	1.34	1.22	1.10	1.00	0.90	0.80	0.75	0.71	0.68

3.3.2. Water Consumption in Coal Electricity Production

Generally, the cooling method in Chinese coal electricity generation can be divided into three categories: closed-loop cooling, open-loop cooling, and air cooling. We picked 12 typical cooling sets for a customized cooling portfolio. Thus, the total water consumption in coal electricity production can be calculated by:

$$W_{\text{electricity production}} = Q \times \sum_{i,j}^{12} w_{ij} \times \gamma_{ij} \tag{10}$$

where Q is the total electricity generated, γ_{ij} represents the percentage of the total electricity generated by each set, w_{ij} represents the water consumption intensity for each set (Table 7) [76].

Table 7. Water consumption coefficient in coal power generation.

Cooling Method	Capacity (MW)	Leading (m³/MWh)	Advanced (m³/MWh)	Base (m³/MWh)
Closed-loop cooling	<300	1.73	1.85	3.20
	300	1.60	1.70	2.70
	600	1.54	1.65	2.35
	1000	1.52	1.60	2.00
Open-loop cooling	<300	0.25	0.30	0.72
	300	0.22	0.28	0.49
	600	0.20	0.24	0.42
	1000	0.19	0.22	0.35
Air cooling	<300	0.30	0.32	0.80
	300	0.23	0.30	0.57
	600	0.22	0.27	0.49
	1000	0.21	0.24	0.42
Average		0.68	0.75	1.21

Due to technological development and changes in national water-saving requirements, the water consumption of coal-fueled power generation has a downward trend [77,78]. The literature [77,78] shows that, the average water consumption factor is 1.34 m³/MWh. Therefore, we assumed that the average water consumption of coal power units would reach 1.21 (m³/MWh) in 2025 and 1.10 (m³/MWh) in 2030 at a rate of 0.024 (m³/MWh) declining per year before 2030. The magnitude of such a declining trend would shrink, with a 0.02 (m³/MWh) reduction per year from 2030 to 2045, and 0.01 (m³/MWh) per year from 2045 to 2050. In 2050, the average water consumption of coal power units will reach the advanced value of 0.75 (m³/MWh). In 2060, coal power units' average water consumption could have reached the advanced value of 0.68 (m³/MWh). The prediction of China's average water consumption factors for coal power generation is shown in Table 6.

3.4. Profits in the Life Cycle of Coal Production

The economic profits of coal electricity production can be calculated through the difference in income and the cost of energy consumed in the life cycle of coal power production, including coal and other energy inputs.

$$\text{Profits}_{\text{life cycle}} = \text{price}_{\text{electricity}} \times Q - Q \times \text{Cost}_{\text{life cycle}} \tag{11}$$

$$\text{Cost}_{\text{life cycle}} = \text{Cost}_{\text{electricity production}} \tag{12}$$

Since the cost of coal mining, washing, and refining are included in the energy price, we only calculate the profits in coal power plants. The energy price for energy inputs and the calculated $\text{Cost}_{\text{electricity production}}$ is shown in Table 8. It is hard to predict the changing trend in raw material and currency inflation. Therefore, we assume the price of all the energy inputs remained unchanged through the years. However, the cost is declining due to the reduction in unit energy consumption. The predicted cost and profits are shown in Table 9.

Table 8. Different energy consumption costs per unit of power generation in coal power plants.

Energy Category	Comprehensive	Coal	Coke	Oil	Gasoline	Kerosene	Diesel	Fuel Oil	Natural Gas	Power
Energy prices (Yuanton) or (Yuan/m^3) or (Yuan/kW·h)		600	2600	4800	5700	3600	4800	3600	3.40	0.45
Energy consumption of coal power generation (gce/kW·h)	351.2	305.5	0.1	0.0	0.1	0.0	0.2	0.0	13.4	32.0
Cost of coal power generation (Yuan/kW·h)	0.397	0.26	0.03	0.00	0.00	0.00	0.00	0.00	0.03	0.08

Table 9. Prediction of energy consumption cost in the future.

Year	2020	2025	2030	2035	2040	2045	2050	2055	2060
Cost of coal power generation (Yuan/kW·h)	0.40	0.39	0.38	0.37	0.36	0.35	0.34	0.33	0.32

3.5. The CCUS Impacts

3.5.1. CCUS Impact on Energy and Water Consumption

The installation of CCUS equipment in coal-fired units will change the structure of the original power generation equipment, cause partial energy loss, increase energy consumption [79], significantly reduce carbon emissions and increase the water consumption of the power generation system. This will also cause water consumption across different types of units to increase by 31–91% [27]. The implementation of CCUS projects consumes water and energy during the process of capturing and storing carbon. The additional energy consumption is around 68.2–85.4 (kgce/t) [80] and the water consumption is around 20 to 40 (m3/t CO_2). There is no clear evidence for the development level of CCUS technologies to improve efficiency. We assume that the intensity of water and energy consumption by the CCUS process will decrease with an average annual rate of 5%, and the results are shown in Table 10.

Table 10. Prediction of the increased water and energy consumption with CCUS.

Year	2020	2025	2030	2035	2040	2045	2050	2055	2060
Increase in energy consumption by adding CCUS (kgce/ton)	76.8	59.4	46	35.6	27.5	21.3	16.5	12.8	9.9
Water consumption increase intensity (m^3/ton)	30	23.2	18	13.9	10.8	8.3	6.4	5	3.9

3.5.2. CCUS Impacts on Economic Profits

The national carbon market was officially established in December 2017. The first batch of about 1700 power generation enterprises was selected to be involved in the carbon emissions trade [65,81]. In 2020, China's average carbon trading price was 28.6 Yuan/ton. There are various predictions on the carbon price, but the price is expected to grow in the future, and the drivers include inflation and, most likely, carbon policy [82]. We excluded the inflation impact under the assumption of no change in energy and electricity prices. An official survey has shown that the carbon price in 2021–2022 is around 50 yuan/ton and is expected to reach 87 and 139 Yuan/ton in 2025 and 2030, respectively [82]. Based on the 25% yearly increase rate between 2020 and 2025 combined with the 10% annual increase rate between 2025 and 2030, we assumed that the increasing annual rate is 9% in 2030–2040, 8% in 2040–2050, and 7% in 2050–2060. The predictions are shown in Table 11. The carbon price would reach 710 Yuan/ton and 1397 yuan/ton in 2050 and 2060. This is lower than the prediction of the World Energy Outlook [83], where the carbon price would reach 250 $ton in advanced economies and 200 $/ton in China in 2050 under a zero-carbon emission world scenario. However, our estimation excludes the impact of inflation, and China's carbon-neutral goal is 2060 rather than 2050. Therefore, based on our prediction, the 2060 carbon price of 1397 yuan/ton or 216 $/ton fits into the 200–250 $/ton range and is believed to be more reasonable for China.

Table 11. Prediction on carbon price in 2020–2060.

Year	2020	2025	2030	2035	2040	2045	2050	2055	2060
Carbon Price (Yuan/ton)	28.6	87	139	214	329	483	710	996	1397

4. Scenarios

4.1. Baseline Scenario

4.1.1. Predicted Electricity Demand in China

The rapid modernization process and socio-economic development in China, combined with the changing consumption pattern, contribute to the fast-growing national electricity demand. The literature has forecasted future electricity demand in China using various methods and models based on key indicators [84]. We selected a reasonable prediction [85], which calculated the average of seven existing models and applied it to the baseline scenario input as shown in Table 12.

Table 12. Prediction of future electricity demand in China.

Year	2020	2025	2030	2035	2040	2045	2050	2055	2060
Electricity demand (trillion kWh)	7.4	8.8	10.2	11.3	12.2	13.1	13.6	14.1	14.4

4.1.2. Predicted Change in the Coal Electricity Share

The proportion of coal-fueled power generation in China dropped from 67.9% to 60.8% in 2015–2020. The coal power share in China would continue to drop, but the declining magnitude is hard to tell. For consistency, we further applied the companion prediction with electricity demand in Xie's study [85] as a baseline scenario input for coal electricity share (Table 13).

Table 13. The baseline scenario for coal electricity share in the future.

Year	2020	2025	2030	2035	2040	2045	2050	2055	2060
Coal electricity share (%)	60.8	49.9	40.9	33.5	27.5	22.5	18.5	15.2	12.4

4.1.3. Predicted Change in the CCUS Implementation in Coal Electricity Production

According to the statistics collected by the Ministry of Science and Technology for CCUS demonstration projects nationwide, since the first CCUS demonstration project in China was put into operation in Shanxi in 2004, there were 38 CCUS projects in operation before 2020, with a total capacity of 5 Mt/year [80]. Under the carbon-neutral target, the overall emission reduction demand of CCUS in China is ~20–408 Mt CO_2 in 2030, ~600–1450 Mt CO_2 in 2050, and ~1–1.82 Gt CO_2 in 2060 [86]. According to the CCUS special report of the International Energy Agency's power operation and maintenance platform, CCUS emission reduction capacity is expected to grow rapidly [87]. The capture scale of CCUS in China's thermal power plants is about 190 Mt CO_2/year by 2030; about 770 Mt

CO_2/year by 2050; and exceeds 1.2 Gt CO_2/year by 2070. Based on the prediction, we deliver a new forecast of carbon mitigation from CCUS in Table 14 and take it as the input for the baseline portfolio.

Table 14. Coal power and national CCUS-related emission reduction demand for baseline scenario, 2020~2060.

Year	2020	2025	2030	2035	2040	2045	2050	2055	2060
From coal power (Mt/Year)	3	20	190	370	520	655	775	880	985
Total carbon mitigation (Mt/Year)	5	9–30	20–408	119–850	370–1300	500–1350	600–1450	800–1650	1000–1820

4.2. Alternative Portfolios and the Difference with the Baseline Scenario

4.2.1. Share of Coal Electricity

The baseline scenario on the contribution of shares changed from coal-based electricity production was based on Xie's study [85]. The prediction is based on an assumed scheme of 18% of change for every five years. The proportion of coal-fueled power generation in China dropped from 67.9% to 60.8% in 2015–2020 and the enhanced efforts of energy transition after 2020. Additionally, there was a 23% decrease in coal-powered electricity in the electricity mix every five years. However, there is a chance that the green energy transition will not go well. In the context of the global removal of coal power, the reduction rate should be faster than that in 2015–2020. Therefore, we assume the "slow" declining rate of coal electricity share in China to be 13% every five years. Three portfolios representing three levels of carbon mitigation were applied in the ATPCC and named "Baseline," "Slow," and "Radical" (Table 15).

Table 15. Scenarios for coal electricity share.

Year	2020	2025	2030	2035	2040	2045	2050	2055	2060
Coal electricity share (Baseline)	60.8	49.9	40.9	33.5	27.5	22.5	18.5	15.2	12.4
Coal electricity share (Slow)	60.8	52.9	46.0	40.0	34.8	30.3	26.4	22.9	20.0
Coal electricity share (Radical)	60.8	46.8	36.0	27.8	21.4	16.5	12.7	9.8	7.5

4.2.2. Carbon Emission Mitigation from CCUS

Based on the baseline scenario of CCUS capture capacity, we further developed the inputs for carbon capture through CCUS for "Slow" and "Radical" as 80% and 120%, respectively, of the baseline CCUS carbon emission capacity. When the two "Slow" scenarios are connected, they reflect a future that is less focused on carbon reduction. Alternatively, two "Radical" scenarios represent the other way around. Thus, the prediction reflects the different levels of carbon-neutral policy constraints. The CCUS capture capacity per scenario is shown in Table 16.

Table 16. Scenario inputs for CCUS carbon capture in 2020–2060.

Year	2020	2025	2030	2035	2040	2045	2050	2055	2060
CCUS capture in coal power plants (Mt/Year) (Baseline)	3	20	190	370	520	655	775	880	985
CCUS capture in coal power plants (Mt/Year) (Slow)	3	16	152	296	416	524	620	704	788
CCUS capture in coal power plants (Mt/Year) (Radical)	3	24	228	444	624	786	930	1056	1182

5. Results

5.1. Scenario Outputs

5.1.1. Baseline Scenario Outputs

The baseline portfolios by 2060 are expected to decrease coal-powered electricity share to 12.4% and falls within the interval between n 20% (claimed by the conservative studies) and 10% (suggested by radical studies). In addition, around one billion tons of carbon could be captured by CCUS applications. Modeled outputs in Table 17 indicate that, in 2060, the two carbon mitigation efforts would reduce total carbon emissions to 0.4 billion tons, which is very close to the carbon-neutral goal. The total energy consumption has a trend of first increasing and then decreasing with a peak year in 2040–2045. Extra energy consumption from the CCUS application would overtake the energy consumption from the life cycle of coal production between 2025 and 2030. The total water consumption has an overall decreasing trend due to technological innovations and reduced coal production. Extra water consumption from CCUS overtakes the water consumption from the life cycle of coal production between 2035–2040. The total revenue in 2060 is the highest among all three scenarios and is 6.4 times the revenue in 2020 for China's coal industry, where 86% of the total revenue comes from the carbon trade through the CCUS implementation. The carbon trade becomes the dominant economic profit in 2040–2045.

Table 17. ATPCC Output of baseline portfolio for China's coal power industry.

Year	2020	2025	2030	2035	2040	2045	2050	2055	2060
Total energy consumption (Gtce)	2.4	3.1	10.4	14.6	15.5	15.0	13.6	12.0	10.3
Energy consumption from CCUS (Gtce)	0.2	1.2	8.7	13.2	14.3	14.0	12.8	11.3	9.8
Total water consumption (billion m^3)	12.7	10.5	11.4	11.0	9.7	8.6	7.3	6.2	5.1
Water consumption from CCUS (billion m^3)	0.1	0.5	3.4	5.1	5.6	5.4	5.0	4.4	3.8
Total carbon emission (Gt)	4.8	4.2	3.7	3.0	2.4	1.8	1.3	0.8	0.4
Revenue from CCUS carbon trade (billion Yuan)	0.1	1.7	26.4	79.2	171.1	316.4	550.3	876.5	1376.0
Total revenue (billion Yuan)	251.2	277.7	328.1	388.4	475.6	610.4	823.1	1127.0	1599.4

5.1.2. Slow Scenario Outputs

In the "Slow" portfolio for coal power industry development, 20% of the coal electricity in the electricity mix will remain in 2060, and around 0.8 (Gt) of carbon will be captured annually through CCUS facilities in coal power plants. The total energy consumption will increase and then reduce, with the peak in 2040–2045 as in the baseline scenario (Table 18). The overall carbon emission will be around 1.5 billion tons in 2060, three times the emissions in the baseline portfolio. The total consumed energy would be 8.7 (Gtce), much less than the baseline portfolio. The extra water consumption of CCUS would overtake the water consumption in the life cycle of coal power production in 2040–2045, a different time from the baseline scenario. The total revenue in China's 2060 coal industry will increase by 481% from 2020. Economic benefits from carbon capture would overtake the electricity in 2045–2050, close to 2050 and fall behind the baseline scenario. The CCUS profits would reach 75% of the total revenue in 2060.

Table 18. ATPCC outputs of "Slow" portfolio for China's coal power industry.

Year	2020	2025	2030	2035	2040	2045	2050	2055	2060
Total energy consumption (Gtce)	2.4	2.9	8.9	12.3	13.0	12.6	11.5	10.1	8.7
Energy consumption from CCUS (Gtce)	0.2	1.0	7.0	10.5	11.4	11.2	10.2	9.0	7.8
Total water consumption (billion m^3)	12.7	11.0	11.7	11.1	9.7	8.5	7.3	6.2	5.2
Water consumption from CCUS (billion m^3)	0.1	0.4	2.7	4.1	4.5	4.4	4.0	3.5	3.1
Total carbon emission (Gt)	4.8	4.5	4.2	3.8	3.3	2.8	2.3	1.9	1.5
Revenue from CCUS carbon trade (billion Yuan)	0.1	1.4	21.1	63.3	136.9	253.1	440.2	701.2	1100.8
Total revenue (billion Yuan)	251.2	293.9	360.5	432.5	522.3	649.0	829.6	1078.6	1461.1

5.1.3. Radical Scenario Outputs

The radical scenario has the least coal electricity share, only 7.5% in 2060, and with the greatest CCUS capacity, 1.2 (Gt) of carbon could be captured through the CCUS application. However, CCUS capacity cannot be fully utilized by 2060 because the zero-carbon emission in coal power plants could have been achieved between 2055 and 2060 (Table 19). We showed that, through the energy transition and the CCUS technology utilization, carbon-neutral goals could be achieved even in coal power plants. The total carbon emission would be very close to zero and thus would lead to the accomplishment of the carbon-neutral national goal. The output of this portfolio has a minor energy consumption but the same pattern through the years as other portfolios. It also accounts for the least overall water consumption. The percentage of water consumption from CCUS would take more than half of the total water consumption between 2030 and 2035. This portfolio will create the least revenue of the three portfolios, but this is the fastest for carbon trade to dominate the incomes of the coal industry, between 2040 and 2045 and close to 2040. The total revenue in 2060 is 5.2 times the revenue in 2020, and the carbon trade would account for 90% of the total economic profits of China's coal industry.

Table 19. ATPCC outputs of "Radical" portfolio for China's coal power industry.

Year	2020	2025	2030	2035	2040	2045	2050	2055	2060
Total energy consumption (Gtce)	2.4	3.2	12.0	17.0	18.1	17.5	15.9	14.0	8.6
Enrgy consumption from CCUS (Gtce)	0.2	1.4	10.5	15.8	17.2	16.7	15.4	13.5	8.2
Total water consumption (billion m^3)	12.7	10.0	11.1	11.0	10.0	8.8	7.6	6.4	4.0
Water consumption from CCUS (billion m^3)	0.1	0.6	4.1	6.2	6.7	6.5	6.0	5.3	3.2
Total carbon emission (Gt)	4.8	3.9	3.2	2.4	1.7	1.0	0.5	0.1	0.0
Revenue from CCUS carbon trade (billion Yuan)	0.1	2.1	31.7	95.0	205.3	379.6	660.3	1051.8	1159.5
Total revenue (billion Yuan)	251.2	260.9	297.3	351.6	442.3	595.2	847.6	1213.3	1294.6

5.2. Tradeoffs between Scenarios

Carbon emission mitigation is the most important goal in the carbon-neutral context. The coupled policy implementation of energy transition and CCUS applications has synergy to improve carbon mitigation. Unintended, complex consequences of these policy implementations also lead to tradeoffs between carbon emission reduction and water/energy consumption. The radical portfolio has the potential to achieve zero-carbon emissions in coal power plants, but with significantly higher energy and water consumption from CCUS application.

It is noted that the radical and baseline portfolios could achieve close to zero-carbon emissions in China's coal industry. Nevertheless, compared to the 2055 and 2060 states in a radical portfolio and the outputs from the baseline scenario, we also found the potential of reducing a large amount of extra water and energy consumption from CCUS with a tradeoff for slightly more carbon emissions. The highest revenues generated from the baseline scenario proved that the relationship between economic profits from China's coal industry and the magnitude of carbon emission reduction is not simply a negative correlation. A win–win situation can be found with high profits and high carbon emission mitigation. Therefore, the optimized equilibrium between two policies regarding the CCUS application and reduced share of coal-generated electricity is never simple and is crucial for the sustainability in the future of China's coal power industry concerning carbon emission mitigation, water/energy consumption, and financial profits.

The baseline and the slow scenario almost account for the same water consumption in 2060. However, the slow scenario has a higher coal share and lower CCUS applications. With the same water constraints, energy consumption in the baseline scenario is much larger than in the slow scenario, but more economic profits would be gained. Therefore, choosing between revenue and energy inputs is crucial for energy-stressed areas. Tough choices must also be made between radical and slow portfolios in water-stressed areas. With relatively similar energy consumption, the radical portfolio accounts for less water consumption and carbon emissions, but lower economic gains.

The radical portfolio has the most economic revenue and accounts for the lowest carbon emissions. Besides that, it illustrates a new way for China's coal power industry to increase revenues. The carbon trading profits only accounted for less than 0.1% of the total revenue in China's coal power industry in 2020. However, it would reach more than 50% before 2050 in each scenario. The carbon-trading profits would account for about

90% of the total revenue in the radical portfolio. This is an inspiring result of China's coal power industry development. One of the critical reasons is that although it is economically beneficial to produce coal electricity, many companies in China's coal industry are in a deficit situation. The reason is that we did not include coal power companies' salaries and administration costs. However, the carbon market and the CCUS implementation could provide a new pathway toward financial profits in this study. With the rapid growth in the amount of captured carbon and the increasing carbon price, the coal industry has the potential to transform the primary revenue source from electricity sales to captured carbon trade.

6. Discussion

6.1. Carbon Mitigation Policies and CEW Nexus in China

Findings in the tradeoff analysis between the portfolios' outputs have important implications for China's carbon neutrality goals. Our results proved the feasibility of carbon neutrality by 2060 in China if there is a zero-carbon-emission possibility in China's coal industry, which has long been considered the most significant contributor to China's carbon emissions [31]. Most of the mitigated carbon emissions come from the decreased contribution of coal-based electricity production to the electricity mix. The second is China's green energy transition policy that converts thermal-electricity-dominated energy systems to renewable-energy-dominated [88]. However, the transformation process may lead to new issues and thus weaken the carbon mitigation efforts. Hydropower accounts for China's largest share of renewable energy and has long been treated as green energy. However, recent studies contested its GHG emissions status and proved it can no longer be considered as a comparatively low-emissions energy source in the Mekong River Basin [89]. The same pattern also was observed in bioenergy efforts. Compared with thermal energy sources, biofuels may account for more carbon emissions [90]. The energy transition could be a solution toward a sustainable carbon-mitigated power system in China, but the featured development energy source must be considered cautiously.

The regulation on reducing coal shares in China's electricity mix is also of concern. The rapid growth of green energy could put the country's power system in a vulnerable situation. The current dominant renewable energy sources, including hydropower, wind parks, and photovoltaic and concentrated solar power plants (PV and CSP) are mostly naturally based. In particular, PV, CSP, and wind power rely on local weather conditions and possess intense space and time fluctuations. Hybrid power generation, including coupled renewable energy sources, is a proper solution but coal energy still needs to sustain the security of an energy system [85].

Coal transportation is a major carbon emission source in the coal power industry. It is due to the long distance between places of coal production and electricity production. Most coal mines are in western China and the power plants are mostly located in eastern China. The reduced coal electricity shares triggered by the green energy transition could relieve the emission stress from coal transportation. It helped water stress related to coal mining and washing where western China has limited water resources. The life cycle of coal electricity accounts for large amounts of water consumption and its water intensity is higher than most other energy resources, except for nuclear power. In this study, we demonstrated that the water consumption in the coal power industry could not compete with CCUS water consumption. This is mainly because we only considered the water consumed by evaporation. A large amount of water withdrawal and use in the coal industry is dissipated back to the environment as polluted water [72]. A "greywater footprint" study is needed for further assessment.

CCUS application in a coal-fueled power plant has been proven crucial and the tradeoffs on extra energy and water consumption should be considered significant even when they are coupled with the implementation of the energy transition. The CCUS had great potential to reduce the extra water consumption by replacing fracturing water with the captured CO_2 in the CCUS application [91]. The energy transition with reduced shares of coal-powered electricity leads to a decreased financial revenue from the coal industry, thus reducing the energy sector's profits nationally. Despite the high cost of electricity in renewable energy sources compared to coal, re-arranging and treating abandoned coal power facilities lead to more investments. However, the revenues from the captured carbon through the CCUS application can fill this gap and have the potential to make more profits. A win–win future can be achieved through the well-organized coupling of CCUS applications and coal power reduction through the economic and environmental energy transition.

6.2. Limitations of the Study

The ATPCC tool was developed with many assumptions, similar to other predictive models in the literature. Some assumptions are due to the lack of data and calculation barriers. We also excluded some parts within the life cycle of coal electricity production. For instance, we only calculated the domestic transportation of imported coal. As mentioned earlier, we only considered the water consumption but not the water withdrawal. We did not track the waste and polluted water treatment in this study. The indirect energy, water consumption, carbon emission, and economic cost were also not considered, including the construction of the power plant, mines, and the CCUS facilities. An enhanced and more comprehensive version of ATPCC could be conducted when more data become available.

Furthermore, the parameters in ATPCC are all in the form of a single value, rather than a range of possibilities. A fuzzy random number method to replace old parameters could be applied in ATPCC if users wish to include uncertainty in the tool. The fuzzy set is commonly used to generate a regular fuzzy number, and it does not refer to one single value but a connected set of possible values. Each potential value has its weight between 0 and 1 [92].

7. Conclusions

The main purpose of this research is to deliver a comprehensive assessment tool to analyze the sustainability of China's coal power industry from a CEW nexus perspective under carbon-neutral mission policy portfolios. The ATPCC developed in this study is a scenario-based life cycle assessment tool for coal-powered electricity production under the two major carbon mitigation policy implementations—energy transition and CCUS application. To the best of our knowledge, the ATPCC is the first study that applied the CEW system for coal-powered electricity production to evaluate the interrelationships of carbon and energy with financial profits. This is also the first study that integrates energy transition and CCUS impacts on the life cycle of coal-powered electricity production in China.

Furthermore, to find sustainable development pathways for China's coal electricity system before 2060 that meet carbon neutrality goals, three portfolios represented different efforts to reduce contributions of coal-generated electricity triggered by the energy transition and carbon captured by CCUS applications.

This study proved the possibility of zero emissions in China's coal electricity industry within the radical portfolio. We also showed that the extra consumption of water and energy from CCUS could create resource stress and would be the dominant consumption source for energy and water in the future. However, economic revenue lost from the reduced production in the coal industry could be compensated from profits in the carbon trade possible from CCUS applications. Revenues from the carbon trade would be the dominant source of profit for China's coal-based power industry in the future.

Comparing the ATPCC outputs across the three portfolios shows that the baseline scenario has the highest energy consumption and economic profits but moderate water consumption and carbon emission. The slow energy transition scenario has the highest water consumption and carbon emissions, but moderate revenues and energy consumption. The radical energy transition scenario has the lowest carbon emissions, water, energy consumption, and financial profit. Therefore, the radical portfolio is the most sustainable strategy from the CEW nexus perspective for China's coal power industry: reduce more coal share and apply as many CCUS facilities as possible. The most profitable scenario is the baseline scenario, balancingthe energy transition and CCUS application.

The major findings from our scenario analysis can be concluded in three aspects:

1. The carbon-neutral state in China's coal-powered electricity industry is achievable with significant efforts on energy transition and CCUS applications.
2. Different implementation levels for the coupling carbon mitigation policies with carbon-neutral synergy would lead to tradeoffs in the CEW nexus and economic profits in China's coal power industry.
3. The tradeoffs would impact the sustainability of China's coal power system development. The tradeoffs are not simply an inverse correlation between one and the other. Sustainability in coal electricity generation in China's future can be achieved by optimizing the energy transition and CCUS applications in different ways to balance carbon emissions, water consumption, energy consumption, and economic profits.

The ATPCC is a straightforward and flexible tool for policymakers to assess CEW tradeoffs. It provides a simple approach to analyzing complex issues with almost all parameters. Policymakers can evaluate the tradeoffs using the tool with only two numerical inputs from different portfolios. While the empirical parameters are based on China's coal industry, they can also be modified for other countries.

Author Contributions: Conceptualization, Y.X., S.R. and J.Q.; methodology, Y.X., J.Q. and G.S.; investigation, Y.X., X.J. and R.Z.; resources, S.R. and X.J.; data curation, X.J. and S.R.; writing—original draft preparation, Y.X., R.Z. and S.R.; writing—review and editing, J.Q., G.S. and X.J.; visualization, Y.X., X.J. and R.Z.; supervision, S.R.; project administration, S.R. and X.J.; funding acquisition, S.R. All authors have read and agreed to the published version of the manuscript.

Funding: This research was funded by the Strategic Consulting Project of Chinese Academy of Engineering (2022-XZ-32).

Institutional Review Board Statement: Not applicable.

Informed Consent Statement: Not applicable.

Data Availability Statement: Not applicable.

Conflicts of Interest: The authors declare no conflict of interest.

References

1. Masson-Delmotte, V.; Zhai, P.; Pirani, A.; Connors, S.L.; Péan, C.; Berger, S.; Caud, N.; Chen, Y.; Goldfarb, L.; Gomis, M.I.; et al. (Eds.) *Climate Change 2021: The Physical Science Basis. Contribution of Working Group I to the Sixth Assessment Report of the Intergovernmental Panel on Climate Change*; Cambridge University Press: London, UK, 2021; *in press*.
2. Salvia, M.; Reckien, D.; Pietrapertosa, F.; Eckersley, P.; Spyridaki, N.A.; Krook-Riekkola, A.; Heidrich, O. Will climate mitigation ambitions lead to carbon neutrality? An analysis of the local-level plans of 327 cities in the EU. *Renew. Sustain. Energy Rev.* **2021**, *135*, 110253. [CrossRef]
3. Altarhouni, A.; Danju, D.; Samour, A. Insurance Market Development, Energy Consumption, and Turkey's CO_2 Emissions. New Perspectives from a Bootstrap ARDL Test. *Energies* **2021**, *14*, 7830. [CrossRef]
4. Zhao, Y.; Su, Q.; Li, B.; Zhang, Y.; Wang, X.; Zhao, H.; Guo, S. Have those countries declaring "zero carbon" or "carbon neutral" climate goals achieved carbon emissions-economic growth decoupling? *J. Clean. Prod.* **2022**, *363*, 132450. [CrossRef]
5. CLP Media Energy Information Research Center. *China Energy Big Data Report (2021)*; CLP Media Energy Information Research Center: Beijing, China, 2021. (In Chinese)
6. Yan, J.; Yang, Y.; Elia Campana, P.; He, J. City-level analysis of subsidy-free solar photovoltaic electricity price, profits and grid parity in China. *Nat. Energy* **2019**, *4*, 709–717. [CrossRef]
7. Meha, D.; Pfeifer, A.; Sahiti, N.; Schneider, D.R.; Yan, J. Sustainable transition pathways with high penetration of variable renewable energy in the coal-based energy systems. *Appl. Energy* **2021**, *304*, 117865. [CrossRef]
8. Chen, J.; Liu, G.; Kang, Y.; Wu, B.; Sun, R.; Zhou, C.; Wu, D. Coal utilization in China: Environmental impacts and human health. *Environ. Geochem. Health* **2014**, *36*, 735–753. [CrossRef] [PubMed]
9. Cui, X.; Hong, J.; Gao, M. Environmental impact assessment of three coal-based electricity generation scenarios in China. *Energy* **2012**, *45*, 952–959. [CrossRef]

10. Vujić, J.; Antić, D.P.; Vukmirović, Z. Environmental impact and cost analysis of coal versus nuclear power: The US case. *Energy* **2012**, *45*, 31–42. [CrossRef]
11. Liu, J.; Wang, K.; Zou, J.; Kong, Y. The implications of coal consumption in the power sector for China's CO_2 peaking target. *Appl. Energy* **2019**, *253*, 113518. [CrossRef]
12. Liu, Z.; Guan, D.; Wei, W.; Davis, S.J.; Ciais, P.; Bai, J.; Peng, S.; Zhang, Q.; Hubacek, K.; Marland, G.; et al. Reduced carbon emission estimates from fossil fuel combustion and cement production in China. *Nature* **2015**, *524*, 335–338. [CrossRef] [PubMed]
13. Cheng, Y.P.; Wang, L.; Zhang, X.L. Environmental impact of coal mine methane emissions and responding strategies in China. *Int. J. Greenh. Gas Control* **2011**, *5*, 157–166. [CrossRef]
14. Wang, J.; Wang, R.; Zhu, Y.; Li, J. Life cycle assessment and environmental cost accounting of coal-fired power generation in China. *Energy Policy* **2018**, *115*, 374–384. [CrossRef]
15. Han, X.; Chen, N.; Yan, J.; Liu, J.; Liu, M.; Karellas, S. Thermodynamic analysis and life cycle assessment of supercritical pulverized coal-fired power plant integrated with No. 0 feedwater pre-heater under partial loads. *J. Clean. Prod.* **2019**, *233*, 1106–1122. [CrossRef]
16. Dong, H.; Liu, Y.; Zhao, Z.; Tan, X.; Managi, S. Carbon neutrality commitment for China: From vision to action. *Sustain. Sci.* **2022**, *accepted/in press.* [CrossRef]
17. Zhang, S.; Chen, W. Assessing the energy transition in China towards carbon neutrality with a probabilistic framework. *Nat. Commun.* **2022**, *13*, 87. [CrossRef]
18. Liu, S.; Peng, G.; Sun, C.; Balezentis, T.; Guo, A. Comparison of improving energy use and mitigating pollutant emissions from industrial and non-industrial activities: Evidence from a variable-specific productivity analysis framework. *Sci. Total Environ.* **2022**, *806*, 151279. [CrossRef]
19. Jiang, K.; Ashworth, P.; Zhang, S.; Liang, X.; Sun, Y.; Angus, D. China's carbon capture, utilization and storage (CCUS) policy: A critical review. *Renew. Sustain. Energy Rev.* **2020**, *119*, 109601. [CrossRef]
20. Yao, X.; Zhong, P.; Zhang, X.; Zhu, L. Business model design for the carbon capture utilization and storage (CCUS) project in China. *Energy Policy* **2018**, *121*, 519–533. [CrossRef]
21. Metz, B.; Davidson, O.; De Coninck, H.C.; Loos, M.; Meyer, L. *IPCC Special Report on Carbon Dioxide Capture and Storage;* Cambridge University Press: Cambridge, UK, 2005.
22. Chu, S. Carbon capture and sequestration. *Science* **2009**, *325*, 1599. [CrossRef]
23. Hasan, M.F.; First, E.L.; Boukouvala, F.; Floudas, C.A. A multi-scale framework for CO_2 capture, utilization, and sequestration: CCUS and CCU. *Comput. Chem. Eng.* **2015**, *81*, 2–21. [CrossRef]
24. Xie, Y.; Hou, Z.; Liu, H.; Cao, C.; Qi, J. The sustainability assessment of CO_2 capture, utilization and storage (CCUS) and the conversion of cropland to forestland program (CCFP) in the Water–Energy–Food (WEF) framework towards China's carbon neutrality by 2060. *Environ. Earth Sci.* **2021**, *80*, 1–17. [CrossRef]
25. Liu, H.J.; Were, P.; Li, Q.; Hou, Z. Worldwide Status of CCUS Technologies and Their Development and Challenges in China. *Geofluids* **2017**, *2017*, 6126505. [CrossRef]
26. Shirkey, G.; Belongeay, M.; Wu, S.; Ma, X.; Tavakol, H.; Anctil, A.; Marquette-Pyatt, S.; Stewart, R.; Sinha, P.; Corkish, R.; et al. An environmental and societal analysis of the US electrical energy industry based on the water–energy nexus. *Energies* **2021**, *14*, 2633. [CrossRef]
27. Li, Q.; Wei, Y.N.; Chen, Z.A. Water-CCUS nexus: Challenges and opportunities of China's coal chemical industry. *Clean Technol. Environ. Policy* **2016**, *18*, 775–786. [CrossRef]
28. Zhai, H.; Rubina, E.S. Carbon capture effects on water use at pulverized coal power plants. *Energy Procedia* **2011**, *4*, 2238–2244. [CrossRef]
29. Newmark, R.L.; Friedmann, S.J.; Carroll, S.A. Water challenges for geologic carbon capture and sequestration. *Environ. Manag.* **2010**, *45*, 651–661. [CrossRef]
30. Yu, L.; Liu, S.; Wang, F.; Liu, Y.; Li, M.; Wang, Q.; Dong, S.; Zhao, W.; Tran, L.-S.P.; Sun, Y.; et al. Effects of agricultural activities on energy-carbon-water nexus of the Qinghai-Tibet Plateau. *J. Clean. Prod.* **2022**, *331*, 129995. [CrossRef]
31. Li, H.; Jiang, H.D.; Dong, K.Y.; Wei, Y.M.; Liao, H. A comparative analysis of the life cycle environmental emissions from wind and coal power: Evidence from China. *J. Clean. Prod.* **2020**, *248*, 119192. [CrossRef]
32. Liang, M.S.; Huang, G.H.; Chen, J.P.; Li, Y.P. Energy-water-carbon nexus system planning: A case study of Yangtze River Delta urban agglomeration, China. *Appl. Energy* **2022**, *308*, 118144. [CrossRef]
33. Lim, X.Y.; Foo, D.C.; Tan, R.R. Pinch analysis for the planning of power generation sector in the United Arab Emirates: A climate-energy-water nexus study. *J. Clean. Prod.* **2018**, *180*, 11–19. [CrossRef]
34. DeNooyer, T.A.; Peschel, J.M.; Zhang, Z.; Stillwell, A.S. Integrating water resources and power generation: The energy–water nexus in Illinois. *Appl. Energy* **2016**, *162*, 363–371. [CrossRef]
35. Chhipi-Shrestha, G.; Hewage, K.; Sadiq, R. Water–energy–carbon nexus modeling for urban water systems: System dynamics approach. *J. Water Resour. Plan. Manag.* **2017**, *143*, 04017016. [CrossRef]
36. Lee, M.; Keller, A.A.; Chiang, P.C.; Den, W.; Wang, H.; Hou, C.H.; Wu, J.; Wang, X.; Yan, J. Water-energy nexus for urban water systems: A comparative review on energy intensity and environmental impacts in relation to global water risks. *Appl. Energy* **2017**, *205*, 589–601. [CrossRef]

37. Trubetskaya, A.; Horan, W.; Conheady, P.; Stockil, K.; Moore, S. A methodology for industrial water footprint assessment using energy–water–carbon nexus. *Processes* **2021**, *9*, 393. [CrossRef]
38. Liu, Y.; Tan, Q.; Han, J.; Guo, M. Energy-Water-Carbon Nexus Optimization for the Path of Achieving Carbon Emission Peak in China considering Multiple Uncertainties: A Case Study in Inner Mongolia. *Energies* **2021**, *14*, 1067. [CrossRef]
39. Zhu, Y.; Zhao, Y.; Li, H.; Wang, L.; Li, L.; Jiang, S. Quantitative analysis of the water-energy-climate nexus in Shanxi Province, China. *Energy Procedia* **2017**, *142*, 2341–2347. [CrossRef]
40. Wang, X.; Zhang, Q.; Xu, L.; Tong, Y.; Jia, X.; Tian, H. Water-energy-carbon nexus assessment of China's iron and steel industry: Case study from plant level. *J. Clean. Prod.* **2020**, *253*, 119910. [CrossRef]
41. Leivas, R.; Laso, J.; Abejón, R.; Margallo, M.; Aldaco, R. Environmental assessment of food and beverage under a NEXUS Water-Energy-Climate approach: Application to the spirit drinks. *Sci. Total Environ.* **2020**, *720*, 137576. [CrossRef]
42. Ma, X.; Zhang, T.; Shen, X.; Zhai, Y.; Hong, J. Environmental footprint assessment of China's ceramic tile production from energy-carbon-water nexus insight. *J. Clean. Prod.* **2022**, *1*, 130606. [CrossRef]
43. Scott, C.A. The water-energy-climate nexus: Resources and policy outlook for aquifers in Mexico. *Water Resour. Res.* **2011**, *47*, W00L04.1–W00L04.18. [CrossRef]
44. Gu, Y.; Dong, Y.N.; Wang, H.; Keller, A.; Xu, J.; Chiramba, T.; Li, F. Quantification of the water, energy and carbon footprints of wastewater treatment plants in China considering a water–energy nexus perspective. *Ecol. Indic.* **2016**, *60*, 402–409. [CrossRef]
45. Yang, X.; Wang, Y.; Sun, M.; Wang, R.; Zheng, P. Exploring the environmental pressures in urban sectors: An energy-water-carbon nexus perspective. *Appl. Energy* **2018**, *228*, 2298–2307. [CrossRef]
46. Yang, X.; Yi, S.; Qu, S.; Wang, R.; Wang, Y.; Xu, M. Key transmission sectors of energy-water-carbon nexus pressures in Shanghai, China. *J. Clean. Prod.* **2019**, *225*, 27–35. [CrossRef]
47. Li, H.; Lin, J.; Zhao, Y.; Kang, J.N. Identifying the driving factors of energy-water nexus in Beijing from both economy-and sector-wide perspectives. *J. Clean. Prod.* **2019**, *235*, 1450–1464. [CrossRef]
48. Li, H.; Zhao, Y.; Zheng, L.; Wang, S.; Kang, J.; Liu, Y.; Shan, Y. Dynamic characteristics and drivers of the regional household energy-carbon-water nexus in China. *Environ. Sci. Pollut. Res.* **2021**, *28*, 55220–55232. [CrossRef]
49. Wang, X.C.; Klemeš, J.J.; Long, X.; Zhang, P.; Varbanov, P.S.; Fan, W.; Dong, X.; Wang, Y. Measuring the environmental performance of the EU27 from the Water-Energy-Carbon nexus perspective. *J. Clean. Prod.* **2020**, *265*, 121832. [CrossRef]
50. Meng, F.; Liu, G.; Chang, Y.; Su, M.; Hu, Y.; Yang, Z. Quantification of urban water-carbon nexus using disaggregated input-output model: A case study in Beijing (China). *Energy* **2019**, *171*, 403–418. [CrossRef]
51. Wang, X.C.; Klemeš, J.J.; Wang, Y.; Dong, X.; Wei, H.; Xu, Z.; Varbanov, P.S. Water-Energy-Carbon Emissions nexus analysis of China: An environmental input-output model-based approach. *Appl. Energy* **2020**, *261*, 114431. [CrossRef]
52. Tian, P.; Lu, H.; Reinout, H.; Li, D.; Zhang, K.; Yang, Y. Water-energy-carbon nexus in China's intra and inter-regional trade. *Sci. Total Environ.* **2022**, *806*, 150666. [CrossRef]
53. Mroue, A.M.; Mohtar, R.H.; Pistikopoulos, E.N.; Holtzapple, M.T. Energy Portfolio Assessment Tool (EPAT): Sustainable energy planning using the WEF nexus approach–Texas case. *Sci. Total Environ.* **2019**, *648*, 1649–1664. [CrossRef]
54. Ifaei, P.; Yoo, C. The compatibility of controlled power plants with self-sustainable models using a hybrid input/output and water-energy-carbon nexus analysis for climate change adaptation. *J. Clean. Prod.* **2019**, *208*, 753–777. [CrossRef]
55. Zhou, N.; Zhang, J.; Khanna, N.; Fridley, D.; Jiang, S.; Liu, X. Intertwined impacts of water, energy development, and carbon emissions in China. *Appl. Energy* **2019**, *238*, 78–91. [CrossRef]
56. Zhao, Y.; Shi, Q.; Qian, Z.; Zheng, L.; Wang, S.; He, Y. Simulating the economic and environmental effects of integrated policies in energy-carbon-water nexus of China. *Energy* **2022**, *238*, 121783. [CrossRef]
57. Liu, X.; Gao, X.; Wu, X.; Yu, W.; Chen, L.; Ni, R.; Zhao, Y.; Duan, H.; Zhao, F.; Chen, L.; et al. Updated hourly emissions factors for Chinese power plants showing the impact of widespread ultralow emissions technology deployment. *Environ. Sci. Technol.* **2019**, *53*, 2570–2578. [CrossRef]
58. Department of Energy Statistics, National Bureau of Statistics. *China Energy Statistics Yearbook 2021*; Department of Energy Statistics, National Bureau of Statistics: Beijing, China, 2021. Available online: http://www.stats.gov.cn/tjsj/ndsj/2021/indexch.htm (accessed on 1 June 2021). (In Chinese)
59. Zhang, K.; Li, Q.S. Strategic research on clean and efficient development and utilization of coal in China under carbon constraint conditions. *Coal Econ. Res.* **2019**, *39*, 10–14. [CrossRef]
60. Ren, S.H.; Xie, Y.C.; Jiao, X.M.; Xie, H.P. Characteristics of Carbon Emissions During Coal Development and Technical Approaches for Carbon Neutral Development. *Adv. Eng. Sci.* **2022**, *54*, 60–68. [CrossRef]
61. Ge, S.R.; Liu, H.T.; Liu, J.L.; Hu, H.S. Energy consumption and energy-saving strategies for coal mine production in China. *J. China Univ. Min. Technol.* **2018**, *47*, 9–14. [CrossRef]
62. Ou, X.; Yan, X.; Zhang, X. Using coal for transportation in China: Life cycle GHG of coal-based fuel and electric vehicle, and policy implications. *Int. J. Greenh. Gas Control* **2010**, *4*, 878–887. [CrossRef]
63. Peng, Y.; Yang, Q.; Wang, L.; Wang, S.; Li, J.; Zhang, X.; Fantozzi, F. VOC emissions of coal-fired power plants in China based on life cycle assessment method. *Fuel* **2021**, *292*, 120325. [CrossRef]
64. Wang, Y.; Zhao, H. The impact of China's carbon trading market on regional carbon emission efficiency. *China Popul. Resour. Environ.* **2019**, *29*, 50–58. (In Chinese)

65. Bao, W.W.; Yang, Y.; Kang, J.N.; Yu, H.P. Thermal Design and Economy Analysis of 1350MW Ultra Super Critical Double Reheat Unit. *Turbine Technol.* **2017**, *59*, 21–25. (In Chinese)

66. Development and Reform Office Climate No.2920. *Guidelines for Accounting Methods and Reporting of Greenhouse Gas Emissions of Chinese Coal Production Enterprises (Trial)*; National Development and Reform Commission: Beijing, China, 2014. (In Chinese)

67. Tu, H.; Liu, C.J. Calculation of carbon dioxide emission from standard coal. *Coal Qual. Technol.* **2014**, *2*, 57–60. (In Chinese)

68. Miller, S.M.; Michalak, A.M.; Detmers, R.G.; Hasekamp, O.P.; Bruhwiler, L.M.; Schwietzke, S. China's coal mine methane regulations have not curbed growing emissions. *Nat. Commun.* **2019**, *10*, 303. [CrossRef]

69. Zhou, A.; Hu, J.; Wang, K. Carbon emission assessment and control measures for coal mining in China. *Environ. Earth Sci.* **2020**, *79*, 1–15. [CrossRef]

70. Sheng, J.; Song, S.; Zhang, Y.; Prinn, R.G.; Janssens-Maenhout, G. Bottom-Up estimates of coal mine methane emissions in China: A gridded inventory, emission factors, and trends. *Environ. Sci. Technol. Lett.* **2019**, *6*, 473–478. [CrossRef]

71. Gao, J.; Guan, C.; Zhang, B. China's CH$_4$ emissions from coal mining: A review of current bottom-up inventories. *Sci. Total Environ.* **2020**, *725*, 138295. [CrossRef]

72. Zhu, Y.; Jiang, S.; Zhao, Y.; Li, H.; He, G.; Li, L. Life-cycle-based water footprint assessment of coal-fired power generation in China. *J. Clean. Prod.* **2020**, *254*, 120098. [CrossRef]

73. Shang, Y.; Hei, P.; Lu, S.; Shang, L.; Li, X.; Wei, Y.; Jia, D.; Jiang, D.; Ye, Y.; Gong, J.; et al. China's energy-water nexus: Assessing water conservation synergies of the total coal consumption cap strategy until 2050. *Appl. Energy* **2018**, *210*, 643–660. [CrossRef]

74. Pan, L.Y.; Liu, P.; Ma, L.; Zheng, L. A supply chain-based assessment of water issues in the coal industry in China. *Energy Policy* **2012**, *48*, 93–102. [CrossRef]

75. Gao, X.; Zhao, Y.; Lu, S.; Chen, Q.; An, T.; Han, X.; Zhuo, L. Impact of coal power production on sustainable water resources management in the coal-fired power energy bases of Northern China. *Appl. Energy* **2019**, *250*, 821–833. [CrossRef]

76. Zhang, C.; Li, J.W. Evaluation of Water Use Efficiency of China's Coal-Fired Units and Analysis of the Effect of Water Quota Improvement. *Energy Found. Grant Proj. Tech. Rep.* **2020**. Available online: https://www.efchina.org/Attachments/Report/repor t-cemp-20201103/%E4%B8%AD%E5%9B%BD%E7%87%83%E7%85%A4%E6%9C%BA%E7%BB%84%E7%94%A8%E6%B0%B4 %E6%95%88%E7%8E%87%E8%AF%84%E4%BC%B0%E5%8F%8A%E7%94%A8%E6%B0%B4%E5%AE%9A%E9%A2%9D%E6 %8F%90%E6%A0%87%E6%95%88%E6%9E%9C%E5%88%86%E6%9E%90.pdf (accessed on 24 April 2022).

77. Li, J.; Zhang, Y.; Deng, Y.; Xu, D.; Xie, K. Water consumption and conservation assessment of the coal power industry in China. *Sustain. Energy Technol. Assess.* **2021**, *47*, 101464. [CrossRef]

78. Zhang, C.; Zhong, L.K.; Wang, J. Decoupling between water use and thermoelectric power generation growth in China. *Nat. Energy* **2018**, *3*, 792–799. [CrossRef]

79. Fennell, P.S. Comparative Energy Analysis of Renewable Electricity and Carbon Capture and Storage. *Joule* **2019**, *3*, 1406–1408. [CrossRef]

80. Zhang, X.; Li, Y.; Ma, Q.; Liu, L.N. Development of Carbon Capture, Utilization and Storage Technology in China. *Eng. Sci.* **2021**, *23*, 070–080. [CrossRef]

81. Yi, L.; He, Q.; Li, Z.P.; Yang, L. Research on the Pathways of Carbon Market Development: International Experiences and Implications for China. *Clim. Chang. Res.* **2019**, *15*, 232–245. Available online: http://www.chinacarbon.info/ (accessed on 24 April 2022). (In Chinese)

82. Slater, H.; De Boer, D.; Qian, G.; Wang, S. *2021 China Carbon Pricing Survey*; China Carbon Forum (CCF): Beijing, China, 2021. (In Chinese)

83. Cozzi, L.; Gould, T.; Bouckart, S.; Crow, D.; Kim, T.Y.; Mcglade, C.; Wetzel, D. *World Energy Outlook 2020*; International Energy Agency: Paris, France, 2020; pp. 1–461.

84. Xiong, J.; Xu, D. Relationship between energy consumption, economic growth and environmental pollution in China. *Environ. Res.* **2021**, *194*, 110718. [CrossRef]

85. Xie, H.P.; Ren, S.H.; Xie, Y.C. Development opportunities of the coal industry towards the goal of carbon neutrality. *J. China Coal Soc.* **2021**, *46*, 1808–1820. (In Chinese) [CrossRef]

86. Cai, B.F.; Li, Q.; Zhang, X.; Cao, C.; Cao, L.; Chen, W.; Chen, Z.; Dong, J.; Fan, J.; Jiang, Y.; et al. *China Carbon Dioxide Capture, Utilization and Storage (CCUS) Annual Report: China CCUS Path Study*; Chinese Academy of Environmental Planning, Institute of Rock and Soil Mechanics, Chinese Academy of Sciences, The Administrative Center for China's Agenda 21: Beijing, China, 2021. (In Chinese)

87. International Energy Agency (IEA). Energy technology perspectives 2020. In *Special Report on Carbon Capture, Utilization and Storage*; IEA: Paris, France, 2020.

88. Zhao, X.; Ma, X.; Chen, B.; Shang, Y.; Song, M. Challenges toward carbon neutrality in China: Strategies and countermeasures. *Resour. Conserv. Recycl.* **2022**, *176*, 105959. [CrossRef]

89. Räsänen, T.A.; Varis, O.; Scherer, L.; Kummu, M. Greenhouse gas emissions of hydropower in the Mekong River Basin. *Environ. Res. Lett.* **2018**, *13*, 034030. [CrossRef]

90. Fan, Y.V.; Klemeš, J.J.; Ko, C.H. Bioenergy carbon emissions footprint considering the biogenic carbon and secondary effects. *Int. J. Energy Res.* **2021**, *45*, 283–296. [CrossRef]

91. Wilkins, R.; Menefee, A.H.; Clarens, A.F. Environmental life cycle analysis of water and CO_2-based fracturing fluids used in unconventional gas production. *Environ. Sci. Technol.* **2016**, *50*, 13134–13141. [CrossRef] [PubMed]
92. Dijkman, J.G.; Van Haeringen, H.; De Lange, S.J. Fuzzy numbers. *J. Math. Anal. Appl.* **1983**, *92*, 301–341. [CrossRef]

Article

Evolution of CCUS Technologies Using LDA Topic Model and Derwent Patent Data

Liangchao Huang [1,2,3], **Zhengmeng Hou** [2,3], **Yanli Fang** [2,3,4,*], **Jianhua Liu** [1,*] and **Tianle Shi** [1]

[1]   Sino-German Research Institute of Carbon Neutralization and Green Development, Zhengzhou University, Zhengzhou 450001, China; liangchao.huang@tu-clausthal.de (L.H.)

[2]   Institute of Subsurface Energy Systems, Clausthal University of Technology, 38678 Clausthal Zellerfeld, Germany

[3]   Research Centre of Energy Storage Technologies, Clausthal University of Technology, 38640 Goslar, Germany

[4]   Sino-German Energy Research Center, Sichuan University, Chengdu 610065, China

*   Correspondence: yanli.fang@tu-clausthal.de (Y.F.); ljh@zzu.edu.cn (J.L.)

Abstract: Carbon capture, utilization, and storage (CCUS) technology is considered an effective way to reduce greenhouse gases, such as carbon dioxide (CO_2), which is significant for achieving carbon neutrality. Based on Derwent patent data, this paper explored the technology topics in CCUS patents by using the latent Dirichlet allocation (LDA) topic model to analyze technology's hot topics and content evolution. Furthermore, the logistic model was used to fit the patent volume of the key CCUS technologies and predict the maturity and development trends of the key CCUS technologies to provide a reference for the future development of CCUS technology. We found that CCUS technology patents are gradually transforming to the application level, with increases in emerging fields, such as computer science. The main R&D institutes in the United States, Europe, Japan, Korea, and other countries are enterprises, while in China they are universities and research institutes. Hydride production, biological carbon sequestration, dynamic monitoring, geological utilization, geological storage, and CO_2 mineralization are the six key technologies of CCUS. In addition, technologies such as hydride production, biological carbon sequestration, and dynamic monitoring have good development prospects, such as CCUS being coupled with hydrogen production to regenerate synthetic methane and CCUS being coupled with biomass to build a dynamic monitoring and safety system.

Keywords: CCUS; LDA topic model; Derwent patent data; technology maturity; R&D

Citation: Huang, L.; Hou, Z.; Fang, Y.; Liu, J.; Shi, T. Evolution of CCUS Technologies Using LDA Topic Model and Derwent Patent Data. *Energies* **2023**, *16*, 2556. https://doi.org/10.3390/en16062556

Academic Editor: Konstantinos Christoforidis

Received: 7 February 2023
Revised: 5 March 2023
Accepted: 7 March 2023
Published: 8 March 2023

1. Introduction

The emission of greenhouse gases, especially carbon dioxide, aggravated global warming and became a major strategic issue that threatens the sustainable development of humanity [1,2]. CCUS technology is an important method to reduce carbon dioxide and to address global climate change, which is of great significance for achieving the goal of carbon neutrality [3,4]. CCUS technology is a carbon dioxide emission reduction and utilization technology, which separates and captures carbon dioxide from various emission sources or the air and transports it to a suitable site for storage or recycling through chemical or biological conversion [5,6]. In recent years, the development of CCUS technology was given high priority worldwide, and the maturity of related technologies is rapidly increasing, showing the development trends of new technologies emerging and energy costs gradually decreasing. The International Energy Agency (IEA) assessed the emission reduction potential of CCUS, which could reduce 6.9×10^9 t of CO_2 per year by 2070 in a sustainable development scenario, accounting for 19.27% of the total emission reduction [7]. The Intergovernmental Panel on Climate Change (IPCC) pointed out that in order to achieve the temperature rise control target of 1.5 °C in 2100, the cumulative global CCUS needs to reduce $5.5 \times 10^{11} \sim 1.0 \times 10^{12}$ t of CO_2 [8]. Currently, the knowledge spillover efficiency of CCUS technology is relatively high in the United States, Europe, Japan, Australia, and New Zealand [9]. Although the research on CCUS technology in China started late, positive results were achieved by the strengthening of basic research, key technology research, and project integration and demonstration. However, compared with the aforementioned international frontier, the individual key technologies and commercial integration levels within CCUS are lagging behind [10–12].

A CCUS technology system covers several key technologies in the following fields: (1) CO_2 capture and transport; (2) CO_2 mineralization; (3) reduction and recycling; (4) biological carbon sequestration; (5) carbon geological storage; and (6) CO_2 enhanced underground resource exploration [13–18]. Previous studies on the evolution of CCUS technology were mostly theoretical analyses, mainly focusing on policies and regulations [19,20], technical routes [21], and life cycle and economic analyses [22]. Some researchers also explored the evolution from a bibliometric perspective [23]. However, only a few studies that analyze the technology evolution process from the perspective of technology topics were based on patent data. In addition, there is a lack of studies that use patent volume to predict the maturity levels and development trends of important CCUS technologies.

The set of words describing the technical details in the patent data text contains the most important information, including the technological innovation. Therefore, based on Derwent patent data, this paper mined the hot technology topics and its heat in CCUS in different years, and thus, a content evolution of CCUS was obtained. The logistic model was applied to fit the patent volume of the key CCUS technologies and predict the levels of the maturity and development trends of the key CCUS technologies. The aim of this paper is to reveal the research status, research focus, and development direction of CCUS technologies and provide the analysis base for the future policies of CCUS technologies.

2. Methodology

2.1. LDA Topic Mining

The LDA topic model is widely used in text topic mining. For semi-structured documents, such as patent documents, the LDA model has the advantages of fast mining, complete coverage, and high accuracy rate. Therefore, this paper applied the LDA topic model to mine and analyze the patent text, which helps to identify the topic words implied in the text, analyze technology evolution, and clarify the trend of technology development.

2.1.1. Data Collection and Pre-Processing

The Derwent Innovation Index (DII) integrates the Derwent World Patents Index (WPI) and the Patents Citation Index to provide patent information worldwide. The patent data in this article come from the Derwent database, based on the patent classification system of the Derwent database and previous research [24]. The searched terms are listed as follows: TS = ("CCS" OR "CCUS" OR "carbon capture" OR "carbon utiliz*" OR "carbon storage" OR "carbon circular utiliz*" OR "carbon sequest*" OR "carbon inject*"OR "CO_2 capture" OR "CO_2 utiliz*" OR "CO_2 storage" OR "CO_2 circular utiliz*" OR "CO_2 sequest*" OR "CO_2 inject*"). The search was conducted on 4 January 2023 based on the above search terms. The searched time dimension was selected as from 2003 to 2022, excluding non-related scientific categories, and a total of 175,922 patent documents were obtained. In addition, during these years, China had the largest number of patents in CCUS.

In order to better mine the text, the patent text was pre-processed by the Python program, which mainly includes the following three steps [25,26]: ① extracting words that can characterize technical information, i.e., title, abstract, and keywords; ② removing common patent words, academic words and stop words such as "what", "why", "and", "before", "use", and other common conjunctions among the top 100 academic words provided by the University of Nottingham [27]; ③ lemmatization (word form reduction) processing of stop words such as "the", "an", "in", "at", etc. The different lexical properties in the words were firstly matched by Wordnet packets. Then, the matching words were unified and lemmatized through the WordNet Lemmatizer program package.

Considering the periodicity of technology evolution, a period stage of 5 years was assumed for topic mining evolution analysis. Thus, 2003–2007 was the first stage, 2008–2012 was the second stage, 2013–2017 was the third stage, and 2018–2022 was the fourth stage. The number of patent documents in each stage and the number of patent terms before and after pre-processing are shown in Table 1. For the convenience of presentation, the total patent data from 2003 to 2022 are expressed as "total text sets", and the patent data of each stage are expressed as "stage *".

Table 1. Number of CCUS technology patents and vocabulary by stage, 2003–2022.

Stage	Number of Patents	Number of Words before Data Cleaning	Number of Words after Data Cleaning
Stage 1 (2003–2007)	14,022	4,914,169	245,302
Stage 2 (2008–2012)	30,153	10,843,401	622,086
Stage 3 (2013–2017)	46,075	16,367,363	988,997
Stage 4 (2018–2022)	85,672	31,873,375	1,925,153
Total text sets	175,922	63,998,308	3,781,538

2.1.2. Parameter Setting

The hyperparameters α and β are the parameters that determine the distribution of topics and words in a document when performing LDA topic mining, respectively. Referring to the relevant study [28], as well as considering the refinement of the topics, the processing speed, and the availability of the results, the parameters $\alpha = 0.5$ and $\beta = 0.1$ were chosen in this paper. The parameter K (optimal number of topics) determines the final number of topics extracted by the LDA model [29]. Perplexity is a widely used method to determine the optimal number of topics. It can be calculated as follows:

$$Perplexity(D_{test}) = \exp\left\{-\frac{\sum_{d=1}^{D} \log p(w_d)}{\sum_{d=1}^{D} N_d}\right\} \tag{1}$$

$$p(w_d) = \prod_{i=1}^{N_d} \sum_{k=1}^{K} p(z_k|d) * p(w_i|z_k) \tag{2}$$

where D_{test} represents the entire text set; D represents the number of documents in the text set; w_d represents the sequence of words in the d document; the non-repetitive vocabulary of the d document is denoted by N_d; $p(w_d)$ is the probability of the word sequence w_d, calculated by multiplying the probabilities of all words in the d document; the i-th word in document d is noted as i; K represents the assumed number of topics; $p(z_k|d)$ represents the probability of matching topic K in a given document; and $p(w_i|z_k)$ represents the probability that topic z_k contains word i.

The optimal number of topics in each stage of the text sets and the total text sets were obtained by Equation (1) and Equation (2), respectively, as shown in Table 2.

Table 2. Optimal number of topics in the text sets at each time stage.

Stage	Stage 1	Stage 2	Stage 3	Stage 4	Total Text Sets
Optimal number of topics	12	13	14	14	24

2.1.3. Topic Mining

The iterative calculation was performed by setting the steps of iterations as 500 according to the above parameters. Output results topic distribution matrix $\theta_{D \times K}$ showed the probability distribution of the K topic in the total text sets, representing the degree of topic contribution to the whole text, as shown in Equation (3).

$$\theta_{D \times K} = \begin{bmatrix} \theta_{1,1} & \theta_{1,2} & \theta_{1,3} & \cdots & \theta_{1,j} & \cdots & \theta_{1,K} \\ \theta_{2,1} & \theta_{2,2} & \theta_{2,3} & \cdots & \theta_{2,j} & \cdots & \theta_{2,K} \\ \theta_{3,1} & \theta_{3,2} & \theta_{3,3} & \cdots & \theta_{3,j} & \cdots & \theta_{3,K} \\ & \vdots & & \ddots & & \vdots & \\ \theta_{i,1} & \theta_{i,2} & \theta_{i,3} & & \theta_{i,j} & \cdots & \theta_{i,K} \\ \vdots & \vdots & \vdots & \cdots & \vdots & \cdots & \\ \theta_{D,1} & \theta_{D,2} & \theta_{D,3} & & \theta_{D,j} & \cdots & \theta_{D,K} \end{bmatrix} \tag{3}$$

Each row in the matrix corresponds to a document, and θ_{ij} presents the probability that the i document belongs to topic j.

2.2. LDA Topic Evolution

2.2.1. Topic Heat Evolution

From the viewpoint of patent topic text, topic heat means the probability of a certain type of patent topic appearing in a fixed period, and a higher probability means a higher topic heat. Therefore, analyzing the changes of topic heat as time could be better for observing the development of the related patents. It can be calculated as

$$Q(Z_{t,k}) = (\sum_{d=1}^{D_t} \theta_{d,k}) / D_t \tag{4}$$

where $Q(Z_{t,k})$ denotes the intensity of topic k in the current time slice t, $\theta_{d,k}$ presents the probability of the topic k in the d document, and D_t indicates the number of documents on time slice t.

2.2.2. Topic Content Evolution

The degree of information similarity between CCUS technical topics was confirmed by calculating the similarity of topics between two close stages with the following formula:

$$sim(D_i, D_j) = \frac{\sum\limits_{g \in D_i} \sum\limits_{k \in D_j} H_{g,k} \times T(w_g) \times T(w_k)}{\sqrt{\sum\limits_{g \in D_i} \sum\limits_{l \in D_i} H_{g,l} \times T(w_g) \times T(w_l)} \cdot \sqrt{\sum\limits_{k \in D_j} \sum\limits_{m \in D_j} H_{k,m} \times T(w_k) \times T(w_m)}} \tag{5}$$

where $sim(D_i, D_j)$ represents the similarity between the two different topics of D_i and D_j, w is the topic vector in the text, H is the topic vector similarity value, and T denotes the topic vector value of term frequency–inverse document frequency (TF–IDF). If the similarity of topics between two stages is $sim > 0.7$, then it is claimed that there is an evolutionary relationship between these two topics.

2.3. Technology Maturity Prediction

Logistic models, also known as the sigmoid curves (S-curves), are widely used to analyze and predict the life cycle of the technology maturity. The corresponding development stage of the life cycle is shown in Figure 1. The general formula for the logistic model can be written as

$$N_t = \frac{k}{1 + \alpha e^{-\beta t}} \tag{6}$$

where N_t is the patent data amount, t is the growth time of the corresponding technology from 10% to 90% of the maximum number of patents, k is the predicted maximum number of patents (saturation) obtained by fitting the number of patents curve, α represents the slope of the curve, and β is the turning point of the slope of the curve from small–large–small, which is also called the midpoint of the S-curve.

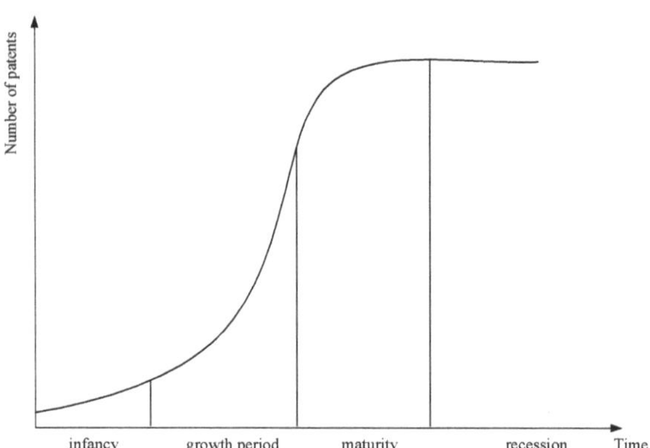

Figure 1. The development stages of the S-curve life cycle.

3. Analysis of the Current State of Technology Evolution

3.1. LDA Topic Mining

According to Equations (1)–(3), the CCUS patents are mined. The selection of topic words adopts the principles of comprehensiveness and accuracy. The topic mining results named by the comprehensive meaning of each topic feature word by querying the relevant patents are listed in Table 3.

Table 3. CCUS technology topic mining results.

Stage	Topic Mining Results
Stage 1	Topic 0: chemical compound, Topic 1: coal, Topic 2: sensor, Topic 3: pipelines, Topic 4: electricity, Topic 5: oxygen-enriched combustion, Topic 6: hydride production, Topic 7: device components, Topic 8: reservoirs, Topic 9: water utilization, Topic 10: condensation process, and Topic 11: microalgae.
Stage 2	Topic 0: microalgae, Topic 1: hydride production, Topic 2: process execution, Topic 3: adsorbent, Topic 4: sensor, Topic 5: reservoirs, Topic 6: water utilization, Topic 7: anti-corrosion technology, Topic 8: chemical compound, Topic 9: oxygen-enriched combustion, Topic 10: pipeline, Topic 11: coal, and Topic 12: electricity.
Stage 3	Topic 0: CO_2 adsorption, Topic 1: economy, Topic 2: CO_2-EOR (CO_2-enhanced oil recovery), Topic 3: CO_2 mineralization, Topic 4: bioenergy, Topic 5: synthesis of organic chemicals, Topic 6: hydride production, Topic 7: membrane separation, Topic 8: alarm system, Topic 9: geological storage, Topic 10: oxygen-enriched combustion, Topic 11: control system, Topic 12: anti-corrosion technology, and Topic 13: water utilization.
Stage 4	Topic 0: CO_2 mineralization, Topic 1: bioenergy, Topic 2: CO_2-EOR, Topic 3: geological storage, Topic 4: biological carbon sequestration, Topic 5: economy, Topic 6: geological utilization, Topic 7: fuel synthesis and preparation, Topic 8: synthesis of organic chemicals, Topic 9: dynamic monitoring, Topic 10: CO_2 separation technology, Topic 11: anti-corrosion technology, Topic 12: oxygen-enriched combustion, and Topic 13: hydride production.
Total text sets	Topic 0: pipelines, Topic 1: oxygen-enriched combustion, Topic 2: CO_2 adsorption, Topic 3: biological carbon sequestration, Topic 4: CO_2-EOR, Topic 5: fuel synthesis and preparation, Topic 6: electricity, Topic 7: geological storage, Topic 8: coal, Topic 9: anti-corrosion technology, Topic 10: alarm system, Topic 11: CO_2 mineralization, Topic 12: dynamic monitoring, Topic 13: water utilization, Topic 14: hydride production, Topic 15: economy, Topic 16: geological utilization, Topic 17: sensor, Topic 18: synthesis of organic chemicals, Topic 19: adsorbent, Topic 20: CO_2 separation technology, Topic 21: reservoirs, Topic 22: bioenergy, and Topic 23: microalgae.

3.2. Measurement Analysis

3.2.1. Time Distribution

The change in the number of patents over time provides a direct observation of the research trends in CCUS, which gives a macro view of the research development. A 3D bar chart of the changes in the number of CCUS patents is shown in Figure 2.

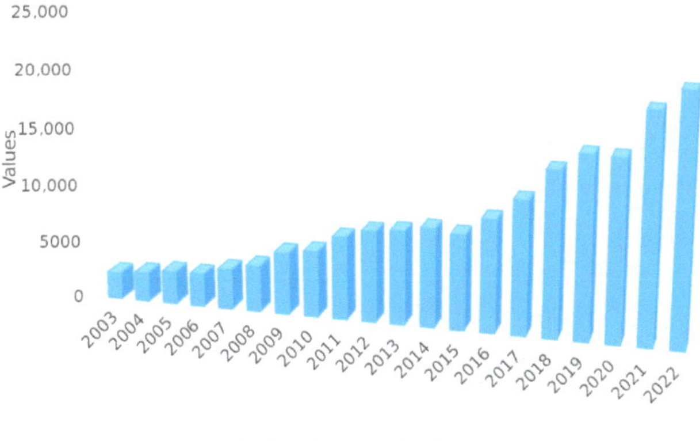

Figure 2. Changes in the number of CCUS patents over years.

As can be seen from Figure 2, the global CCUS number of patents was maintaining a continuous growth trend. In 2016, the growth of number of patents obviously accelerated due to the signing of the Paris Agreement at the end of 2015. In 2018, the IPCC's Special Report on Global Warming of 1.5 °C was released, which pointed out the significance of CCUS technology in addressing global climate change [8]. As a result, 2018 became the year with the highest growth rate of CCUS patents. In 2020, due to the sudden outbreak of the coronavirus, field experiments on related technologies were restricted, resulting in a decline in the number of CCUS patents that year. As the situation improved, the number of patents picked up and continued to grow in 2021.

3.2.2. Research Field Distribution

Analyzing the distribution of the research results in different fields can help to show the current research directions in the field. The top 10 fields involved in CCUS were selected by classifying the subject field of the patent text data, and thus, a corresponding radar map of the global CCUS subject field distribution of these top 10 fields was obtained. The radar maps of the top 10 fields in 2003 and 2022 are displayed in Figure 3.

In Figure 3, the 2003 and 2022 CCUS patents cover roughly the same fields. In 2003, the number of patent applications in chemistry ranked first and those in engineering ranked second, while the number of patent applications in 2022 in engineering ranked first and those in chemistry ranked second. This indicates that the technology patents of CCUS gradually transformed to the application level. Different from in 2003, in 2022, the field of computer science was newly added to the top 10 and rose to eighth place. This shows that the technologies related to computer science (e.g., sensor, dynamic monitoring) in CCUS patents are developing at a high speed, and more attention is being paid to safety and the economy in the implementation of CCUS technologies.

In summary, the core technologies of CCUS are concentrated in the fields of engineering, chemistry, and instruments instrumentation. At the same time, it is also important to focus on emerging areas, such as computer science, and strengthen the development of these areas, which would have a propulsive effect on CCUS technology.

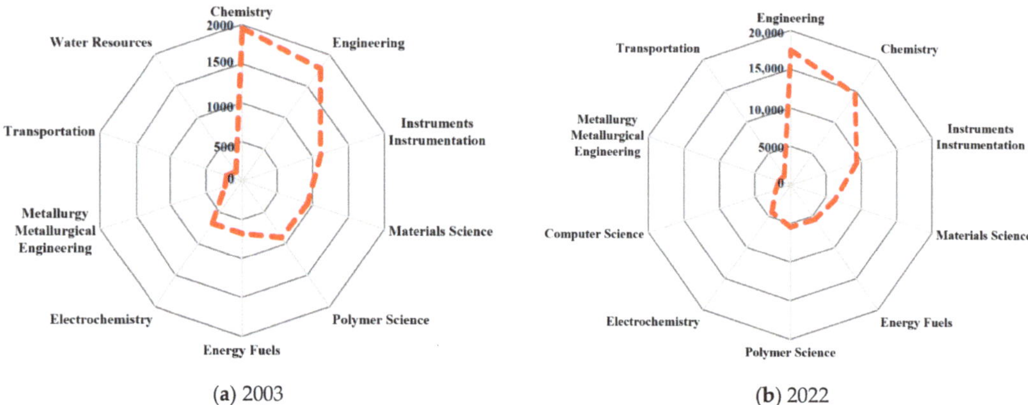

(**a**) 2003   (**b**) 2022

Figure 3. Global radar map of CCUS subject fields in 2003 and 2022.

3.2.3. Comparison of International R&D Institutes

Among the top 100 R&D institutions with CCUS technology patent applications from 2003 to 2022, China accounted for the highest proportion (52%), followed by Japan (16%), the United States (15%), Korea (7%), Germany (6%), France (5%), and Saudi Arabia (2%). The statistics on the top 15 innovative entities in CCUS patent applications and the number of patents in the US, Europe, Japan, Korea, and China are shown in Table 4.

Table 4. Top 15 patent entities and their patent numbers in CCUS patent applications in China and in US, Europe, Japan, and Korea.

Chinese Patent Entities	Number of Patents	US, Europe, Japan, and Korea Patent Entities	Number of Patents
China Petroleum Chem Corp	1324	Lg Chem Ltd. (KR)	701
Univ Tsinghua	537	General Electric Co. (US)	546
Univ Zhejiang	516	Basf Se (DE)	491
Univ Cent South	506	Posco (KR)	458
Cas Dalian Chem Physical Inst	469	Toyota Jidosha Kk (JP)	439
Univ Xian Jiaotong	461	Air Liquide Sa (FR)	397
Univ Tianjin	446	Samsung Electronics Co., Ltd. (JP)	379
Univ South China Technology	419	Bosch Gmbh Robert (DE)	379
Univ Dalian Technology	416	Toshiba Kk (JP)	357
State Grid Corp China	389	Jfe Steel Corp (JP)	332
Petrochina Co., Ltd.	385	Int Business Machines Corp (US)	322
Univ Beijing Sci Technology	392	Shell Int Res Mij Bv (US)	312
Univ Kunming Sci Technology	347	Nippon Steel Corp (JP)	310
Univ Southeast	341	L Oreal Sa (FR)	307
Sinopec Corp	335	Toray Ind Inc. (JP)	301

As can be seen from Table 4, there are certain differences in the institutes of authorization between China and the US, Europe, Japan, and Korea. By 2022, the top 15 innovation entities with the highest number of patent applications in the US, Europe, Japan, and Korea are all listed companies. Among the top 15 innovation entities in China, only 4 are listed companies, and the others are universities and university-affiliated research institutes.

In the US, Europe, Japan, and Korea, CCUS technology R&D is more in line with market demand after market research, so CCUS technology can be better applied and commercialized, which promotes the continuous development of the CCUS industry. At the same time, the patent institutes of the US, Europe, Japan, and Korea are larger in scale, and the products derived from core technology occupy a higher share of the international market with a stronger competitiveness. In China, universities occupy a large proportion of the CCUS patent authorization market. The orientation of universities themselves is education and scientific research, which deviates from the purpose of enterprises pursuing market interests through innovation. From this perspective, China's CCUS development is not as close to the market as that of the US, Europe, Japan, and Korea. While using universities to strengthen basic research, China should promote industrial development and break through technological blockades.

3.3. CCUS Technology Topic Heat Evolution Analysis

Based on the topic mining results of the CCUS total text sets, the intensity values of each topic in different years are calculated according to Equation (4). A topic evolution river diagram is drawn as time according to the change trend, as shown in Figure 4, where T_i represents the i topic, and $i = 1,2,3, \ldots ,23$.

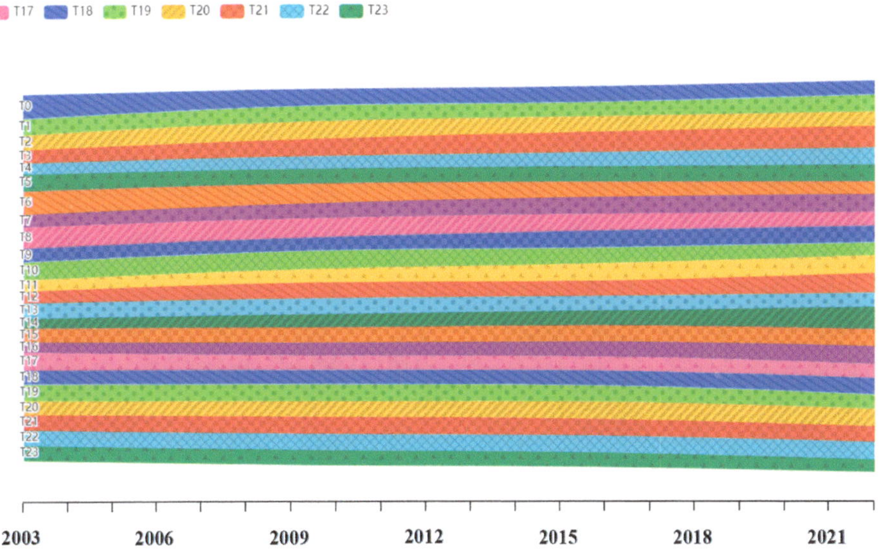

Figure 4. River map of the evolution of CCUS patent topic heat.

In the river map, the wider the width of a certain river is, the greater the intensity of the corresponding topic is. This can be seen from the CCUS patent topic evolution river map. Pipelines (T0), electricity (T6), coal (T8), alarm systems (T10), sensors (T17), adsorbents (T19), and microalgae (T23) are the basic technologies for the pre-development of CCUS. Their heat tends to decrease over time. Two reasons cause this situation. On the one hand, as the corresponding technology reaches maturity, the heat naturally decreases. On the other hand, some of the basic technologies evolve into new technologies.

Biological carbon sequestration (T3), CO_2-EOR (T4), geological storage (T7), anti-corrosion technology (T9), CO_2 mineralization (T11), dynamic monitoring (T12), hydride production (T14), economy (T15), geological utilization (T16), synthesis of organic chemicals (T18), CO_2 separation technology (T20), and bioenergy (T22) are the key technologies for the current development of CCUS. The heat of these aforementioned technologies is increasing over time. Most of them are CO_2 utilization and storage technologies, while CO_2 capture technologies are less hot. Hydride production (T14), biological carbon sequestration (T3), dynamic monitoring (T12), geological utilization (T16), geological storage (T7), and CO_2 mineralization (T11) are also key technologies with a relatively high heat.

The evolution of oxygen-enriched combustion (T1) and fuel synthesis and preparation (T5) are two topics that could be classified as having three phases. The heat of these two topics was relatively high in the first stage, and then it decreased in the second and third stages because the related technologies were at a bottleneck. In the fourth stage, due to a breakthrough in technological bottlenecks, such as oxy-fuel combustion costs and energy intensity, the heat of these topics gradually increased. Regarding CO_2 adsorption (T2), water utilization (T13), and reservoirs (T21), the heat of these three topics was upward-declining, which indicates that a certain technical bottleneck was broken in the first stage, and then the technologies tended to mature and evolve to higher-level technologies.

3.4. Analysis of the Evolution of Technical Subject Matter Content in CCUS

In accordance with the topic correlation mentioned above, the similarity of the topics at each stage, as calculated by Equation (5), was presented in the form of a Sankey diagram. For the purpose of differentiation, the topics from the first stage to the fourth stage are labeled "A, B, C and D", as shown in Figure 5.

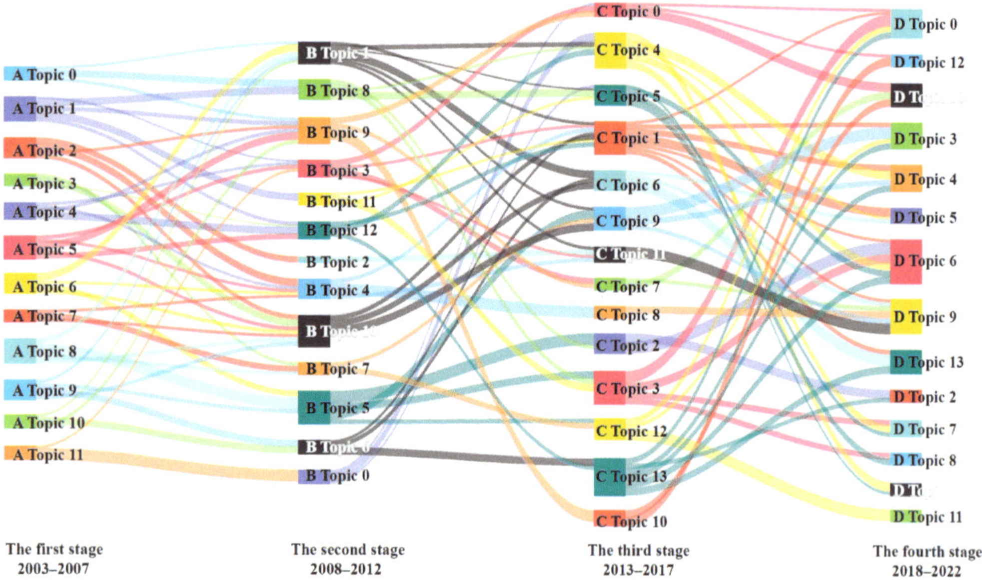

Figure 5. CCUS patent topic evolution river diagram.

With the development of global carbon neutrality, countries made large-scale innovations in the field of CCUS technology, evolving from the basic technical level to the application level. The focus of R&D changed over time. At present, CCUS technologies are beginning to be applied and commercialized, with more emphasis placed on breaking energy technology bottlenecks and reducing costs, such as CO_2-EOR, biological carbon sequestration, CO_2 mineralization, and other application-oriented technologies.

The basic CCUS technologies, such as pipelines, sensors, microalgae, electricity, coal, and adsorbents were developed well at the beginning. However, with continuous improvement and subsequent maturity, technologies are gradually transforming into geological storage, dynamic monitoring, bioenergy, geological utilization, and CO_2 separation technology.

The related topic heat for carbon dioxide capture technology shows a decreasing trend. In contrast, the technologies of carbon dioxide storage and utilization technology and CCUS management are on the rise. The key areas of CO_2 storage and utilization technologies mainly are bio-utilization (e.g., bioenergy and biological carbon sequestration), geological utilization (e.g., CO_2-EOR/EGR), chemical utilization (e.g., hydride production, etc.), and geological storage (e.g., aquifers or oil and gas formations). In addition, the key area of CCUS management technology is mainly dynamic monitoring.

4. Predictive Analysis of Key Technologies' Evolution in CCUS

Six technologies with the highest heat are selected for analysis in this section. They are hydride production (T14), biological carbon sequestration (T3), dynamic monitoring (T12), geological utilization (T16), geological storage (T7), and CO_2 mineralization (T11). The number of patents of each technology from 2003 to 2022 was searched by using the Derwent Innovation Index (DII), as shown in Figure 6.

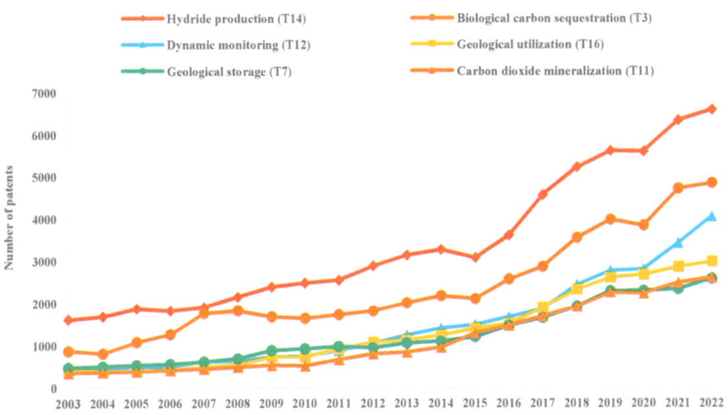

Figure 6. Number of patents for key CCUS technologies from 2003 to 2022.

From Figure 6, it can be seen that the number of patents for the selected six key technologies of CCUS grew significantly from 2003 to 2022, even though the number of patents grew relatively slowly between 2003 and 2015. The rate of growth greatly accelerated in 2016 after the Paris Agreement was signed at the end of 2015. From the perspective of the number of patents for each key technology, the number of patents of hydride production (T14) and biological carbon sequestration (T3) was relatively high, and the growth rate remained unabated since 2020. With respect to the topics of dynamic monitoring (T12), geological utilization (T16), geological storage (T7), and CO_2 mineralization (T11), the number of patents is approximate. However, the growth rate of dynamic monitoring (T12) is explosive after 2020, due to the development of computer science and the improvement of safety and economic production awareness.

Based on the logistic model, the number of patents for Topic 14, Topic 3, Topic 12, Topic 16, Topic 7, and Topic 11 from 2003 to 2022 were fitted by Origin software, and thus, the corresponding field logistic curve fitting plots and topic fitting statistical parameters were obtained. Figure 7a–f represent the fitted curves of six selected key CCUS technologies for the number of patents. The curves before 2022 represent the existing number of patents, and the curves after 2022 represent the forecasted number of patents.

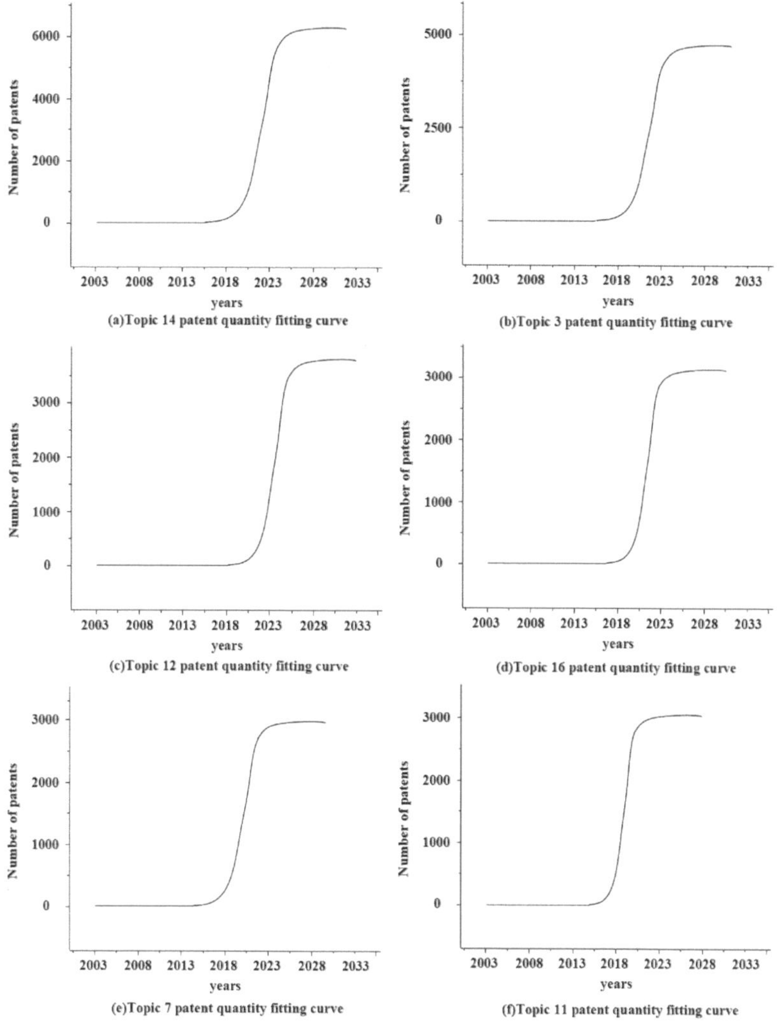

Figure 7. Fitted curve of six key CCUS technologies based on the number of patents.

The degree of freedom of the fitting curve statistics was set as 22. The adjusted values of R^2 were 0.99184, 0.99615, 0.99827, 0.99652, 0.98949, and 0.99549, which indicates that the individual topic curves fit the number of patents well. Based on Figure 7, the development stages of the key CCUS technologies were determined, and the results are shown in Table 5. The infancy period of the six key CCUS technologies lasted for a long time, and the development speed increased slowly. Around 2016, various technologies entered the growth stage, and the number of patents for related technologies began to increase significantly. After the growth period, related technologies enter the forecasted mature period. At this stage, the technology accumulation is sufficient, and part of the research starts to be transformed into application. Therefore, the growth rate of the number of patents is relatively slow at this stage, even though it is still growing. Compared with the other three technologies, the maturity periods for the topics of hydride production (T14), biological carbon sequestration (T3), and biological carbon sequestration (T3) come late, so the related topics should be the key technologies for future development. Just as CCUS is coupled with hydrogen production to regenerate synthetic methane, CCUS is coupled with biomass to build a dynamic monitoring and safety system. After the maturity period, the number of relevant patents reaches saturation, which is caused by the wide use in commerce. Then, the development of the matured technology faces three possibilities. One possibility is that the heat of technology research begins to decline, and the related technology is washed. Another possibility is that the related technology is replaced by new technology, while the third possibility is that some patented technology fails to be commercialized.

Table 5. Development stages of key CCUS technologies.

Topic	Infancy Period	Growth Period	Mature Period
hydride production (T14)	2003–2015	2016–2023	2024–2027
biological carbon sequestration (T3)	2003–2015	2016–2022	2023–2026
dynamic monitoring (T12)	2003–2017	2018–2024	2025–2028
geological utilization (T16)	2003–2016	2017–2021	2022–2025
geological storage (T7)	2003–2013	2014–2020	2021–2025
CO_2 mineralization (T11)	2003–2014	2015–2019	2020–2024

5. Conclusions

Firstly, in this paper, we captured a total of 175,922 patents related to CCUS technology from 2003 to 2022 by using Derwent patent data. Secondly, the LDA topic model was used to mine the technical topics of the CCUS patents. Then, the evolution of global CCUS technology was analyzed from three aspects: measurement analysis, technical topic heat, and technical topic content evolution. Finally, the logistic model was applied to fit the patent volume of the key CCUS technologies and predict the degrees of maturity freedom and development trends of the key CCUS technologies. The specific conclusions are as follows:

(1) Due to the signing of the Paris Agreement, the outbreak node of CCUS technology patents was in 2016. In the key fields of research, CCUS technology patents are gradually transformed to the application level, while the emerging fields, such as computer science are on the rise. Regarding the main R&D institutes, the US, Europe, Japan, and South Korea are led by enterprises, while China is dominated by universities and research institutes. Although the total numbers of patents and R&D institutes in China are the highest in the world, there is still a long way for CCUS technologies to go in China to be applied and commercialized.

(2) The transformation of CCUS technologies can be summarized in two steps. Firstly, CCUS technologies gradually transform from the basic level of pipelines, electricity, coal, and sensors to the technical level, such as CO_2 adsorption, water utilization, and reservoirs. Then, hydride production, biological carbon sequestration, CO_2-EOR, and CO_2 mineralization transform to the application level.

(3) The trend of the topic related to carbon capture technology is declining. The trends of technologies for carbon dioxide storage and utilization technology and CCUS management technology are on the rise. The topics of hydride production, biological carbon sequestration, dynamic monitoring, geological utilization, geological storage, and CO_2 mineralization are relatively hot, and the related technologies are the key technologies of CCUS.

(4) The key technologies, such as geological utilization, geological storage, and CO_2 mineralization are matured. New technologies would merge to drive the development of CCUS in the aforementioned fields. Technologies such as hydride production, biological carbon sequestration, and dynamic monitoring are in the growth period, each with a bright development prospect. Just as CCUS is coupled with hydrogen production to regenerate synthetic methane, CCUS is coupled with biomass to build a dynamic monitoring safety system.

Author Contributions: Conceptualization, Z.H.; investigation, Y.F. and J.L.; methodology, J.L. and T.S.; supervision, Z.H.; data curation, L.H. and T.S.; writing—original draft, L.H. and Y.F.; writing—review and editing, Z.H. and J.L. All authors have read and agreed to the published version of the manuscript.

Funding: The Henan Institute for Chinese Development Strategy of Engineering & Technology (Grant No. 2022HENZDA02), the Science & Technology Department of Sichuan Province Project (Grant No. 2021YFH0010), and the Soft Science Major Project of Henan Province (Grant No. 212400410002).

Data Availability Statement: Data are available in a publicly accessible repository.

Conflicts of Interest: The authors declare no conflict of interest.

References

1. Liu, J.; Shi, T.; Huang, L. A Study on the Impact of Industrial Restructuring on Carbon Dioxide Emissions and Scenario Simulation in the Yellow River Basin. *Water* **2022**, *14*, 3833. [CrossRef]
2. Jorgenson, A.K.; Fiske, S.; Hubacek, K.; Li, J.; McGovern, T.; Rick, T.; Schor, J.B.; Solecki, W.; York, R.; Zycherman, A. Social science perspectives on drivers of and responses to global climate change. *Wiley Interdiscip. Rev. Clim. Chang.* **2019**, *10*, e554. [CrossRef]
3. Xie, Y.; Hou, Z.; Liu, H.; Cao, C.; Qi, J. The sustainability assessment of CO_2 capture, utilization and storage (CCUS) and the conversion of cropland to forestland program (CCFP) in the Water–Energy–Food (WEF) framework towards China's carbon neutrality by 2060. *Environ. Earth Sci.* **2021**, *80*, 468. [CrossRef]
4. Ali, S.A.; Mulk, W.U.; Ullah, Z.; Khan, H.; Zahid, A.; Shah, M.U.H.; Shah, S.N. Recent Advances in the Synthesis, Application and Economic Feasibility of Ionic Liquids and Deep Eutectic Solvents for CO2 Capture: A Review. *Energies* **2021**, *15*, 9098. [CrossRef]
5. IEA. About CCUS. Available online: https://www.iea.org/reports/about-ccus (accessed on 5 January 2023).
6. Hou, Z.; Luo, J.; Xie, Y.; Wu, L.; Huang, L.; Xiong, Y. Carbon Circular Utilization and Partially Geological Sequestration: Potentialities, Challenges, and Trends. *Energies* **2023**, *16*, 324. [CrossRef]
7. IEA. Energy Technology Perspectives 2020. Available online: https://www.iea.org/reports/energy-technology-perspectives-2020 (accessed on 5 January 2023).
8. IPCC. Global Warming of 1.5 °C. Available online: https://www.ipcc.ch/sr15/ (accessed on 5 January 2023).
9. Bae, J.; Chung, Y.; Lee, J.; Seo, H. Knowledge spillover efficiency of carbon capture, utilization, and storage technology: A comparison among countries. *J. Clean. Prod.* **2020**, *246*, 119003. [CrossRef]
10. Sun, L.; Dou, H.; Li, Z.; Hu, Y.; Hao, X. Assessment of CO_2 storage potential and carbon capture, utilization and storage prospect in China. *J. Energy Inst.* **2018**, *91*, 970–977. [CrossRef]
11. Zhang, X.; Li, Y.; Ma, J.; Liu, L. Development of Carbon Capture, Utilization and Storage Technology in China. *Strateg. Study CAE* **2021**, *23*, 70–80. [CrossRef]
12. You, K.; Yu, Y.; Cai, W.; Liu, Z. The change in temporal trend and spatial distribution of CO_2 emissions of China's public and commercial buildings. *Build. Environ.* **2023**, *229*, 109956. [CrossRef]
13. Hills, C.D.; Tripathi, N.; Carey, P.J. Mineralization technology for carbon capture, utilization, and storage. *Front. Energy Res.* **2020**, *8*, 142. [CrossRef]
14. Xiong, Y.; Hou, Z.; Xie, H.; Zhao, J.; Tan, X.; Luo, J. Microbial-mediated CO_2 methanation and renewable natural gas storage in depleted petroleum reservoirs: A review of biogeochemical mechanism and perspective. *Gondwana Res.* **2022**; *in press*. [CrossRef]
15. Luo, J.; Hou, Z.; Feng, G.; Liao, J.; Haris, M.; Xiong, Y. Effect of Reservoir Heterogeneity on CO_2 Flooding in Tight Oil Reservoirs. *Energies* **2022**, *15*, 3015. [CrossRef]
16. Nocito, F.; Dibenedetto, A. Atmospheric CO_2 mitigation technologies: Carbon capture utilization and storage. *Curr. Opin. Green Sustain. Chem.* **2020**, *21*, 34–43. [CrossRef]
17. Ahmed, D.S.; El-Hiti, G.A.; Yousif, E.; Ali, A.A.; Hameed, A.S. Design and synthesis of porous polymeric materials and their applications in gas capture and storage: A review. *J. Polym. Res.* **2018**, *25*, 1–21. [CrossRef]
18. Ali, S.A.; Shah, S.N.; Shah, M.U.H.; Younas, M. Synthesis and performance evaluation of copper and magnesium-based metal organic framework supported ionic liquid membrane for CO_2/N_2 separation. *Chemosphere* **2023**, *311*, 136913. [CrossRef]
19. Jiang, K.; Ashworth, P.; Zhang, S.; Liang, X.; Sun, Y.; Angus, D. China's carbon capture, utilization and storage (CCUS) policy: A critical review. *Renew. Sustain. Energy Rev.* **2020**, *119*, 109601. [CrossRef]
20. Zhang, H. Regulations for carbon capture, utilization and storage: Comparative analysis of development in Europe, China and the Middle East. *Resour. Conserv. Recycl.* **2021**, *173*, 105722. [CrossRef]
21. Zhang, X.; Fan, J.L.; Wei, Y.M. Technology roadmap study on carbon capture, utilization and storage in China. *Energy Policy* **2013**, *59*, 536–550. [CrossRef]
22. Zhang, Z.; Wang, T.; Blunt, M.J.; Anthony, E.J.; Park, A.H.A.; Hughes, R.W.; Webley, P.A.; Yan, J. Advances in carbon capture, utilization and storage. *Appl. Energy* **2020**, *278*, 115627. [CrossRef]
23. Jiang, K.; Ashworth, P. The development of Carbon Capture Utilization and Storage (CCUS) research in China: A bibliometric perspective. *Renew. Sustain. Energy Rev.* **2021**, *138*, 110521. [CrossRef]

24. Pires, E.A.; Ribeiro, N.M.; Quintella, C.M. Sistemas de Busca de Patentes: Análise comparativa entre Espacenet, Patentscope, Google Patents, Lens, Derwent Innovation Index e Orbit Intelligence. *Cad. De Prospecção* **2020**, *13*, 13. [CrossRef]
25. Qi, Y.; Peng, W.; Xiong, N.N. The effects of fiscal and tax incentives on regional innovation capability: Text extraction based on python. *Mathematics* **2020**, *8*, 1193. [CrossRef]
26. Setchi, R.; Spasić, I.; Morgan, J.; Harrison, C.; Corken, R. Artificial intelligence for patent prior art searching. *World Pat. Inf.* **2021**, *64*, 102021. [CrossRef]
27. AcademicVocabulary. Available online: http://www.nottingham.ac.uk/alzsh3/acvocab/wordlists.htm (accessed on 12 January 2023).
28. Jelodar, H.; Wang, Y.; Yuan, C.; Feng, X.; Jiang, X.; Li, Y.; Zhao, L. Latent Dirichlet allocation (LDA) and topic modeling: Models, applications, a survey. *Multimed. Tools Appl.* **2019**, *78*, 15169–15211. [CrossRef]
29. Sbalchiero, S.; Eder, M. Topic modeling, long texts and the best number of topics. Some Problems and solutions. *Qual. Quant.* **2020**, *54*, 1095–1108. [CrossRef]

Article

Hydromechanical Impacts of CO_2 Storage in Coal Seams of the Upper Silesian Coal Basin (Poland)

Maria Wetzel [1], Christopher Otto [1,*], Min Chen [2], Shakil Masum [2], Hywel Thomas [2], Tomasz Urych [3], Bartłomiej Bezak [4] and Thomas Kempka [1,5]

[1] GFZ German Research Centre for Geosciences, 14473 Potsdam, Germany; kempka@gfz-potsdam.de (T.K.)
[2] Geoenvironmental Research Centre, School of Engineering, Cardiff University, Cardiff CF24 3AA, UK
[3] Central Mining Institute, 40-166 Katowice, Poland
[4] Polska Grupa Górnicza S.A., 40-039 Katowice, Poland
[5] University of Potsdam, Institute of Geosciences, 14476 Potsdam, Germany
* Correspondence: christopher.otto@gfz-potsdam.de

Abstract: Deep un-mineable coal deposits are viable reservoirs for permanent and safe storage of carbon dioxide (CO_2) due to their ability to adsorb large amounts of CO_2 in the microporous coal structure. A reduced amount of CO_2 released into the atmosphere contributes in turn to the mitigation of climate change. However, there are a number of geomechanical risks associated with the commercial-scale storage of CO_2, such as potential fault or fracture reactivation, microseismic events, cap rock integrity or ground surface uplift. The present study assesses potential site-specific hydromechanical impacts for a coal deposit of the Upper Silesian Coal Basin by means of numerical simulations. For that purpose, a near-field model is developed to simulate the injection and migration of CO_2, as well as the coal-CO_2 interactions in the vicinity of horizontal wells along with the corresponding changes in permeability and stresses. The resulting effective stress changes are then integrated as boundary condition into a far-field numerical model to study the geomechanical response at site-scale. An extensive scenario analysis is carried out, consisting of 52 simulation runs, whereby the impacts of injection pressures, well arrangement within two target coal seams as well as the effect of different geological uncertainties (e.g., regional stress regime and rock properties) is examined for operational and post-operational scenarios. The injection-induced vertical displacements amount in maximum to 3.59 cm and 1.07 cm directly above the coal seam and at the ground surface, respectively. The results further demonstrate that neither fault slip nor dilation, as a potential consequence of slip, are to be expected during the investigated scenarios. Nevertheless, even if fault integrity is not compromised, dilation tendencies indicate that faults may be hydraulically conductive and could represent local pathways for upward fluid migration. Therefore, the site-specific stress regime has to be determined as accurately as possible by in-situ stress measurements, and also fault properties need to be accounted for an extensive risk assessment. The present study obtained a quantitative understanding of the geomechanical processes taking place at the operational and post-operational states, supporting the assessment and mitigation of environmental risks associated with CO_2 storage in coal seams.

Keywords: CO_2 storage; horizontal well; coal swelling; regional stress regime; fault reactivation; hydromechanical simulation; environmental risk assessment

Citation: Wetzel, M.; Otto, C.; Chen, M.; Masum, S.; Thomas, H.; Urych, T.; Bezak, B.; Kempka, T. Hydromechanical Impacts of CO_2 Storage in Coal Seams of the Upper Silesian Coal Basin (Poland). *Energies* **2023**, *16*, 3279. https://doi.org/10.3390/en16073279

Academic Editor: Shu Tao

Received: 15 March 2023
Revised: 24 March 2023
Accepted: 4 April 2023
Published: 6 April 2023

1. Introduction

An essential key technology to reduce the amount of carbon dioxide (CO_2) released into the atmosphere and mitigate the effects of climate change is the safe and permanent storage of CO_2 in the deep geologic subsurface [1]. Anthropogenic CO_2 can be injected in porous formations, e.g., deep saline aquifers, depleted oil and gas reservoirs, but also coal beds. Deep un-mineable coal deposits are viable reservoirs for geological time periods, due to their high adsorption potential and large pore volumes [2], related to their characteristic internal structure [3]. Coal beds are naturally fractured sedimentary rocks, which generally consist of a well-defined fracture network (cleats), separating distinct blocks of the coal matrix, which comprises large amounts of macro- and micropores (Figure 1b). Moreover, coal seams are located in areas that are close to major CO_2 emissions sources [4,5], what would reduce costs of CO_2 transport from the source to the storage site. Whether there are also economic incentives to use coal seams as storage formations remains a topic of future research, as previous studies have only considered CO_2 storage in coal beds in combination with ECBM (Enhanced Coal Bed Methane Recovery) [6,7].

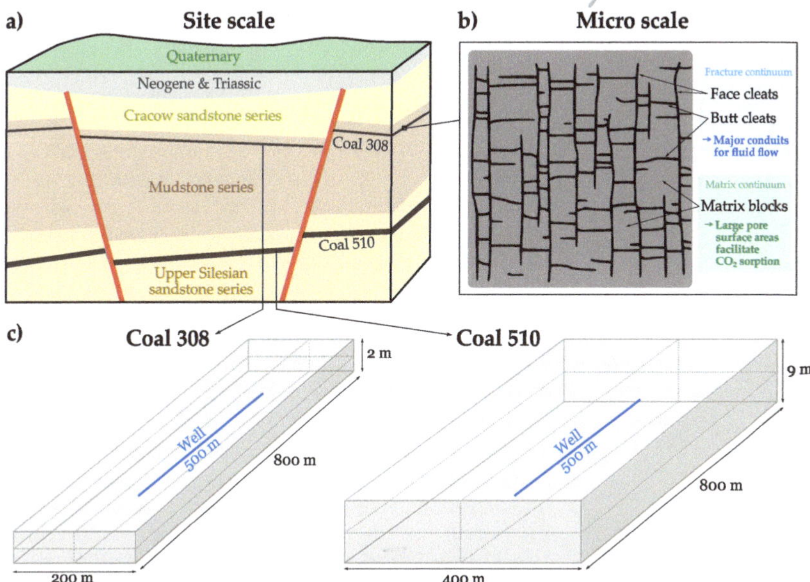

Figure 1. (**a**) Schematic illustration of the characteristic lithological sequence including the two selected target coal seams. (**b**) Typical coal microstructure consisting of a cleat network and a matrix, whereby both are considered in the near-field model by a dual porosity approach. (**c**) Dimensions of the near-field model domain simulating CO_2 injection and coal-CO_2 interactions in the borehole vicinity.

Under subsurface conditions, the injected CO_2 is partly present as a free phase and partly immobilised by adsorption to the coal matrix, what causes coal swelling. The adsorption-induced swelling of the coal matrix constricts the fluid flow pathways, leading to a reduction in permeability, and can result in a considerable loss of injectivity in the well bore vicinity [8–10]. Coal swelling constitutes a major challenge, as experienced in several demonstration projects related to CO_2 injection into coal beds [11–13]. Fang et al. [14] developed a dynamic matrix- and fracture-permeability model to simulate the CO_2-ECBM process, which takes into account the effects of pressure relief, CO_2 pressure injection, matrix shrinkage, and swelling on permeability changes. In general, coal swelling is almost proportional to the amount of CO_2 adsorbed in its matrix [3]. The capacity for CO_2 storage depends on the coal rank and operational conditions [15], such as temperature, pressure and coal moisture content [16,17]. Using horizontal wells could substantially enlarge the coal-CO_2 contact area, and establish a good connection to the cleats as dominant flow pathways [18]. Many numerical simulation studies investigated the potential of using horizontal wells in the context of ECBM [19–23], whereas their field applications are rarely reported [24]. Hence, in order to deliver sustained rates of CO_2 injection and overcome the challenge posed by coal swelling, this study investigates a horizontal well configuration.

A number of hydromechanical risks is associated with the storage of CO_2, such as fault or fracture reactivation, cap rock integrity and ground surface movement. An injection of CO_2 into the geological subsurface always induces a pressure build-up, whereby also small changes in reservoir pressure alter the pre-injection stress state and result in geomechanical effects. Even small pore pressure changes may result in microseismic events due to rock heterogeneities, and represent a risk of generating new fractures, what could in turn affect caprock integrity. Larger changes of the recent stress field may reactive critically stressed faults, which then may act as preferential leakage pathways for a buoyancy-driven migration of CO_2 into shallower aquifers. To address these geomechanical risks, numerical models can be used to predict the mechanical response related to an injection of CO_2 [25–27].

In the present study, the hydromechanical impact of a prospective commercial-scale CO_2 storage is assessed for a coal deposit. Therefore, a near-field model is used to simulate the injection and migration of CO_2 as well as the coal-CO_2 interactions for the vicinity of horizontal wells, along with the corresponding changes in permeability and stress. The resulting effective stress changes are then integrated as a boundary condition into a far-field numerical model to study the geomechanical response at site-scale. The far-field model comprises all relevant lithostratigraphic units as well as the exact geometries of the local fault system. This serves as basis for an extensive scenario analysis investigating the effect of different geological uncertainties (e.g., regional stress regime and rock properties). The main aim of the present work is to obtain a quantitative understanding of the geomechanical processes taking place at the operational and post-operational states to assess and mitigate environmental risks, such as ground surface uplift and a fault reactivation.

2. Materials and Methods

2.1. Geology of the Study Site

The prospective commercial-scale storage site is located in the central part of the Upper Silesian Coal Basin. The lithostratigraphy in the study area includes Quaternary (Pleistocene) and Neogene (Miocene) sediments, lying unconformable on top of Triassic deposits and the Upper Carboniferous. The productive coal-bearing Carboniferous strata consist of the Cracow sandstone series, the Mudstone series as well as the Upper Silesian sandstone series [28].

Two coal seams are found to be suitable for CO_2 storage: coal seams 308 and 510 (Figure 1a). Coal seam 308 is situated in the Mudstone series and represents a well studied seam, which has been exploited for decades. It is located at an average depth of 405 m and gently dips towards ESE, whereby its mean thickness is about 2 m. Coal seam 510 is part of the Upper Silesian sandstone series and lies at a considerably greater depth of 1250 m below the ground surface. It gently dips towards SSW and exhibits, with in average 9 m, the highest thickness of all seams in this area [29].

Several steep dipping major faults separate the study area into individual tectonic blocks, whereby each block exhibits a relatively homogeneous internal structure. The tectonic block of interest forms a trench structure, whereby its lateral extent considerably decreases with depth. The dominant strike directions for the four major faults limiting the considered fault block are SW–NE and SE–NW. Moreover, there is a network of minor faults with offsets of up to several meters, which have to be considered for the assessment of fault reactivation and as potential CO_2 leakage pathways.

2.2. Near-Field Model Simulating CO_2 Injection and Adsorption-Induced Coal Swelling

For the near-field model domain, a coupled hydromechanical model is developed to assess CO_2 flow within the target coal seams and coal-CO_2 interaction along with the corresponding changes in permeability and stress. Numerical simulations are performed using the parallel finite element-based simulator COMPASS (COde for Modelling PArtially Saturated Soils), which has been developed by Thomas et al. (e.g., [30,31]). The governing equations of flow, deformation and the adsorption/desorption kinetics are included with appropriate couplings to simulate fluid flow through an elastically deformable fractured rock.

The characteristic internal structure of coal is conceptualised using a dual continuum model. The cleat network constitutes the primary flow path, whereas the coal matrix provides available adsorption sites for CO_2 (Figure 1b). The fracture system and porous matrix are treated as distinct continua, and flow interactions between both continua are integrated using mass exchange. The following general assumptions are made: (i) CO_2 adsorption occurs in the matrix continuum only, (ii) coal is a dry, homogeneous, isotropic and dual poroelastic medium, (iii) strain is small, (iv) coal is isothermal, and (v) Darcy's law is applied for gas flow in fractures. The reliability of the proposed model has been extensively evaluated by numerous validation studies. The reader is kindly referred to [18,31,32] for all details on the mathematical model implementation and verification benchmarks. These studies demonstrated that simulated gas transport and coal-CO_2 interactions are consistent with the respective experimental results.

Two models are set up for the coal seams 308 and 510 at the study site, whereby model dimensions vary accordingly. For both coal seams, the total model length is 800 m, of which 500 m corresponds to the length of the horizontal well (Figure 1c). The model height relates to the respective average thickness of the coal seams and amounts to 2 m and 9 m for seam 308 and 510, respectively. The model width is based on the anticipated maximal CO_2 migration distances, and amounts to 200 m and 400 m for coal seams 308 and 510, respectively. Due to the underlying symmetry, only one quarter of the domain is simulated. The model domain is discretised using tetrahedral elements. A zero-flux boundary for gas flow is applied at the external domain boundaries. For coal deformation, a constant volume condition is applied. It is assumed that there is no CO_2 present in the model domain prior to the injection. Figure 2 exhibits the adsorption isotherm and adsorption-induced swelling strain based on van Wageningen et al. [33], which are considered for the numerical simulations. Further model parameters and the corresponding references are listed in Table 1.

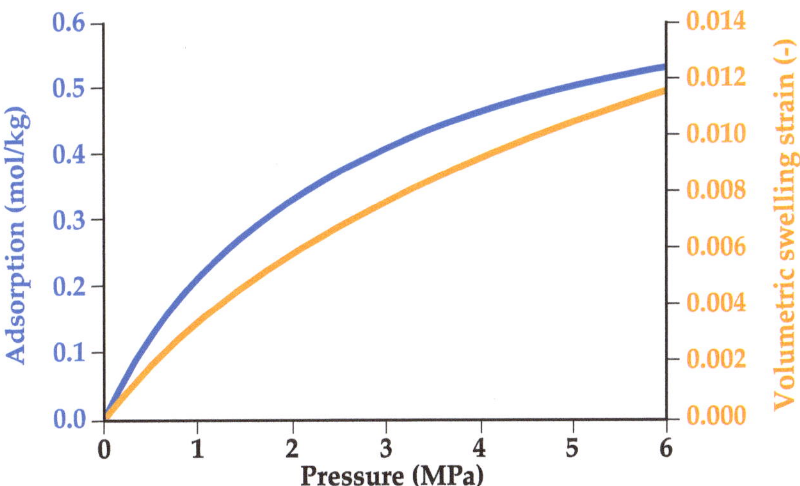

Figure 2. CO_2 adsorption isotherm (blue line) and adsorption-induced swelling strain (orange line) used in the simulations [33].

Three injection scenarios are simulated for respective coal seam, whereby different constant injection pressures are assigned to the horizontal well. These pressures are related to the hydrostatic pressures at the respective seam depths, assuming that the maximum pressure increase is below the 1.5-fold of the hydrostatic pressure. Consequently, constant injection pressures of 0.5 MPa, 1.0 MPa and 1.5 MPa are assumed for coal seam 308.

The scenarios for the deeper coal seam 510 are oriented on the hypothetical hydrostatic pressure increase of the scenarios for seam 308 to maintain comparability between the results. Thus, constant injection pressures of 1.5 MPa, 3.0 MPa and 4.5 MPa used at the horizontal well in coal seam 510. The resulting total amounts of injected CO_2 for the respective scenarios are listed in Table 2. The injection operation is simulated for a time period of one year (operational state), afterwards CO_2 injection is stopped. To evaluate the effectiveness of long-term storage as well as CO_2 migration and the evolution of adsorption-induced coal swelling, the total simulation time is extended to 15 years (post-operational state).

Table 1. Hydrochemical parameters of the near-field model.

Parameter		Value	Reference
Intrinsic permeability (mD)	Seam 308	1.0	[34]
	Seam 510	0.3	[35]
Initial fracture porosity (-)	Seam 308	0.015	[34]
	Seam 510	0.006	[35]
Matrix porosity (-)		0.04	[36]
Gas viscosity (Pa·s)		1.84×10^{-5}	[37]
Unstressed fracture compressibility (MPa^{-1})		0.029	[38]
Matrix elastic modulus (GPa)		11.5	[39]
Langmuir pressure constant (MPa)		2.6	[33]
Langmuir volume constant (mol/kg)		0.76	[33]

Table 2. Injection and effective pressures as well as the amount of injected CO_2 per single well for the six near-field scenarios.

Coal Seam	Constant Injection Pressure (MPa)	Pressure Increase in Relation to Respective Hydrostatic Pressure (-)	Amount of Injected CO_2 (t)	Maximum Effective Pressure Change (MPa)	
				1 year	15 years
308	1.5	1.36	2018	15.5	7.3
	1.0	1.24	1239	11.8	6.3
	0.5	1.12	560	7.2	4.4
510	4.5	1.37	22,139	20.9	9.4
	3.0	1.29	12,551	17.1	7.6
	1.5	1.12	5002	11.2	5.2

2.3. Far-Field Model Assessing the Commercial-Scale Hydromechanical Impact

A hydromechanical model is implemented by means of the numerical simulator FLAC^{3D} [40], aiming at capturing the geomechanical response and evaluating risks of a prospective CO_2 injection at field-scale. The model consists of five individual stratigraphic units and two coal seams, which are simplified using the average dipping angle, azimuth, and mean depth for each geological layer. The thickness of all units varies within the model, except for the coal seams, which have a constant thickness of 2 m and 9 m, respectively (Figure 3).

The local fault system, consisting of four major and five minor faults, is implemented using their entire complex geometry. In order to localise zones of potentially reduced mechanical integrity, and thus potential fluid migration pathways at reactivated fault segments, fault zones are implemented into the hydromechanical model as 67,655 ubiquitous joint elements. Ubiquitous joints resemble weak zones within a geomechanical model, whereby dip angles and directions of faults may be individually assigned to each model element at the fault plane. The faults cut the entire Carboniferous strata, while Neogene, Triassic and Quaternary units remain unaffected. Fault displacements are incorporated to consider the effect of rock property changes along the faults. A linear-elastic constitute law is assigned to all lithological layers in the numerical model, whereas the Mohr–Coulomb failure model is used for the ubiquitous joint elements. Fault properties are assigned with zero cohesion, a friction angle of 20° and a dilation angle of 10° to maintain conservative assumptions but not to underestimate potential shear and tensile failure [41]. The resulting numerical model, with the applied simplifications, accurately captures the characteristic geometry of the lithological units and especially the complex fault system (Figure 4).

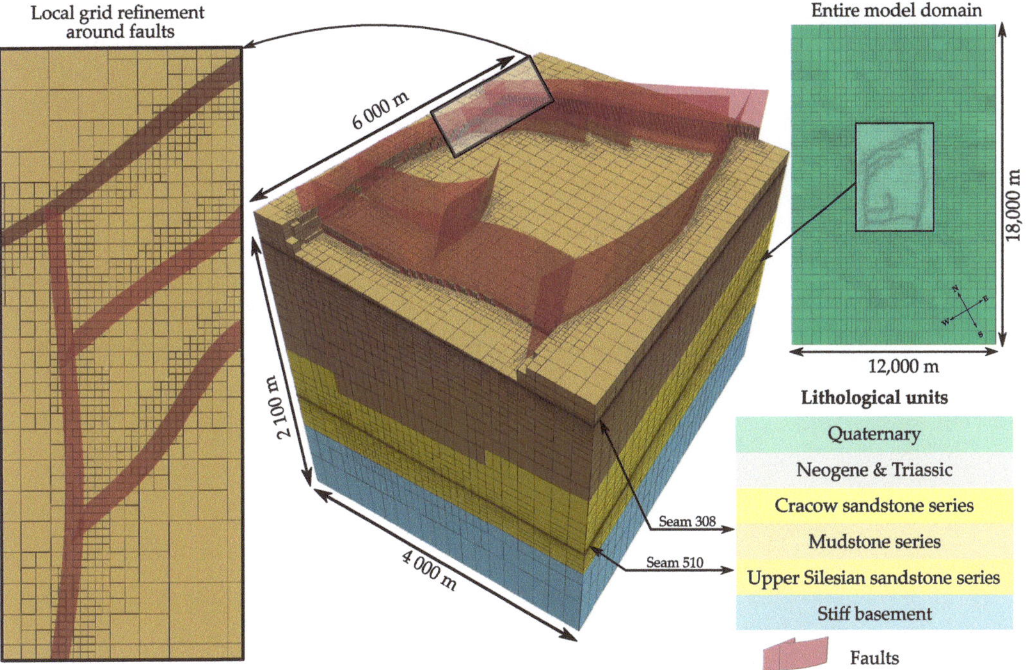

Figure 3. Numerical model of the potential storage site. The entire model domain has the three-fold size of the tectonic block to avoid mechanical impacts on the lateral model boundary conditions. The thickness of all layers, aside from the coal seams, varies within the model as a result of the varying dip directions and angles of the lithostratigraphic top surfaces.

The model has an extent of 4000 m × 6000 m and is discretised with a lateral resolution of 200 m. The size of the entire model domain (12,000 m × 18,000 m) corresponds to the three-fold of the area of interest to avoid impacts of the applied mechanical boundary conditions on the simulation results as discussed in [42]. Local grid refinements by a factor of three are applied to the elements in the near-field model domain and fault vicinity to accurately account for local hydromechanical effects (Figure 3). Fixed velocities perpendicular to the model bottom and the lateral boundaries are applied, while the model top is defined to remain unconstrained. The model is discretised by 500,000 elements, with grid element sizes decreasing towards the target coal seams and increasing towards the model boundaries.

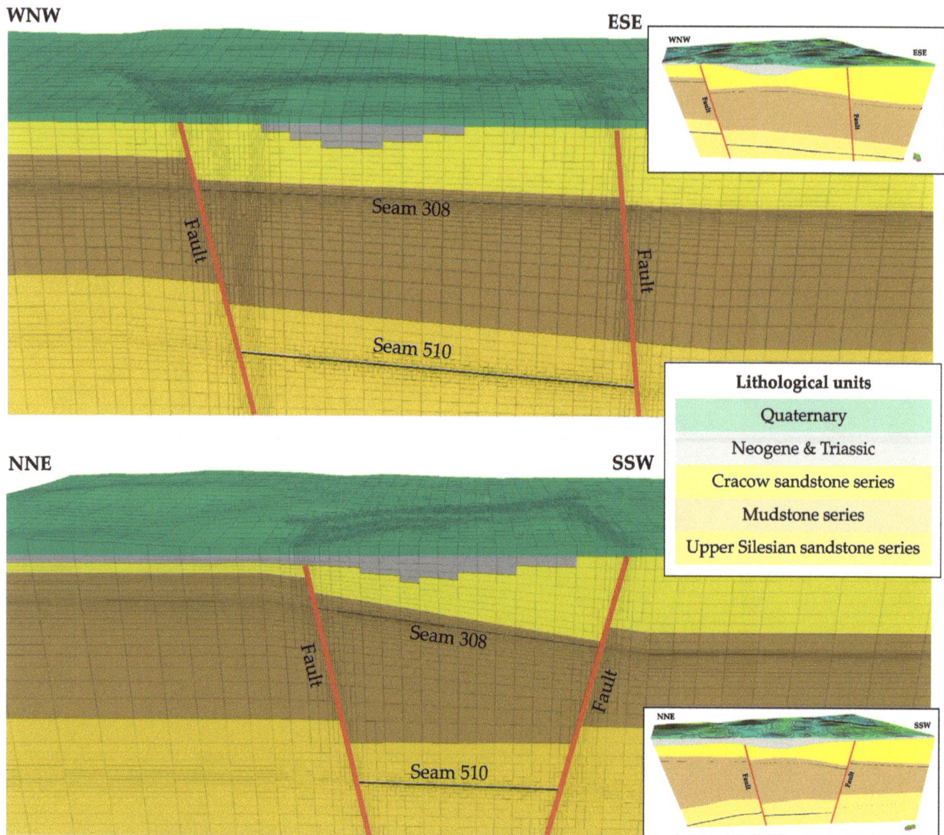

Figure 4. Comparisons of WNW-ESE (**above**) and NNE-SSW (**bottom**) cross sections of the numerical model with the respective cross-sections of the structural geologic model (illustrated cut-outs on the right).

In order to analyse potential site-specific hydromechanical impacts of the prospective CO_2 storage operation, rock parameter uncertainties as well as the regional stress regime are assessed as geological uncertainties. The effect of mechanical rock properties is investigated in two scenarios: a soft and a stiff one (Table 3). The stiffer parameter for the Carboniferous strata refer to averaged data from the Wieczorek mine [41,43], whereas the soft parameters are derived from average values of a few dozen borehole samples from a local coal deposit [44]. Geomechanical parameters of the Quaternary, Neogene, Triassic and the stiff basement are assumed based on their lithologies. Elastic properties of the two coal seams are assigned according to the near-field model with Young's moduli of 1.8 GPa [36] and 2.2 GPa [44] for coal seam 308 and 510, respectively. These values are maintained constant, since changing elastic moduli of the coal would result in different simulated effective stress changes within the near-field model, what would require dedicated near-field simulations to incorporate the effect of coal property changes.

Table 3. Geomechanical parameters used for the undertaken scenario analysis.

Property		Quaternary	Neogene and Triassic	Cracow Sandstone Series	Mudstone Series	Coal (308)	Upper Silesian Sandstone	Coal (510)	Stiff Basement
Young's modulus	Soft	2	2	3.3	3.6	1.8	5.9	2.2	80
(GPa)	Stiff	5	5	7.5	8.1	1.8	8.4	2.2	80
Poisson's ratio (-)		0.4	0.25	0.21	0.26	0.3	0.23	0.3	0.27
Tensile strength (MPa)		0.0	1.0	2.0	2.6	0.6	2.6	0.6	3.8
Cohesion (MPa)		0.0	2.0	14.9	18.1	3.8	14.0	3.8	10.9
Friction angle (°)		35	31	31	31	31	31	31	31
Density (kg/m^3)		2100	2200	2115	2418	1388	2378	1388	2465

Additionally, the effect of the regional stress regime is assessed. For the Upper Silesian Coal Basin in Poland, most of hydraulic fracturing tests and focal mechanism data indicate the predominance of a strike-slip stress regime with a maximum horizontal stress orientation in NW-SE direction and a S_{Hmax} azimuth of 167° [45], whereby S_{Hmax} was reported to be 1.22 S_V. By contrast, the analysis of local focal mechanisms around the local coal deposit indicates a predominant normal faulting regime, whereby the azimuth of the maximum horizontal stress orientation is 79° [44]. Both stress regimes are investigated, whereby the vertical stress gradient results from the gravitational load of the overburden. The equilibrated mechanical state of the numerical model serves as starting data set for the respective scenario analysis.

2.4. Workflow to Couple Near- and Far-Field Simulations

Since so far not one numerical simulator is able to assess the complex processes of mechanics, adsorption and multiphase flow across several spatial scales, numerical simulations are performed at both scales separately. The effects of interactions between fluid flow, stress, and gas adsorption in the near-field domain are then implemented within the far-field model to study hydromechanical effects due to adsorption-induced coal swelling and pore pressure increase.

Therefore, a robust and adaptive coupling workflow is developed, which is able to flexibly transfer data between different simulation programs and model grids. Near-field simulations are performed with the numerical modelling code COMPASS, which employs tetrahedral grids for model discretisation. The hydromechanical far-field model is generated with the geomechanical simulator FLAC^{3D} and uses hexahedral elements. Effective stress changes have been chosen as coupling parameter. Due to fixed displacement boundary conditions in the near-field models, effective stresses considerably increase, mainly due to the high surface stresses resulting from the CO_2 adsorption. Thus, transferring and implementing effective stress changes enables to investigate hydromechanical effects of the CO_2 storage operation in the far-field.

At first, the FLAC^{3D} mesh of the near-field model domain is refined by a factor of three to accurately transfer the near-field simulation results. A Python-based workflow extracts and calculates effective stress changes for the operational and post-operational time step in the near-field and interpolates them to the refined far-field mesh, where effective stress changes are assigned as internal boundary condition. Thus, results for each near-field injection domain are replicate within the target coal seams for several injection wells in the far-field model.

2.5. Injection Well Arrangement

In order to store a preferably high amount of CO_2, as many horizontal wells as possible are placed within target coal seams, whereby the total number of panels is an important factor regarding previous economic considerations. An essential criterion to minimise the risk of fault reactivation is the safety distance to the known geological fault zones. It must be chosen carefully in order to not negatively affect the operation and become a concern for health, saftey and the environment. Based on the calculated migration distance of the CO_2 from the near-field simulations, a minimum distance of at least 200 m between the horizontal injection blocks (marked in blue in Figure 5) and all nine faults is considered.

Thus, regarding the predefined well length of 500 m, 10 and 12 horizontal wells can be placed within the upper and lower seams, respectively. For coal seam 308, an additional limiting factor exists, since most of the Eastern parts of the coal seam are already mined. Thus, only the Western part of coal seam 308 is taken into consideration for CO_2 storage by the present study. Figure 5 shows the resulting injection well arrangement investigated in the numerical simulations: 10 panels can be applied, considering a near-field model domain of 200 m × 800 m. The horizontal extent of coal seam 510 within the tectonic block is smaller and comprises an area of about 2400 m × 4000 m. Due to the respective dip direction, only three of the five minor faults cut the coal seam and must be considered in the well design. Thus, 12 panels can be realised for the investigated near-field domain of 400 m × 800 m lateral size.

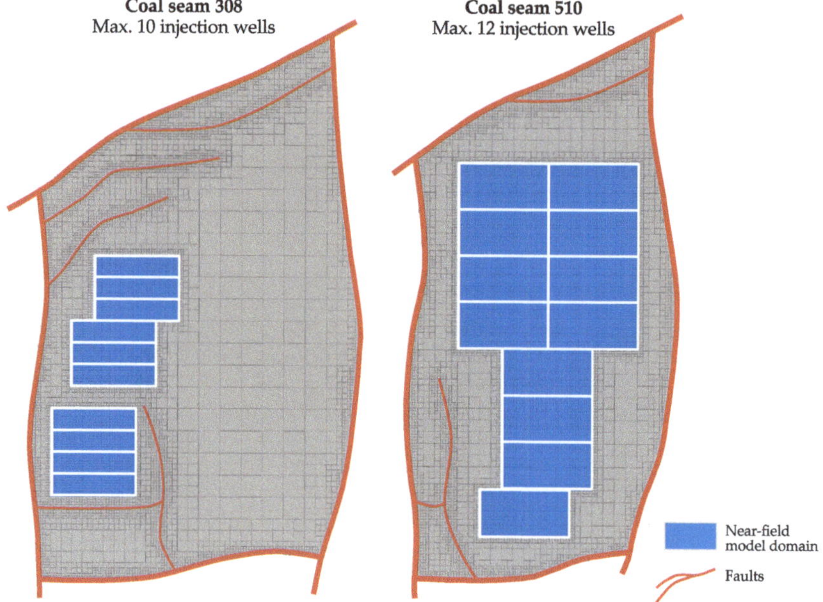

Figure 5. Injection well arrangement for coal seams 308 and 510 with 10 and 12 panels, respectively. The near-field model domain (blue) has an extent of 200 m × 800 m and 400 m × 800 m for seams 308 and 510, respectively.

3. Results

3.1. Effective Stress Changes in Near-Field Simulations

Two representative simulations were selected from the near-field simulations to assess the hydromechanical impact of CO_2 injection for certain stress states: (i) an operational state (1 year), which corresponds to the time of the injection stop, and thus the maximum pressure increase, and (ii) a post-operational state (15 years). Especially changes in effective stress are evaluated, since it has been defined as coupling parameter between the near- and far-field simulations. The changes in effective stress partly result from pore pressure changes due to the injection, and partly from adsorption-induced coal swelling.

For the upper coal seam 308, the highest effective stress change of up to 15.5 MPa occurs for the scenario simulating a constant injection pressure of 1.5 MPa at a simulation time of 1 year. After 15 years of simulation, stresses are relieved and distributed in the near surroundings of the coal seam, since CO_2 further migrates after the injection stop. The maximum effective stress increase within the near-field model is still about 7.3 MPa (Figure 6). In this scenario, 2018 t of CO_2 are injected per panel (Table 2).

Coal seam 510 is located at a considerably greater depth, resulting in significantly higher injection pressures. The highest effective stress increase amounts to 20.9 MPa for the scenario with a constant injection pressure of 4.5 MPa. After 15 years, the observed maximum effective stress increase within the near-field model is still about 9.4 MPa (Figure 6). The amount of injected CO_2 is with 22,139 t per panel, more than ten times higher than that for coal seam 308. This is related to the higher density of CO_2 at the given depths, as well as the higher thickness (factor of 4.5) of coal seam 510. The maximum effective stress and the spatial expansion of the pressure increase is considerably lower for scenarios assuming lower injection pressures, as shown in Figure 6.

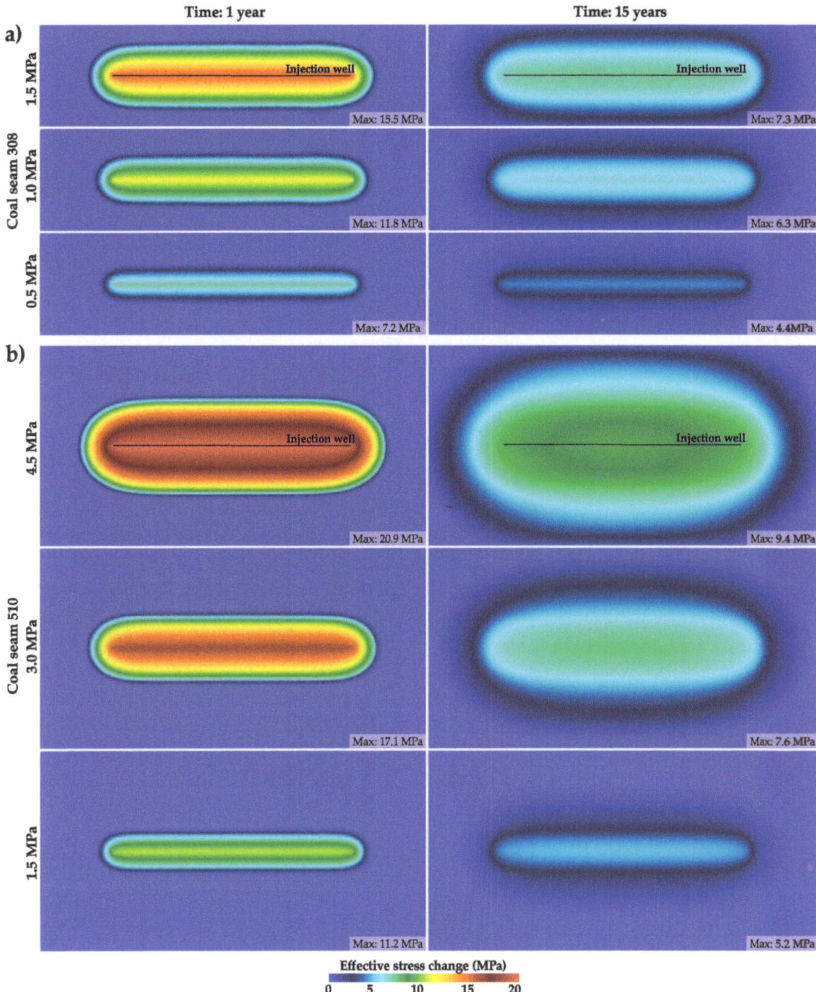

Figure 6. Effective stress changes in coal seams (**a**) 308 and (**b**) 510 for the investigated scenarios at simulation times of 1 year (operational state) and 15 years (post-operational state).

3.2. Vertical Displacements in the Far-Field Simulations

An extensive scenario analysis, consisting of 52 simulations, is conducted to obtain a quantitative understanding of the mechanical processes taking place during the operational and post-operational states. Moreover, the impact of different geological uncertainties, such as the regional stress regime and rock parameter uncertainties are assessed. The elaborated injection well arrangement enables to assess the maximum effect of a CO_2 storage operation at the potential commercial-scale site for the 10 and 12 injection panels in coal seam 308 and 510, respectively. A direct hydromechanical effect of CO_2 storage is the occurrence of vertical displacements in the rock mass due to considerable changes in the stress field. The results discussed in the following refer to the scenarios considering soft rock properties to present the maximum hydromechanical effects, whereby the impact of rock properties is separately analysed in Section 3.3.

3.2.1. CO_2 Injection into Seam 510

Maximum effective pressure changes, and thus vertical displacements occur for a constant injection pressure of 4.5 MPa at a simulation time of 1 year. At this time, the effective pressure change amounts to 20.9 MPa, leading to a maximal vertical displacement of 3.59 cm directly above the coal seam. Due to effective stress changes and the swelling of the coal seam, the rock mass is displaced both, above and below the coal seam, whereby the displacements below the seam are lower due to the higher overburden stresses, and limited ability to displace rock mass into that direction (Figure 7). The dip of the coal seam of 3.1° towards the South is visible within the lateral distribution of vertical displacements, whereby the displacements are slightly higher for the shallower panels. The high displacements directly above the injection well strongly decrease in vertical direction due to the distribution of the stresses, so that the displacements are less localised and occur over a broader area. As Figure 8b,c illustrate, high displacements are also laterally limited to the close vicinity of the injection wells, and strongly decrease within a distance of decametre. Within the Mudstone series, the vertical displacement amounts to circa 1 cm in maximum (Figure 7), and the effect of each single well is not any more recognisable. At the ground surface, the simulated maximum displacement amounts to 0.99 cm. It occurs in the central part of the investigated fault block and slightly decreases towards its boundaries, whereby an area within a distance of 3500 m × 2400 m exhibits a ground surface uplift of more than 0.5 cm. The symmetric structure of the surface displacement is related to the relatively even distribution of horizontal injection wells within coal seam 510.

At the simulation time of 15 years, the maximal vertical displacements directly above the seam are considerably lower, since the CO_2 is migrated away from the wells and effective stress distribution in the near-field model domain is less localised (Figure 8). The maximum displacements amount to 1.97 cm and 0.89 cm above coal seam 510 and at the ground surface, respectively. In this scenario, 265,674 t of CO_2 are stored within the 12 panels in coal seam 510.

Lower injection pressures of 3.0 MPa and 1.5 MPa result in lower vertical displacements of in maximum 2.64 cm and 1.47 cm above the coal seam, respectively. Moreover, the lower the injection pressure, the lower the difference in vertical displacements between the investigated operational (1 year) and post-operational states (15 years) at the ground surface (Figure 7). The maximum displacements as well as the amounts of injected CO_2, depending on the injection pressure for all scenarios, can be found in the Appendix A (Table A1).

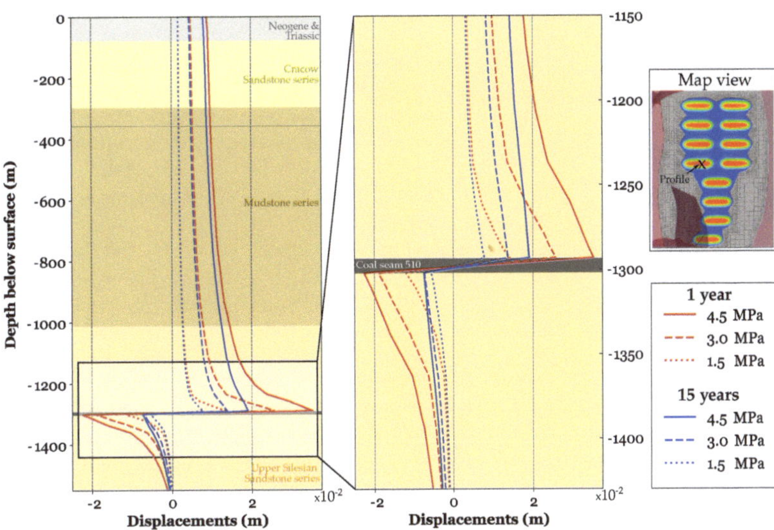

Figure 7. Vertical profile from the model top through one of the horizontal injection panels of coal seam 510 exhibiting the simulated vertical displacements for constant injection pressures of 4.5 MPa, 3.0 MPa and 1.5 MPa during the operational (1 year) and post-operational state (15 years).

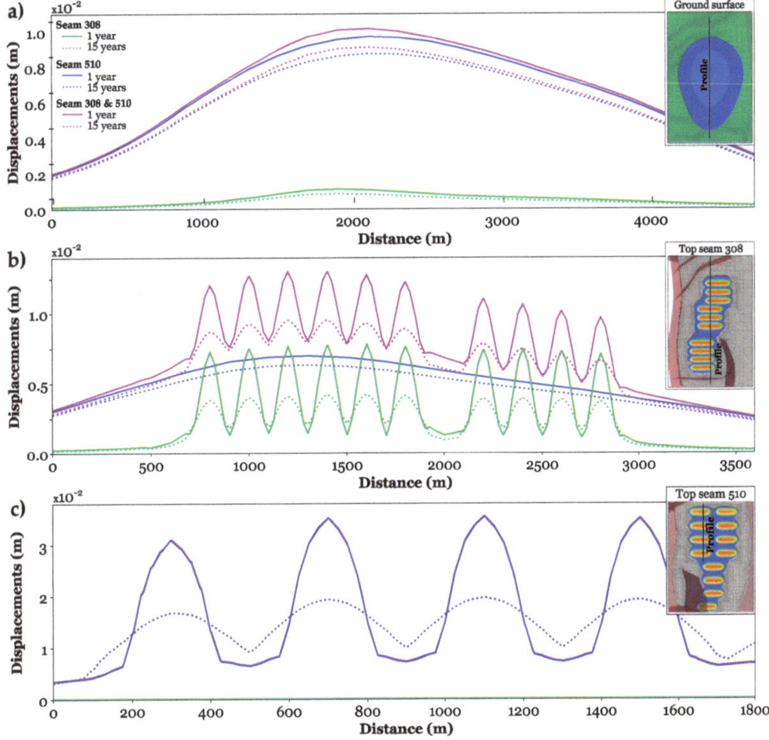

Figure 8. Vertical displacements (**a**) at the ground surface, (**b**) the top of coal seam 308, and (**c**) seam 510 for a CO_2 injection into seam 308, 510 and both seams simultaneously considering the operational and post-operational scenarios.

3.2.2. CO_2 Injection into Seam 308

For all scenarios considering an injection into coal seam 308, the vertical displacements are relatively low compared to those dedicated to seam 510. Maximum effective stress changes occur for the operational state simulating 1.5 MPa constant injection pressure in each of the 10 horizontal wells. Hence, the resulting maximum effective pressure change of 15.5 MPa leads to a vertical displacement of 0.77 cm, which occurs in the centre of each near-field injection panel (Figure 8). At the model top, the ground surface uplift is in maximum 0.34 cm.

At the post-operational state (15 years), the maximal vertical displacements are considerably lower, since the CO_2 is migrated away from the wells and the effective stress distribution in the near-field model domain is less localised. In this phase, the maximum displacements are 0.42 cm and 0.25 cm above the coal seam and at the ground surface, respectively. In this scenario, 20,185 t CO_2 are stored within the 10 panels in coal seam 308.

3.2.3. CO_2 Injection into Both Seams Simultaneously

In order to evaluate the maximum hydromechanical effects of the CO_2 storage operation at the selected site, four scenarios are simulated, considering CO_2 injection into both coal seams simultaneously. Therefore, the maximum elaborated injection pressures of 1.5 MPa and 4.5 MPa for coal seam 308 and 510 are applied, respectively. Figure 8 illustrates that a CO_2 injection into both seams leads to the highest vertical displacements of up to 1.55 cm compared to 0.77 cm and 1.08 cm for the injection operation solely targeting seams 308 and 510, respectively. The maximum ground surface uplift for an injection operation using both seams is 8% higher, with 1.07 cm, compared to the scenario simulating an injection into coal seam 510, only. Moreover, the area exhibiting vertical displacements of more than 0.5 cm is slightly larger with 2500 m × 3000 m, especially in the Western part of the model, where CO_2 is injected into seam 308 (Figure 9).

Regarding the changes in rock mass uplift between the operational and post-operational scenario, a clear tendency can be observed: the lower the injection pressure, the lower the difference in ground surface uplift between the operational (1 year) and post-operational states (15 years). For a simultaneous injection into both coal seams, the difference between the operational and post-operational scenarios reduces from 45% above the deeper seam 510 to 23.9% at the shallower seam 308, and 12.1% at the model top. Thus, the hydromechanical impact of CO_2 injection is the highest within the immediate surrounding of the injection well, where large and localised effective stress changes occur.

For the simulated storage operation in both coal seams, the injected amount of CO_2 is the highest with 285,858 t, whereas the major part of 93% or 265,673 t is injected into coal seam 510, while only 7% or 20,185 t can be stored in coal seam 308. This is related to the higher seam thickness and depth below ground surface, and consequently the higher volume of coal seam 510 and higher density of CO_2.

3.3. *Effects of Rock Parameter Uncertainties*

Elastic rock properties of the main lithological units are varied to assess the effects of parameter uncertainties based on the data available for the study area [43,44]. In the soft rock scenario, the elastic moduli are about half as high as the moduli of the stiff rock scenario (Table 3). Only the properties of the two coal seams are maintained constant, consistent with the near-field simulations.

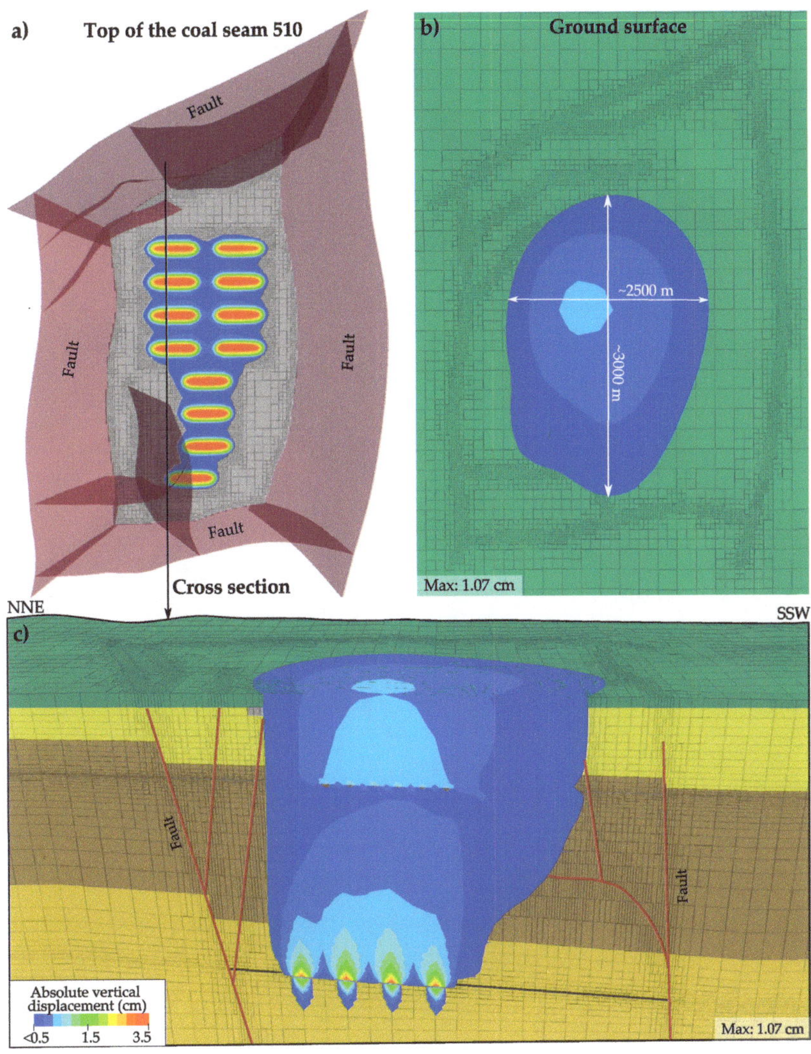

Figure 9. Simulated vertical displacements in (**a**) map view above coal seam 510, (**b**) at the ground surface and (**c**) along a NNE-SSW trending cross section as for an injection in both coal seams with the respective maximum injection pressures for the operational state (1 year).

The vertical displacements are slightly lower in all scenarios, where stiff rock properties are applied (Figure 10). Nevertheless, the differences in vertical displacements amount to 6.4% and 7.3% in maximum above the coal seams 510 and 308, respectively, and are thus relatively small. At the ground surface, the differences are even lower and range between 1.5% and 6.3% for coal seams 510 and 308, respectively. The difference in percentage terms for the vertical displacement above the respective seam decrease with increasing injection pressures and simulation times (Table 4).

Since the integrated effective stress changes lead to coal swelling in the respective target seam, the properties of the surrounding rock mass affect the vertical displacements marginally, only. Nevertheless, elastic properties of the coal seam itself can considerably influence the coal swelling behaviour, and thus the vertical displacements. Due to the previously undertaken extensive field and laboratory investigations, mechanical properties of the coals in the target area are well known, and thus not constitute a systemic uncertainty.

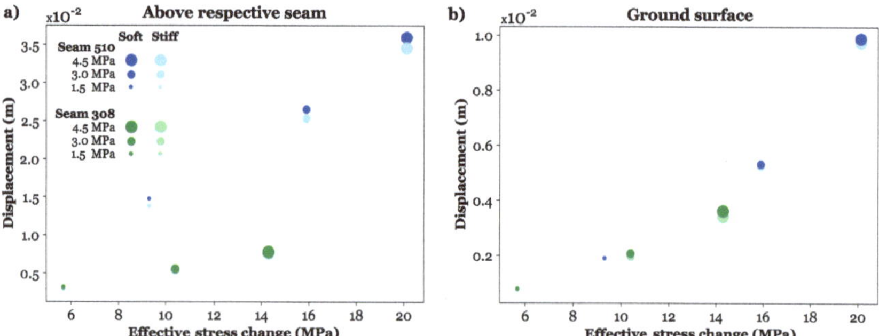

Figure 10. Maximum vertical displacements (**a**) above the respective coal seams as well as (**b**) the ground surface compared for the scenarios with soft and stiff rock properties for the operational state (1 year).

Table 4. Percentage deviation in maximum displacements between the soft and stiff rock property scenarios.

	Constant Injection Pressure (MPa)	Max. Displacements for the Stiff Property Scenario (%)			
		Above Coal Seam		At Ground Surface	
		1 year	15 years	1 year	15 years
Coal seam 308	1.5	−2.5	−1.7	−6.1	−5.5
	1.0	−3.4	−2.5	−6.2	−5.9
	0.5	−7.3	−4.6	−6.3	−6.3
Coal seam 510	4.5	−3.9	−3.9	−1.5	−1.4
	3.0	−4.6	−3.7	−1.5	−1.5
	1.5	−6.4	−4.9	−1.5	−1.5

3.4. Stress Field and Fault Reactivation Potential

Besides ground surface uplift, a potential risk of CO_2 injection includes the potential reactivation of pre-existing geologic faults due to spatial and temporal changes in the recent stress field. Reactivated faults may then act as preferential leakage pathways for a buoyancy-driven upward migration of CO_2. If a pore pressure increase accompanying fluid injection reactivates faults or fault segments, essentially depends on the initial stress state. Therefore, two different stress regimes are assessed as geological uncertainty for the storage site: (i) a normal faulting regime, which has been reported for the local coal deposit, and (ii) a strike-slip regime, which is characteristic for the Upper Silesian Coal Basin. The nine implemented faults present in the study area are examined in the scope of a comprehensive fault slip and dilation tendency analysis. Fault slip tendency is the ratio of resolved shear stress to resolved normal stress at a surface [46], whereby fault dilation tendency is defined by the stress acting normal to a given surface [47]. They describe the likelihood of a plane to slip and to dilate at a certain applied stress state [48]. Both parameters range between 0 and 1, whereby higher values indicate a higher tendency for mechanical failure.

Initial slip and dilation tendencies of the fault system considering a normal faulting regime ($S_V > S_{Hmax} > S_{hmin}$) are shown in Figure 11a,b. In this scenario, slip tendencies are generally very low and do not exceed 0.05 in most parts of the fault system. Only the deeper parts of the W-E trending minor faults, as well as the SE fault exhibit slightly higher slip tendencies of up to 0.08. Regarding the dilation tendency, several faults of the system are initially nearly-critically or critically stressed for dilation. Especially the NE and SW faults, and parts of the NW fault exhibit high values of more than 0.85 (Figure 11b). The high dilation tendencies for the assessed normal faulting regime are an indicator for the presence of a conduit behaviour at the respective faults, potentially enabling fluid flow along the fault planes.

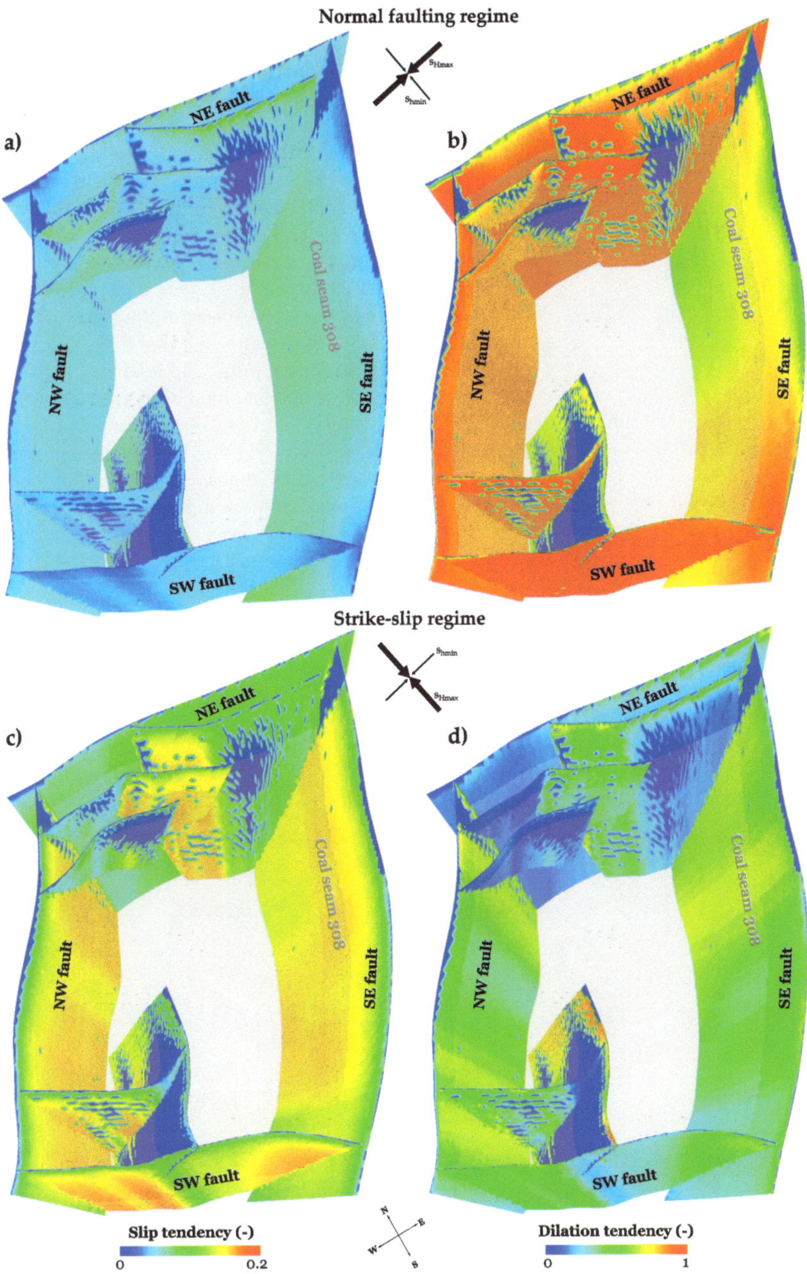

Figure 11. (**a**) Simulated slip and (**b**) dilation tendencies for the fault system of the study area for a normal faulting regime, as well as (**c**) slip and (**d**) dilation tendencies for a predominant strike-slip regime.

Initial slip and dilation tendencies of the fault system considerably vary for a strike-slip regime ($S_{Hmax} > S_V > S_{hmin}$) and are displayed in Figure 11c,d. In this scenario, slip tendencies are more than twice the slip tendencies of those of a normal faulting regime, but still relatively low with maximum values of 0.18. There are local variations due to the undulating nature of the faults and the orientation of their fault surfaces, with higher slip tendencies in the deeper parts of the SE and NW faults as well as fault segments striking NW-SE. The dilation tendencies in the strike-slip regime are considerably lower than those in the normal faulting regime. The maximum dilation tendency of 0.92 is only archived in some parts of the NE-SW-trending minor fault in the Southern part of the tectonic block. The SE and NW faults as well as fault segments trending NW-SE show medium dilation tendencies of around 0.5, whereas the NE fault and the W-E-trending minor faults show low dilation tendencies, indicating the presence of a sealing behaviour which prevents fluid migration along the respective fault planes.

Mechanical failure of the ubiquitous joint elements occurs in none of the 52 assessed scenarios. Thus, only the scenarios exhibiting the highest effective stress changes (simultaneous injection into coal seams 308 and 510) are discussed in the following. For the considered normal faulting regime, slip tendency is changing slightly for the operational state. Most of the faults exhibit a minor decrease in slip tendency, especially in the shallower parts, with up to 9% in maximum (Figure 12a). Only along the NW-SE trending parts of the minor faults as well as along the SW fault, a slight increase in slip tendency by 13.5% in maximum is observed. Changes in fault dilation tendencies are even lower, whereby the highest deviations occur in the shallower parts of the faults (Figure 12b). In most parts of the faults, the dilation tendency is slightly reduced by 6.8% in maximum. Only in the NE-SE trending parts of the minor faults, dilation tendency increases by 5.8% in maximum.

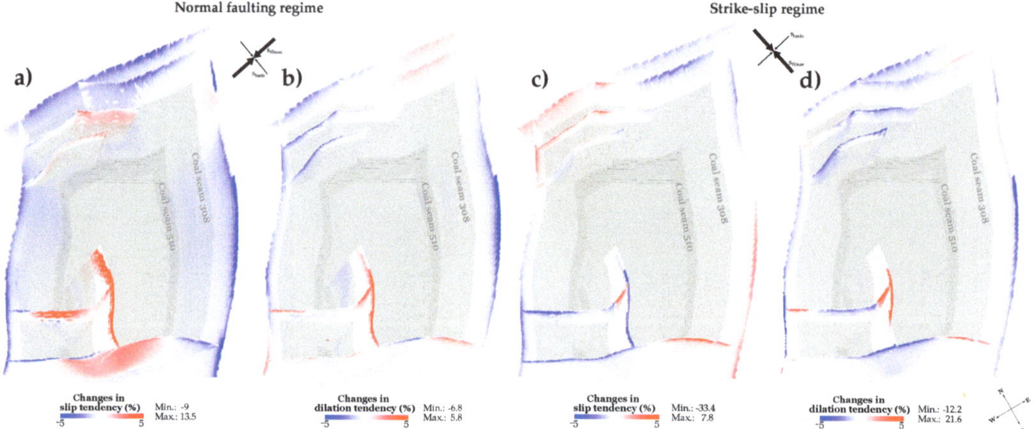

Figure 12. Simulated changes in (**a**) slip and (**b**) dilation tendencies at the operational state (1 year) for a normal faulting regime and a CO_2 storage within both coal seams. (**c**) Slip and (**d**) dilation tendencies for a strike-slip regime.

For the considered strike-slip regime, slip and dilation tendencies mainly change in the shallower parts of the faults, which located above the coal seam 308 (Figure 12c). The impact of the simulated CO_2 injection on the stress regime is still relatively low, with slight changes in slip and dilation tendencies, only. The values for slip tendency in the fault segments in the NE and SW parts of the fault block are reduced by up to 33.4%, whereas values for the fault segments in the SE and NW parts increase by 7.8% in maximum. Nevertheless, even if the initial slip tendencies are higher for the considered strike-slip regime, absolute values are still relatively low. Fault dilation tendencies decrease mostly in the shallower parts of the faults, with 12.2% in maximum (Figure 12d). Some segments of the minor faults in the Southern part of the fault block show an increase in dilation tendency by up to 21.5%. However, the initial values in these areas are relatively low.

Neither fault slip nor dilation, as a potential consequence of slip, arise in the extensive scenario analysis. Consequently, fault reactivation is not expected at any time of CO_2 injection in the study area considering the applied injection well arrangement, injection pressures and the given model parametrisation.

4. Discussion

In this study, hydromechanical impacts of a potential CO_2 storage in coal beds have been assessed for a commercial-scale operation in the Upper Silesian Coal Basin. Specifically, bandwidths of vertical displacements and fault reactivation depending on the local stress field were investigated as part of a sensitivity analysis, consisting of 52 scenarios. The undertaken analysis allows for obtaining a quantitative understanding of the mechanical processes taking place during the operational and post-operational states. Thereby, these predictive numerical models can increase the safety and sustainability of CO_2 storage in coal seams.

The near-field model was employed to simulate CO_2 flow within coal seams and coal-CO_2 interactions, along with the corresponding changes in permeability and stress. Therefore, a dual porosity modelling approach was applied to represent fluid flow and coal-CO_2 interactions within the characteristic microstructure of coal. Although this approach is in general able to express different characteristics of fractures and the rock matrix, conventional dual porosity models are particularly appropriate for reservoirs with highly connected small-scale fractures. For large-scale fractures, hydraulic fractures and highly-localised anisotropy, the dual porosity modelling approach may lead to inaccuracies [23,49]. More attempts are required for an accurate representation of multiscale fractures in fluid flow. In this study, the coal deformation was assumed to be linearly elastic, however, recent work by Chen et al. [50] indicates that nonlinear deformation of fractured coal can also affect the stress field of coal seams. Thus, the nonlinearity of coal deformation should be taken into account in future studies.

The injected amount of CO_2 per single panel of 22,139 t in maximum for one year of operation is low compared to the annual emissions of power plants, which are in the range of several Millions of tons CO_2. Nevertheless, an injection of CO_2 in deep unmineable coal seams can be used to increase the efficiency of coal bed methane recovery [51,52], due to the higher adsorption affinity of CO_2 compared to methane [53]. Chećko et al. [51] simulated a CO_2 injection into a coal seam of the Upper Silesian Coal Basin combined with enhanced coal bed methane (ECBM) recovery at a depth comparable to that in this study. During a time period of one year, 1,954,213 sm^3 (equivalent to 3547 t) of CO_2 have been injected via one single horizontal well. When additionally shrinking and swelling of the coal matrix was considered within their model, the aggregate amount of injected CO_2 decreased to 625,000 sm^3 (equivalent to 1134 t). Taking into account that respective coal seam 405 has a significantly lower thickness of 3 m compared to 9 m for seam 510, the injected amount for one year is in a comparable range to the value reported in this paper. Chećko et al. [51] further showed that a simultaneous injection of CO_2 can result in a 50% increase in coal bed methane extraction. Nevertheless, due to the observed reduction in coal permeability, they concluded that hydraulic fracturing is required increase the methane recovery efficiency.

Due to the comparable low aggregate amount of 285,858 t CO_2 in maximum, which has been injected via the 22 panels, also the geomechanical impact was limited. The maximum vertical displacements amounted to 3.59 cm and 1.07 cm at the coal seam top and the ground surface, respectively, and no reactivation of the faults zones was to be expected under the given conditions. Nevertheless, the main system-controlling parameters were identified. As expected, the strain increment was proportional to the pressure increase, and thus the injection rate. Magnitude and spatial distribution of displacements further depended on site-specific characteristics, such as seam geometry (e.g., thickness and dip). The simulation results further demonstrated that even considerably lower mechanical properties of the surrounding rock mass exhibited a marginal effect on ground surface uplift, only. Nevertheless, elastic properties of the coal seam itself could considerably influence the coal swelling behaviour, and thus the vertical displacements. Due to the previously undertaken extensive field and laboratory investigations, mechanical properties of the respective coals itself are well known, and thus do not constitute a systemic uncertainty.

The extensive slip and dilation tendency analysis of the fault system in the study area exhibited a fluid conduit behaviour in case of a normal faulting regime for two major faults. Nevertheless, this tendency could not be observed for a strike-slip regime (Section 3.4). Hence, a major uncertainty at potential storage sites is the local stress field, since even if fault integrity is not compromised, faults might be initially hydraulically conductive and could represent local pathways for upward fluid migration. This illustrates the importance in determining the local stress regime as accurately as possible, e.g., via in-situ stress measurements, such as borehole logging, leak-off tests and the recognition of high energy tremors via local focal mechanisms. The results of such analyses can be used to identify the areas and individually known faults of highest reactivation potentials, and thereby guide the operational monitoring at field scale.

Moreover, site-specific modelling assessments of prospective storage sites need to include the entire fault system, whereby especially detailed fault geometries should be integrated. Nevertheless, especially larger faults generally represent highly complex zones, as demonstrated by tectonic field-work studies, e.g., [54–56]. Thus, faults zones are composed of various fault segments, Riedel shears, multiple fault strands, dilatational jogs or relay ramps [57]. Parts of these particular fault elements can be sub-seismic, and thus not considered in initial modelling studies. However, they can become extremely relevant within the context of subsurface utilisation, since they represent potential fluid migration pathways in case of reactivation. In this context, also fault properties and their spatial and temporal variability are one of the major unknowns in field-scale hydromechanical simulations. Therefore, the complexity of the fault system and fault parametrisation need to be assessed as part of the extensive risk assessment.

5. Summary and Conclusions

The presented work evaluated the potential hydromechanical effects of CO_2 storage in coal seams at a prospective storage site in the Upper Silesian Coal Basin in Poland. In order to examine the complex interacting processes of mechanics, adsorption and multiphase flow with the required accuracy, numerical simulations were performed at near-field and far-field scale. While near-field models of the well bore vicinity have been applied to simulate CO_2 injection, adsorption and resulting permeability and stress evolution, the far-field model assessed commercial-scale hydromechanical impacts. Especially, ground surface uplift and the potential reactivation of fault zones were investigated, since faults could act as preferential leakage pathways for the buoyancy-driven upward migration of CO_2. In order to flexibly integrate near-field simulation results as internal boundary conditions within the far-field model, a workflow was developed using the effective stress changes as coupling parameter.

The numerical model of the study site, with a spatial extent of 4000 m × 6000 m, comprises five main lithostratigraphic units, two target coal seams as well as the exact geometries of the four major and five minor faults. Two coal seams have been selected as potential storage formations, due to their lateral abundance and sufficient thickness: (i) coal seam 308 has a average thickness of 2 m and is located at an average depth of 405 m, (ii) coal seam 510 is situated at an average depth of 1250 m below the ground surface and has an average thickness of 9 m. A horizontal well configuration, considering a well length of 500 m was examined for injection aiming to enlarged coal-CO_2 contact area.

In order to assess the impacts of the injection pressures, well arrangement as well as the effect of different geological uncertainties (regional stress regime and rock properties) for the operational and post-operational states, a comprehensive scenario analysis was conducted, consisting of 52 hydromechanically coupled simulation runs. The highest impact was observed for a simultaneous CO_2 injection in both coal seams at constant injection pressures of 1.5 MPa and 4.5 MPa for coal seams 308 and 510, respectively. The resulting maximum vertical displacements amounted to 3.59 cm, 1.27 cm and 1.07 cm directly above the coal seams 510, 308 and at the ground surface, respectively, whereby an area of circa 2500 m × 3000 m exhibited surface uplift of more than 0.5 cm. Hydromechanical impacts of CO_2 injection in coal were highest in the immediate surrounding of the injection area, where large and localised effective stress changes occured. Moreover, the larger the distance to the injection location, the smaller the differences between the operational and post-operational states, whereby the differences in ground surface uplift amounted to 10.1%. The storage of 285,858 t of CO_2 in maximum was simulated at the prospective commercial-scale site, whereby the major part of 93% or 265,673 t was stored into the deeper coal seam 510, and only 7% or 20,185 t into the shallower seam 308. These differences are related to the considerably higher CO_2 density at the depth of coal seam 510, as well as the storage formation volume.

Soft and stiff rock property scenarios were examined, whereby elastic moduli in the soft rock scenario are about 50% of those used in the stiff one. Nevertheless, the differences in vertical displacements above the seams were relatively small with 6.4% and 7.3% in maximum for coal seams 510 and 308 in the operational state, respectively. Hence, the properties of the surrounding rock mass marginally affected the vertical displacements resulting from the pore pressure increase experienced due to CO_2 injection.

Moreover, the local fault system present in the study area was examined by means of a comprehensive fault slip and dilation tendency analysis, considering two different stress regimes as geological uncertainty: (i) a strike-slip regime, which is characteristic for the Upper Silesian Coal Basin, as well as (ii) a normal faulting regime, which is reported for the local coal deposit. For the investigated normal faulting regime, slip tendencies were generally very low and did not exceed 0.05 and 0.18 for the normal faulting and strike-slip regime, respectively. Nevertheless, the dilation tendency exhibited high values over 0.85 for the normal faulting regime, indicating potential fluid conduit characteristics of the respective faults, likely enabling CO_2 leakage. By contrast, the dilation tendencies were considerably lower in case of a strike-slip regime. Failure of the ubiquitous joint elements representing the fault plane did not occur at any time of CO_2 injection considering the applied well design, injection pressures and given parametrisation. Hence, neither fault slip nor dilation, as a potential consequence of the pore pressure increase, were to be expected during the operational or post-operational states, and fault integrity was not compromised at any time of the simulated CO_2 storage operation.

Even if fault integrity was not affected, faults may be initially hydraulically conductive and could represent local pathways for upward migration of CO_2 or displaced brine. Hence, the local stress regime has to be determined as accurately as possible by in-situ stress measurements. Thereby, regions with the highest fault reactivation potentials or conductive fault segments can be identified to guide the geophysical monitoring during the operational and post-operational phases of the project. The conducted sensitivity analysis enabled to obtain a quantitative understanding of the important mechanical processes during the operational and post-operational phases.

In conclusion, application of numerical models led to specific new insights into hydromechanical processes at a potential CO_2 storage site. The coupling of CO_2 flow near- and geomechanical far-field modeling approach were proven appropriate for determining hydromechanical impacts at commercial-scale operation. Further, elaborated simulation concept is suitable to identify and quantitatively assess and mitigate these impacts to ensure operational safety. Thereby, the predictive numerical models can contribute to increasing the safety and sustainability of CO_2 storage in coal seams. Based on the results presented in this study, assuming that a certain injection pressure (below the 1.5-fold of the hydrostatic pressure) is not exceeded during site operation at the investigated area, it can be concluded that:

- Vertical surface displacements occur at a tolerable extent with no structural damage of surface infrastructure expected.
- Parameter variations of rock strength have no significant influence on the hydromechanical impacts of the CO_2 storage operation, if conservative rock properties are taken into consideration.
- Fault reactivation is unlikely and can be further reduced by increasing the amount of in-situ stress measurements to identify regions with high fault reactivation potentials and conductive fault segments.
- Post-operational impact are unlikely due to the degrading pressure in the reservoir formation.

The presented site-specific findings will contribute to the assessment of potential environmental risks at other CO_2 storage sites in coal deposits if CO_2 storage in coal seams becomes economically competitive. The proposed methodology for model coupling across different scales and hydromechanical impact assessment is directly applicable to any coal deposit worldwide.

Author Contributions: Conceptualisation, M.W., T.K. and M.C.; methodology, M.W., T.K. and M.C.; formal analysis, M.W., T.K. and M.C.; investigation, M.W. and M.C.; resources, H.T. and T.K.; data curation, T.U. and B.B.; writing—original draft preparation, M.W. and M.C.; writing—review and editing, C.O. and T.K.; visualisation, M.W.; supervision, T.K. and H.T.; project administration, S.M. All authors have read and agreed to the published version of the manuscript.

Funding: The research was conducted as part of the "Establishing a Research Observatory to Unlock European Coal Seams for Carbon Dioxide Storage (ROCCS)" project. The ROCCS project has received funding from the Research Fund for Coal and Steel under Grant Agreement No. 899336. In addition, supported within the funding programme "Open Access Publikationskosten" Deutsche Forschungsgemeinschaft (DFG, German Research Foundation)—Project Number 491075472. The financial support is gratefully acknowledged.

Data Availability Statement: The data and software will be made available on specific user requests.

Conflicts of Interest: The contact author has declared that none of the authors has any competing interests.

Appendix A

Table A1. All scenarios of the investigated sensitivity analysis regarding initial conditions, maximum displacements and the respective maximum amount of stored CO_2.

Coal Seam	Stress Regime	Rock Properties	Injection Pressure (MPa)	Time (years)	Max. Displacement above Seam (10^{-2} m)	Max. Displacement at Surface (10^{-2} m)	Amount of CO_2 (t)
510	SS	soft	4.5	1	3.59	0.99	265,674
510	SS	soft	4.5	15	1.97	0.89	265,674
510	SS	soft	3	1	2.64	0.53	150,613
510	SS	soft	3	15	1.43	0.50	150,613
510	SS	soft	1.5	1	1.47	0.19	60,026
510	SS	soft	1.5	15	0.82	0.18	60,026
510	SS	stiff	4.5	1	3.45	0.97	265,674
510	SS	stiff	4.5	15	1.89	0.88	265,674
510	SS	stiff	3	1	2.52	0.52	150,613
510	SS	stiff	3	15	1.37	0.49	150,613
510	SS	stiff	1.5	1	1.37	0.19	60,026
510	SS	stiff	1.5	15	0.78	0.18	60,026
510	NF	soft	4.5	1	3.59	0.99	265,674
510	NF	soft	4.5	15	1.97	0.89	265,674
510	NF	soft	3	1	2.64	0.53	150,613
510	NF	soft	3	15	1.43	0.50	150,613
510	NF	soft	1.5	1	1.47	0.19	60,026
510	NF	soft	1.5	15	0.82	0.18	60,026
510	NF	stiff	4.5	1	3.45	0.97	265,674
510	NF	stiff	4.5	15	1.89	0.88	265,674
510	NF	stiff	3	1	2.52	0.52	150,613
510	NF	stiff	3	15	1.37	0.49	150,613
510	NF	stiff	1.5	1	1.37	0.19	60,026
510	NF	stiff	1.5	15	0.78	0.18	60,026
308	SS	soft	1.5	1	0.77	0.34	20,185

Table A1. *Cont.*

Coal Seam	Stress Regime	Rock Properties	Injection Pressure (MPa)	Time (years)	Max. Displacement above Seam (10^{-2} m)	Max. Displacement at Surface (10^{-2} m)	Amount of CO_2 (t)
308	SS	soft	1.5	15	0.42	0.25	20,185
308	SS	soft	1	1	0.55	0.19	12,392
308	SS	soft	1	15	0.34	0.17	12,392
308	SS	soft	0.5	1	0.31	0.07	5599
308	SS	soft	0.5	15	0.21	0.08	5599
308	SS	stiff	1.5	1	0.76	0.36	20,185
308	SS	stiff	1.5	15	0.41	0.26	20,185
308	SS	stiff	1	1	0.53	0.21	12,392
308	SS	stiff	1	15	0.33	0.18	12,392
308	SS	stiff	0.5	1	0.29	0.08	5599
308	SS	stiff	0.5	15	0.20	0.08	5599
308	NF	soft	1.5	1	0.77	0.34	20,185
308	NF	soft	1.5	15	0.42	0.25	20,185
308	NF	soft	1	1	0.55	0.19	12,392
308	NF	soft	1	15	0.34	0.17	12,392
308	NF	soft	0.5	1	0.31	0.07	5599
308	NF	soft	0.5	15	0.21	0.08	5599
308	NF	stiff	1.5	1	0.76	0.36	20,185
308	NF	stiff	1.5	15	0.41	0.26	20,185
308	NF	stiff	1	1	0.53	0.21	12,392
308	NF	stiff	1	15	0.33	0.18	12,392
308	NF	stiff	0.5	1	0.29	0.08	5599
308	NF	stiff	0.5	15	0.20	0.08	5599
308 and 510	SS	soft	1.5 and 4.5	1	3.59	1.07	285,858
308 and 510	SS	soft	1.5 and 4.5	15	1.97	0.94	285,858
308 and 510	NF	soft	1.5 and 4.5	1	3.59	1.07	285,858
308 and 510	NF	soft	1.5 and 4.5	15	1.97	0.94	285,858

SS—Strike-slip regime, NF—Normal faulting regime.

References

1. IPCC. *Special Report: Carbon Dioxide Capture and Storage;* Technical Report; Cambridge University Press: Cambridge, UK, 2005.
2. Masoudian, M.S. Multiphysics of carbon dioxide sequestration in coalbeds: A review with a focus on geomechanical characteristics of coal. *J. Rock Mech. Geotech. Eng.* **2016**, *8*, 93–112. [CrossRef]
3. White, C.M.; Smith, D.H.; Jones, K.L.; Goodman, A.L.; Jikich, S.A.; LaCount, R.B.; DuBose, S.B.; Ozdemir, E.; Morsi, B.I.; Schroeder, K.T. Sequestration of carbon dioxide in coal with enhanced coalbed methane recovery—A review. *Energy Fuels* **2005**, *19*, 659–724. [CrossRef]
4. Sun, L.; Dou, H.; Li, Z.; Hu, Y.; Hao, X. Assessment of CO_2 storage potential and carbon capture, utilization and storage prospect in China. *J. Energy Inst.* **2018**, *91*, 970–977. [CrossRef]
5. Vangkilde-Pedersen, T.; Anthonsen, K.L.; Smith, N.; Kirk, K.; Neele, F.; van der Meer, B.; Le Gallo, Y.; Bossie-Codreanu, D.; Wojcicki, A.; Le Nindre, Y.M.; et al. Assessing European capacity for geological storage of carbon dioxide—The EU GeoCapacity project. *Energy Procedia* **2009**, *1*, 2663–2670. [CrossRef]
6. Hong, W.Y. A techno-economic review on carbon capture, utilisation and storage systems for achieving a net-zero CO_2 emissions future. *Carbon Capture Sci. Technol.* **2022**, *3*, 100044. [CrossRef]
7. Dutta, P.; Zoback, M.D. CO_2 sequestration into the Wyodak coal seam of Powder River Basin—Preliminary reservoir characterization and simulation. *Int. J. Greenh. Gas Control* **2012**, *9*, 103–116. [CrossRef]
8. van Bergen, F.; Pagnier, H.; Krzystolik, P. Field experiment of enhanced coalbed methane-CO_2 in the upper Silesian basin in Poland. *Environ. Geosci.* **2006**, *13*, 201–224. [CrossRef]
9. Steadman, E.N.; Anagnost, K.K.; Botnen, B.W.; Botnen, L.A.; Daly, D.J.; Gorecki, C.D.; Harju, J.A.; Jensen, M.D.; Peck, W.D.; Romuld, L.; et al. The Plains CO_2 Reduction (PCOR) partnership: Developing Carbon management options for the central interior of North America. *Energy Procedia* **2011**, *4*, 6061–6068. [CrossRef]
10. Sheng, Y.; Benderev, A.; Bukolska, D.; Eshiet, K.I.I.; da Gama, C.D.; Gorka, T.; Green, M.; Hristov, N.; Katsimpardi, I.; Kempka, T.; et al. Interdisciplinary studies on the technical and economic feasibility of deep underground coal gasification with CO_2 storage in bulgaria. *Mitig. Adapt. Strateg. Glob. Chang.* **2016**, *21*, 595–627. [CrossRef]
11. Fujioka, M.; Yamaguchi, S.; Nako, M. CO_2-ECBM field tests in the Ishikari Coal Basin of Japan. *Int. J. Coal Geol.* **2010**, *82*, 287–298. [CrossRef]
12. van Bergen, F.; Krzystolik, P.; van Wageningen, N.; Pagnier, H.; Jura, B.; Skiba, J.; Winthaegen, P.; Kobiela, Z. Production of gas from coal seams in the Upper Silesian Coal Basin in Poland in the post-injection period of an ECBM pilot site. *Int. J. Coal Geol.* **2009**, *77*, 175–187. [CrossRef]
13. Van Bergen, F.; Winthaegen, P.; Pagnier, H.; Krzystolik, P.; Jura, B.; Skiba, J.; van Wageningen, N. Assessment of CO_2 storage performance of the Enhanced Coalbed Methane pilot site in Kaniow. *Energy Procedia* **2009**, *1*, 3407–3414. [CrossRef]

14. Fang, H.; Sang, S.; Liu, S. Establishment of dynamic permeability model of coal reservoir and its numerical simulation during the CO_2-ECBM process. *J. Pet. Sci. Eng.* **2019**, *179*, 885–898. [CrossRef]

15. Busch, A.; Gensterblum, Y. CBM and CO_2-ECBM related sorption processes in coal: A review. *Int. J. Coal Geol.* **2011**, *87*, 49–71. [CrossRef]

16. Cai, Y.; Pan, Z.; Liu, D.; Zheng, G.; Tang, S.; Connell, L.; Yao, Y.; Zhou, Y. Effects of pressure and temperature on gas diffusion and flow for primary and enhanced coalbed methane recovery. *Energy Explor. Exploit.* **2014**, *32*, 601–619. [CrossRef]

17. Pone, J.D.; Halleck, P.M.; Mathews, J.P. Sorption capacity and sorption kinetic measurements of CO_2 and CH_4 in confined and unconfined bituminous coal. *Energy Fuels* **2009**, *23*, 4688–4695. [CrossRef]

18. Masum, S.A.; Chen, M.; Hosking, L.J.; Stańczyk, K.; Kapusta, K.; Thomas, H.R. A numerical modelling study to support design of an in-situ CO_2 injection test facility using horizontal injection well in a shallow-depth coal seam. *Int. J. Greenh. Gas Control* **2022**, *119*, 103725. [CrossRef]

19. Ridha, S.; Pratama, E.; Ismail, S. Performance assessment of CO_2 sequestration in a horizontal well for enhanced coalbed methane recovery in deep unmineable coal seams. *Chem. Eng. Trans.* **2017**, *56*, 589–594.

20. Ren, J.; Zhang, L.; Ren, S.; Lin, J.; Meng, S.; Ren, G.; Gentzis, T. Multi-branched horizontal wells for coalbed methane production: Field performance and well structure analysis. *Int. J. Coal Geol.* **2014**, *131*, 52–64. [CrossRef]

21. Connell, L.D.; Pan, Z.; Camilleri, M.; Meng, S.; Down, D.; Carras, J.; Zhang, W.; Fu, X.; Guo, B.; Briggs, C.; et al. Description of a CO_2 enhanced coal bed methane field trial using a multi-lateral horizontal well. *Int. J. Greenh. Gas Control* **2014**, *26*, 204–219. [CrossRef]

22. Gentzis, T.; Bolen, D. The use of numerical simulation in predicting coalbed methane producibility from the Gates coals, Alberta Inner Foothills, Canada: Comparison with Mannville coal CBM production in the Alberta Syncline. *Int. J. Coal Geol.* **2008**, *74*, 215–236. [CrossRef]

23. Sams, W.; Bromhal, G.; Jikich, S.; Odusote, O.; Ertekin, T.; Smith, D. *Simulating Carbon Dioxide Sequestration/ECBM Production in Coal Seams: Effects of Permeability Anisotropies and Other Coal Properties*; SPE 84423; Society of Petroleum Engineers: Richardson, TX, USA, 2003.

24. Zhang, J.; Feng, Q.; Zhang, X.; Hu, Q.; Wen, S.; Chen, D.; Zhai, Y.; Yan, X. Multi-fractured horizontal well for improved coalbed methane production in eastern Ordos basin, China: Field observations and numerical simulations. *J. Pet. Sci. Eng.* **2020**, *194*, 107488. [CrossRef]

25. Liu, G.; Smirnov, A.V. Carbon sequestration in coal-beds with structural deformation effects. *Energy Convers. Manag.* **2009**, *50*, 1586–1594. [CrossRef]

26. Connell, L.D.; Detournay, C. Coupled flow and geomechanical processes during enhanced coal seam methane recovery through CO_2 sequestration. *Int. J. Coal Geol.* **2009**, *77*, 222–233. [CrossRef]

27. Chen, Z.; Liu, J.; Elsworth, D.; Connell, L.D.; Pan, Z. Impact of CO_2 injection and differential deformation on CO_2 injectivity under in-situ stress conditions. *Int. J. Coal Geol.* **2010**, *81*, 97–108. [CrossRef]

28. Kedzior, S. Accumulation of coal-bed methane in the south-west part of the Upper Silesian Coal Basin (southern Poland). *Int. J. Coal Geol.* **2009**, *80*, 20–34. [CrossRef]

29. Smoliński, A.; Rompalski, P.; Cybulski, K.; Chećko, J.; Howaniec, N. Chemometric study of trace elements in hard coals of the upper Silesian Coal Basin, Poland. *Sci. World J.* **2014**, *2014*, 234204. [CrossRef]

30. Thomas, H.R.; Rees, S.W.; Sloper, N.J. Three-dimensional heat, moisture and air transfer in unsaturated soils. *Int. J. Numer. Anal. Methods Geomech.* **1998**, *22*, 75–95. https://doi.org/10.1002/(SICI)1096-9853(199802)22:2<75::AID-NAG909>3.0.CO;2-K.

31. Chen, M.; Masum, S.A.; Thomas, H.R. 3D hybrid coupled dual continuum and discrete fracture model for simulation of CO_2 injection into stimulated coal reservoirs with parallel implementation. *Int. J. Coal Geol.* **2022**, *262*, 104103. [CrossRef]

32. Chen, M.; Masum, S.; Sadasivam, S.; Thomas, H. Modelling anisotropic adsorption-induced coal swelling and stress-dependent anisotropic permeability. *Int. J. Rock Mech. Min. Sci.* **2022**, *153*, 105107. [CrossRef]

33. Van Wageningen, W.F.; Wentinck, H.M.; Otto, C. Report and modeling of the MOVECBM field tests in Poland and Slovenia. *Energy Procedia* **2009**, *1*, 2071–2078. [CrossRef]

34. Konecny, P.; Kozusnikova, A. Influence of stress on the permeability of coal and sedimentary rocks of the Upper Silesian basin. *Int. J. Rock Mech. Min. Sci.* **2011**, *48*, 347–352. [CrossRef]

35. Pagnier, H.; van Bergen, F.; Krzystolik, P.; Skiba, J.; Jura, B.; Hadro, J.; Wentink, P.; De-Smedt, G.; Kretzschmar, H.J.; Fröbel, J.; et al. Reduction of CO_2 Emission by Means of CO_2 Storage in Coal Seams in the Silesian Coal Basin of Poland; RECOPOL (Final Report); 2006. Available online: https://ieaghg.org/docs/General_Docs/Reports/2006-10%20Final%20RECOPOL%20Report.pdf (accessed on 3 April 2023).

36. Majewska, Z.; Majewski, S.; Zietek, J. Swelling of coal induced by cyclic sorption/desorption of gas: Experimental observations indicating changes in coal structure due to sorption of CO_2 and CH_4. *Int. J. Coal Geol.* **2010**, *83*, 475–483. [CrossRef]

37. Chen, M.; Hosking, L.J.; Sandford, R.J.; Thomas, H.R. Dual porosity modelling of the coupled mechanical response of coal to gas flow and adsorption. *Int. J. Coal Geol.* **2019**, *205*, 115–125. [CrossRef]

38. Reeves, S.R.; Taillefert, A. Reservoir Modeling for the Design of the RECOPOL CO_2 Sequestration Project, Poland; Topical Report, DOE Contract No. DE-FC26-00NT40924; 2002. Available online: https://www.adv-res.com/Coal-Seq_Consortium/ECBM_Sequestration_Knowledge_Base/Coal-Seq%20Topical%20Reports/Topical%20Report%20-%20RECOPOL%20Design%20Modelling.pdf (accessed on 3 April 2023).

39. Hol, S.; Spiers, C.J. Competition between adsorption-induced swelling and elastic compression of coal at CO_2 pressures up to 100 MPa. *J. Mech. Phys. Solids* **2012**, *60*, 1862–1882. [CrossRef]

40. Itasca Consulting Group, Inc. *FLAC3D Software Version 5.01. User's Manual. Advanced Three-Dimensional Continuum Modelling for Geotechnical Analysis of Rock, Soil and Structural Support*; Itasca Consulting Group, Inc.: Minneapolis, MN, USA, 2014.

41. Otto, C.; Kempka, T.; Kapusta, K.; Stańczyk, K. Fault reactivation can generate hydraulic short circuits in underground coal gasification—New insights from regional-scale thermo-mechanical 3D modeling. *Minerals* **2016**, *6*, 101. [CrossRef]

42. Kempka, T.; Nielsen, C.M.; Frykman, P.; Shi, J.Q.; Bacci, G.; Dalhoff, F. Coupled hydro-mechanical simulations of CO_2 storage supported by pressure management demonstrate synergy benefits from simultaneous formation fluid extraction. *Oil Gas Sci. Technol.* **2015**, *70*, 599–613. [CrossRef]

43. Chećko, J. Analysis of geological, hydrogeological and mining conditions in the area of the developed georeactor located in KWK Wieczorek mine. *Prz. Górn.* **2013**, *69*, 37–45.

44. Dubinski, J.; Stec, K.; Bukowska, M. Geomechanical and tectonophysical conditions of mining-induced seismicity in the Upper Silesian Coal Basin in Poland: A case study. *Arch. Min. Sci.* **2019**, *64*, 163–180. [CrossRef]

45. Jarosiński, M. Recent tectonic stress field investigations in Poland: A state of the art. *Geol. Q.* **2006**, *50*, 303–321.

46. Morris, A.; Ferrill, D.A.; Henderson, D.B. Slip-tendency analysis and fault reactivation. *Geology* **1996**, *24*, 275–278. http://dx.doi.org/10.1130/0091-7613(1996)024<0275:STAAFR>2.3.CO.

47. Ferrill, D.A. Stressed rock strains groundwater at Yucca Mountain, Nevada. *GSA Today* **1999**, *9*, 1–8.

48. Hobbs, B.E.; Means, W.D.; Williams, P.F. *An Outline of Structural Geology*; Wiley: New York, NY, USA, 1976; Volume 25.

49. Moinfar, A.; Varavei, A.; Sepehrnoori, K.; Johns, R.T. Development of a coupled dual continuum and discrete fracture model for the simulation of unconventional reservoirs. In Proceedings of the SPE Reservoir Simulation Symposium, The Woodlands, TX, USA, 18–20 February 2013; Volume 2, pp. 978–994. [CrossRef]

50. Chen, M.; Masum, S.A.; Thomas, H.R. Three-dimensional cleat scale modelling of gas transport processes in deformable fractured coal reservoirs. *Gas Sci. Eng.* **2023**, *110*, 204901. [CrossRef]

51. Chećko, J.; Urych, T.; Magdziarczyk, M.; Smolinski, A. Research on the processes of injecting CO_2 into coal seams with CH_4 recovery using horizontal wells. *Energies* **2020**, *13*, 416. [CrossRef]

52. Wong, S.; Gunter, W.D.; Mavor, M.J. Economics of CO_2 sequestration in coalbed methane reservoirs. In Proceedings of the SPE/CERI Gas Technology Symposium 2000, GTS 2000, Calgary, AB, Canada, 3–5 April 2000. [CrossRef]

53. Godec, M.; Koperna, G.; Gale, J. CO_2-ECBM: A review of its status and global potential. *Energy Procedia* **2014**, *63*, 5858–5869. [CrossRef]

54. Price, N.J.; Cosgrove, J.W. *Analysis of Geological Structures*; Cambridge University Press: Cambridge, UK, 1990.

55. Van der Zee, W. *Dynamics of Fault Gouge Development in Layered Sand-Clay Sequences*; Shaker Verlag GmbH: Dueren, Germany, 2002.

56. Koledoye, B.A.; Aydin, A.; May, E. A new process-based methodology for analysis of shale smear along normal faults in the Niger Delta. *AAPG Bull.* **2003**, *87*, 445–463. [CrossRef]

57. Ligtenberg, J.H. Detection of fluid migration pathways in seismic data: Implications for fault seal analysis. *Basin Res.* **2005**, *17*, 141–153. [CrossRef]

energies

Article

Exploring Public Attitudes and Acceptance of CCUS Technologies in JABODETABEK: A Cross-Sectional Study

Charli Sitinjak [1,2,*], **Sitinjak Ebennezer** [3] **and Józef Ober** [4]

[1] Faculty of Psychology, Esa Unggul University, Jakarta 11510, Indonesia

[2] Centre for Research in Psychology and Human Well-Being (PSiTra), Faculty of Social Sciences and Humanities, Universiti Kebangsaan Malaysia, Bangi 43600, Selangor, Malaysia

[3] Petroleum Engineering, Balikpapan College of Oil and Gas Technology, Kalimantan 76127, Indonesia; sitinjakebennezer@gmail.com

[4] Department of Applied Social Sciences, Faculty of Organization and Management, Silesian University of Technology, Roosevelta 26-28, 41-800 Zabrze, Poland; jozef.ober@polsl.pl

* Correspondence: p112562@siswa.ukm.edu.my

Citation: Sitinjak, C.; Ebennezer, S.; Ober, J. Exploring Public Attitudes and Acceptance of CCUS Technologies in JABODETABEK: A Cross-Sectional Study. *Energies* **2023**, *16*, 4026. https://doi.org/10.3390/en16104026

Academic Editors: Michael Zhengmeng Hou, Hejuan Liu, Cheng Cao, Liehui Zhang and Yulong Zhao

Received: 20 April 2023
Revised: 7 May 2023
Accepted: 9 May 2023
Published: 11 May 2023

Abstract: One of the most essential elements of environmental protection is an appropriate policy towards carbon capture, utilisation, and storage (CCUS). On the one hand, these technologies are being dynamically developed. Still, on the other hand, we often encounter social resistance to change and new technologies, which is one of the main barriers to their implementation. This research examined public acceptance and awareness of Indonesia's CCUS technologies. Five hundred respondents completed an online survey representing Jakarta, Bogor, Depok, Bekasi, and Tangerang. The study found that the respondents had more favourable feelings towards carbon capture and utilisation (CCU) than CO_2 capture and storage (CCS), perceiving CCU as more innovative, necessary, cost-effective, secure, environmentally friendly, and beneficial to regional and national economies than CCS. However, in Indonesia, most respondents did not embrace the development of CCUS technology due to a lack of knowledge and fear, which can lead to violence. The results indicate that an individual's awareness of perceived risks and the ability to safeguard the environment are crucial to their acceptance of CCUS technology. These findings contribute to understanding the public perception of CCUS technologies in Indonesia and can help to develop effective communication strategies to improve public understanding and acceptance of CCUS initiatives.

Keywords: CO_2; CCU; CCS; carbon capture; carbon utilisation; carbon storage; public acceptance; cross-sectional study

1. Introduction

The issue of the public understanding and acceptance of carbon capture, storage, and utilisation (CCUS) technologies has attracted the attention of governments, actors in the energy industry, and academics. It is widely acknowledged that stakeholders and the general public can substantially influence investment and location decisions for CO_2 capture, storage, and utilisation facilities, as well as CO_2-derived products [1]. As a result, understanding the attitudes and perceptions of the public towards CCUS technologies is crucial to the design and implementation of effective strategies for public participation in potential CCUS initiatives in Europe [2]. Given the increasing urgency of addressing climate change by reducing carbon emissions and the potential role of CCUS technologies in achieving this goal, this is particularly relevant.

During the last two decades, social researchers have paid considerable attention to investigating public attitudes towards CCUS technologies [3,4]. This body of work has included studies exploring the degree of public awareness and understanding of CCS projects and the factors that contribute to the support or opposition of CCUS technologies in various contexts and populations. In reviewing the available studies, the general trend suggests moderate public acceptance of CCUS technologies [5]. These findings have important implications for policymakers, industry stakeholders, and researchers, as they suggest that efforts to promote the widespread implementation of CCUS technologies may be met with varying levels of public support and opposition [6]. More research is needed to explore the factors that underpin these attitudes and identify effective strategies for engaging the public in conversations around CCUS technologies.

Several studies [1,3,4] have examined the public's perceptions of CCU and CCS and found that CCU is generally viewed more favourably than CCS. Furthermore, among the numerous aspects of CCUS technologies, CO_2 storage receives the lowest level of public acceptance [7]. However, it is important to note that the configuration of the application case can influence public attitudes towards these technologies. For example, a study by Loukouzas et al. [8] found that the general public preferred scenarios combining CCS with bioenergy over those that involve shale gas, underground coal gasification, or heavy industry. Similarly, a second study [9] found that the general public preferred CCS in conjunction with bioenergy and heavy industry over coal-fired power plants. When examining public perceptions and attitudes towards CCUS technologies, it is crucial to consider the contexts in which they are implemented [10].

In addition to examining the general public's attitudes towards CCUS technologies, several studies have investigated whether individuals who are likely to be directly affected by adjacent CCS installations have different perceptions of the technology than the general public. The results of these investigations have been contradictory. Some studies [11,12] have found that CCS scepticism is higher in areas where CCS storage may impact residents. However, a more recent study involving five countries (Canada, the Netherlands, Norway, the United States, and the United Kingdom) by [13] found that local samples exhibited the same or even higher levels of acceptance for CCS than the general population. These discrepancies in the findings illustrate the difficulty in understanding public perceptions of CCUS technologies and suggest that the local context and other factors may play a role in determining attitudes towards these technologies.

The public acceptance of CCUS technology is influenced by its perceived risks and advantages [10,14]. CO_2 leaks or blowouts, induced seismicity, and local repercussions on property values or tourism are only some of the potential concerns [15]. There is also some doubt as to whether or not CCS is a long-term solution to keep polluting companies around. On the contrary, CCS is primarily lauded for its potential to aid in the fight against global warming. CCS is also believed to provide local economic advantages and ease the transition to a decarbonised society [8,9,14].

The level of trust people have significantly influences public opinion on CCUS technologies among relevant stakeholders [16,17]. Public confidence can be increased through effective communication with stakeholders, especially when the message is seen to align with the interests of those involved [15]. Although the links between these components have not been demonstrated, some researchers have also investigated how affect and prior attitudes play a role in determining public perceptions of CCUS technology [11,14,18]. These results highlight the importance of thinking about the various elements that affect the way people see CCUS technologies. Effective public engagement initiatives might benefit from policymakers and industry stakeholders talking to the public and learning about their concerns and goals [19].

The CCUS Project has been trying to figure out why the reception of CCUS projects varies. This research aims to contribute to this cause by investigating the general public's familiarity with and attitude towards CCUS technology in Indonesia. Except for a few notable exceptions [20,21], the social acceptance of CCUS technology has received less attention in these countries in recent years [22,23].

Based on the above review [1–23], three questions can be formulated, the answers to which will help fill the identified research gap:

- To what extent does the public know, understand, value, and accept CCS and CCU technologies?

- How does the public perceive and respond to CCS and CCU technologies?
- What are the determinants of people's willingness to adopt CCS and CCU technologies?

The successful deployment of CCUS technologies requires first gaining a thorough understanding of how the public views these innovations. The public's understanding, perception of benefits and costs, and general evaluation and acceptability can help policymakers and industry stakeholders build more effective public participation programmes. Reporting on social acceptance sheds light on how acceptance differs between different research populations, but identifying important individual determinants of acceptance can also help to target specific communities and resolve their concerns.

The understanding and acceptance of carbon capture, storage, and utilisation (CCUS) technologies by the general public have been a major concern for governments, industry actors, and academics. It is widely acknowledged that stakeholders and the general public can have a significant impact on investment and location decisions for CO_2 capture, storage, and utilisation facilities, as well as CO_2-derived products. Therefore, it is essential to understand public attitudes and perceptions towards CCUS technologies to develop and implement effective public engagement strategies in Indonesia for potential CCUS initiatives. Given the increasing urgency of addressing climate change by reducing carbon emissions and the potential role of CCUS technologies in achieving this goal, this is particularly relevant.

Social scientists have devoted substantial amounts of attention to analysing public attitudes towards CCUS technologies. This research included studies examining the level of public awareness and understanding of CCS initiatives, as well as the factors that contribute to the support or opposition of CCUS technologies in diverse contexts and populations. The available studies suggest that public adoption of CCUS technologies is generally moderate. These findings have significant implications for policymakers, industry stakeholders, and researchers, as they suggest that efforts to promote the widespread adoption of CCUS technologies may be met with variable degrees of public support and opposition. Therefore, additional research is required to investigate the factors underlying these attitudes and to identify effective strategies to involve the public in discussions about CCUS technologies.

1.1. CCS and CCU for Indonesia

Indonesia is a developing nation with abundant fossil fuel resources, and fossil fuels will continue to be the primary energy source for decades to come. In an effort to mitigate climate change, the Indonesian government has committed to reducing greenhouse gas emissions from fossil fuel consumption. Figure 1 shows the poor air quality in Indonesia as a result of greenhouse gas emissions.

Indonesia has initiated research related to CCUS as a means to address the issue of climate change and reduce CO_2 emissions. However, it should be noted that this particular programme addresses the function of natural carbon dioxide sequestration, as opposed to carbon dioxide conversion. The concept of carbon capture and utilisation (CCU) technology involves the transformation of carbon dioxide (CO_2) into a range of valuable products. Usman investigated the synthesis of various chemical compounds, such as carboxylates, carbonates, carbamates, isocyanates, and polymeric materials, from CO_2 [22] for their potential applications in electrochemistry and photochemistry. According to Jiutian et al. [23], the activation of CO_2 and its co-reactants can produce valuable chemicals using high-energy salts or catalysts.

The utilisation of fossil fuels has demonstrated a notable increase, rising from 53.4 million tonnes of oil equivalent (Mtoe) in 1990 to 154.93 Mtoe in 2013. During this period, the proportion of oil consumption ranged from 50% to 60%. Under the NEP scenario, the proportion of oil in the energy mix is projected to decrease to 25–30%, while the aggregate consumption of fossil fuels is expected to rise to 690 Mtoe by 2050. The escalation in the use of fossil fuels is expected to be accompanied by a corresponding increase in carbon dioxide emissions [18]. The number of CO_2 emissions in 1990 was recorded at 133.9 Mtoe, which remained constant at 133.9 Mtoe in 2013. According to projections, global energy consumption is expected to reach 1000.6 and 2065.98 Mtoe in 2030 and 2050, respectively. According to LEMIGAS [18], the Indonesian government projected that the depleted oil and gas fields located in Kutai, Tarakan, and the South Sumatra region have the potential to store approximately 640 Mtoe of CO_2.

Figure 1. Effect of greenhouse gas emissions in JABODETABEK. Source: https://www.thejakartapost.com, accessed on 2 May 2023.

According to Rakhiemah [24], the most appealing alternative for CCS in Indonesia is the combination of CO_2 storage with EOR. This is due to the possibility of generating additional revenue from oil production, which can counterbalance the expenses associated with CCS. However, it is anticipated that this alternative will not meet the CO_2 reduction objective for extended-term CCS, as the utilisation of oil generated from EOR results in CO_2 emissions. Moreover, implementing carbon capture and storage (CCS) in the power industry can alleviate carbon dioxide emissions and accelerate industrialisation by facilitating the construction of additional power plants to meet the country's electrification objectives.

1.2. Carbon Capture and Storage (CCS) and Carbon Capture and Utilisation (CCU) Technologies

Carbon capture and storage (CCS) entails extracting carbon dioxide from industrial emissions and then storing it permanently in geological formations. The technique is commonly used in various industries, such as cement, steel, power generation, and chemical production. In contrast, carbon capture and utilisation (CCU) is part of a broader set of carbon recycling applications describing the reuse of captured carbon either directly (e.g., to fertilise greenhouses in beverages) or as an ingredient in new products (e.g., concrete, fuel, and chemicals). CCUs can replace the use of additional fossil fuels, thereby reducing emissions. This method can be considered a removal if carbon is removed from the atmosphere and remains in a closed circle for decades or centuries (e.g., when incorporated into cement building materials).

CCS technology is utilised to regulate the emission of CO_2 trapped from various processes, such as precombustion, post-combustion, and oxy-fuel combustion. The stages of a CCS project can be categorised into four phases: CO_2 capture, CO_2 transportation, CO_2 injection, and post-injection of CO_2 [25,26].

In the short term, storing CO_2 in geological locations such as deep saline formations, depleted oil or gas reservoirs, deep unmineable coal seams, and shale formations can reduce CO_2 emissions [27], as shown in Figure 2. Compared to pure CCS technology, CCUS technology focusses more on using captured CO_2, while sequestration (S) is secondary.

CCUS technology can reduce the cost of sequestration and provide benefits by enhancing the production of hydrocarbons or heat energy. Therefore, it has gained popularity in recent years. Depending on the purpose of CO_2 injection, several related technologies have been developed, including Enhanced Oil Recovery (EOR), Enhanced Coalbed Methane Recovery (ECBM), Enhanced Gas Recovery (EGR), Enhanced Shale Gas Recovery (ESG), and Enhanced Geothermal Systems (EGSs).

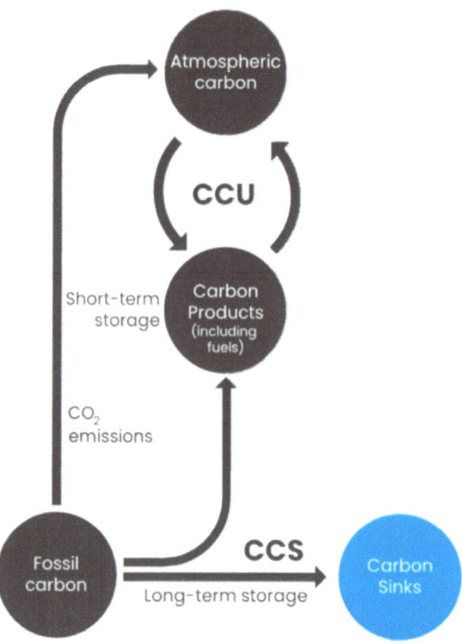

Figure 2. CCS and CCU process. Source: own elaboration.

The engineering projects for CCS and CCUS technologies are highly complex and require extensive research in various engineering and science disciplines, including geology, geoengineering, geophysics, environmental engineering, mathematics, and computer science, for their successful implementation. The site selection process for these projects is crucial to their success and must consider safety, the economy, the environment, and public acceptance at all levels of operation, including country-wide, basin-wide, regional, or sub-basin levels [28,29]. Although CCS and CCUS technologies have similarities in their site selection process, they will induce different physical and chemical responses in underground porous or fractured rock formations, depending on existing hydrological (H), thermal (T), mechanical (M), and chemical (C) fields [30,31].

In the following sections, we examine the key study variables in each population: knowledge, attitudes, perceptions, preferences, acceptance, and trust. CCUS technologies will be the focus of our attention. We will also compare and contrast these factors based on age, sex, and socioeconomic status. The variables that influence the spread of CCU and CCS systems will be the subject of future research. In the last section, we discuss and conclude this study by pointing out the theoretical and practical implications of the research conducted.

2. Materials and Methods

2.1. Participants

The JABODETABEK region was selected as the focus of this study due to its unique geographical location and high levels of tourism, resulting in increased CO_2 emissions. The region is also a hub for economic and social development, with a rapidly growing population and an increasing demand for energy. Therefore, understanding public attitudes towards CCUS technologies in this region is critical for developing effective climate change mitigation strategies.

The region is located on the northwest coast of Java Island and includes the cities of Jakarta, Bogor, Depok, Bekasi, and Tangerang. It is a popular tourist destination and attracts millions of visitors annually, making it one of the largest contributors to greenhouse gas emissions in the region. Furthermore, the JABODETABEK region is experiencing rapid economic and social growth, making it an excellent benchmark for evaluating the social acceptance of CCUS technologies.

A total of 500 people were included in this study, with 100 people from each of the five cities. The people of Jakarta, Bogor, Depok, Bekasi, and Tangerang were the pool from which the participants were drawn. The research team used an online form to find all participants at least 18 years of age.

2.2. Procedure

The research team devised a web-based survey to examine public sentiments about carbon capture and utilisation systems in the JABODETABEK region, which is known for its high tourist activity and consequently high CO_2 emissions. Due to its accelerated growth and potential as a benchmark for the social acceptance of CCUS, this region was selected as a unique case study.

Before the questionnaire was distributed to the participants, it underwent an ethical evaluation and was approved by the Ethics Secretariat of the JEP UKM Department (reference number: UKM/PPI/111/JEP-2023-113). The survey was then emailed to 6000 randomly selected individuals within the JABODETABEK region. The participants had the option to complete the survey at their leisure.

Before starting the survey, participants were informed of the research objectives and asked for their permission. The survey was completed by 500 people from five communities in the region (Jakarta, Bogor, Depok, Bekasi, and Tangerang). The response rate of 8.3% is within the average range for email-distributed surveys.

2.3. Measures

To gauge public opinion on CCUS technologies in Jakarta, Bogor, Depok, Bekasi, and Tangerang, this study used a questionnaire designed by the research team. To construct a prediction model for the adoption of CCUS, this survey was created to gauge the general public's familiarity with and opinion on CCUS technologies. Items from earlier studies on public acceptance of CCUS technologies in various countries were included in the questionnaire, along with items designed by the research team to examine the many dimensions of public acceptance of energy technology.

Baseline questions, CCS/CCU details, awareness and overall assessment, attribute beliefs, acceptance, preferred alternatives, and trust were only some of the topics covered in the questionnaire's several sections. Two paragraphs detailed the essential characteristics of CCS/CCU technologies in the second portion of the questionnaire. Sections ii, iii, and iv of the questionnaire were the meat and potatoes of the study, since they contained most of the items aimed at gauging the level of knowledge of the respondents, emotions, perceptions of benefits and drawbacks, and general dispositions towards them. To assess the reliability of the aforementioned questionnaire, Cronbach's coefficient alpha was used, the results of which indicated a good level of reliability (attitude $\alpha = 0.862$; beliefs $\alpha = 0.750$; perceived benefit/cost $\alpha = 0.854$; trust $\alpha = 0.755$; emotion $\alpha = 0.883$; social norm $\alpha = 0.766$; cognitive dissonance $\alpha = 0.722$).

2.4. Data Analysis

The structural equation modelling (SEM) approach was used to evaluate the data obtained from this investigation. Structural equation modelling (SEM) analysis investigated the connections between study variables: awareness, general evaluation of CCUS technologies, affect, perception of qualities, benefits and costs, acceptance, preference over other technologies, and trust. The SEM analysis also allowed us to determine the most important individual factors that determine the acceptability of technologies such as CCU and CCS. In addition to structural equation modelling, the study variables were cross-analysed by age, sex, urban/rural residency, and income using descriptive statistics and bivariate correlations. The calculations in the statistical analysis were performed using the Statistical Package for the Social Sciences (SPSS) version 26.

3. Results

3.1. Demographic Characteristics

Participants in the current study were selected using a convenience sampling strategy. A total of 500 people were included in the study, with men making up 49.8% and women accounting for 50.2 percent. The ages ranged widely, with the highest percentage (18.2%) falling in the "young adults" bracket. Those between the ages of 51 and 60 made up the next largest group (17%), followed by those between the ages of 21 and 30 (17.2%), demonstrating a wide spread of ages among the participants.

Most of the respondents said their annual income was less than IDR 5,500,000. Participants with incomes between IDR 2,500,000 (13.6%) and IDR 2,500,000–3,500,000 (10.6%) were the most numerous. These results indicate that the majority of the people who participated in the survey had a modest income. Most of the participants had completed high school (21.2% of the total) or college (21.4%). Less than twenty-one percent of the survey population reported having no formal education. Most of the participants (20%) and those who leased (21.4%) had their own homes, while only a minority (17.8%) lived in family homes. These results indicate that the study sample included people from various backgrounds in terms of housing (see Table 1).

3.2. Level of Awareness towards CCUS Technology

Next, we aimed to assess the respondents' level of awareness regarding CCUS technology. To achieve this, we provided information on CCUS technology and its potential to mitigate climate change and asked participants to evaluate CCU and CCS as alternative solutions to address climate change. We provided comprehensive information about CCUS in an online form created for participants. To measure the participants' awareness of CCUS, we used a 5-point Likert scale, ranging from 1 (very poor) to 5 (very good).

Table 2 and Figure 3, presented above, provide an overview of the responses from the participants regarding their awareness of CCUS technology. The results indicate that most of the participants had a low level of knowledge of CCUS technology. Specifically, more than half of the participants in each area rated their awareness as "very poor" or "poor". The highest percentage of participants who rated their awareness as "very poor" were in Bekasi (47%), followed by Tangerang (33%), Jakarta (27%), Depok (28%), and Bogor (37%). These findings suggest a significant lack of awareness among participants about CCUS technology and its potential to mitigate climate change. More efforts are needed to raise awareness and promote education on this important topic in the areas studied.

3.3. Investigating Fear and Interest Levels among Respondents

As part of our study on CCUS technologies, we sought to dig deeper into how people respond emotionally to this innovative technology. To do so, we asked our participants to rate to what degree the CCUS information fact sheet triggered their fear and interest (see Table 3).

Subjective reactions to CCUS technologies in Jakarta, Bogor, Depok, Bekasi, and Tangerang are shown in the table below. There appears to be a range of feelings towards CCUS technologies among the participants, as evidenced by the results. Participants in Tangerang were more likely to report "very fear" about CCUS technology, while those in Depok were less likely to have such feelings.

Table 1. Sample demographics. Source: own elaboration.

Demographic Factors	Frequency	Percentage	Demographic Factors	Frequency	Percentage
Gender			Income		
Male	249	49.8	<IDR 2,500,000	68	13.6
Female	251	50.2	2,500,000–3,500,000	53	10.6
Age			3,501,000–4,500,000	56	11.2
18–20	91	18.2	4,501,000–5,500,000	51	10.2
21–30	86	17.2	5,501,000–6,500,000	53	10.6
31–40	83	16.6	7,501,000–8,500,000	47	9.4
41–50	82	16.4	8,501,000–9,500,000	75	15
51–60	85	17	9,501,000–10,500,000	49	9.8
>60 years old	73	14.6	>IDR 10,501,000	48	9.6
Educational Status			Residence		
No formal education	105	21	Family house	89	17.8
Primary education/junior high school	87	17.4	Private house	100	20
Senior high school	106	21.2	Rent	107	21.4
Bachelor	95	19	Contract	104	20.8
Postgraduate	107	21.4	Boarding house	100	20
Working Status					
Unemployed	75	15			
A contract in the private sector	77	15.4			
Permanent position in the private sector	87	17.4			
Civil servant	73	14.6			
Self-employed	98	19.6			
Contract in the government	90	18			

Table 2. Level of awareness. Source: own elaboration.

Area	Participants	% of Total Participants	Very Poor (1)	Poor (2)	Moderate (3)	Good (4)	Very Good (5)
Jakarta	220	0.44	60	55	66	11	28
Bogor	65	0.13	24	16	5	14	6
Depok	80	0.16	22	23	24	6	5
Bekasi	85	0.17	40	23	6	10	6
Tangerang	50	0.1	20	13	23	5	7

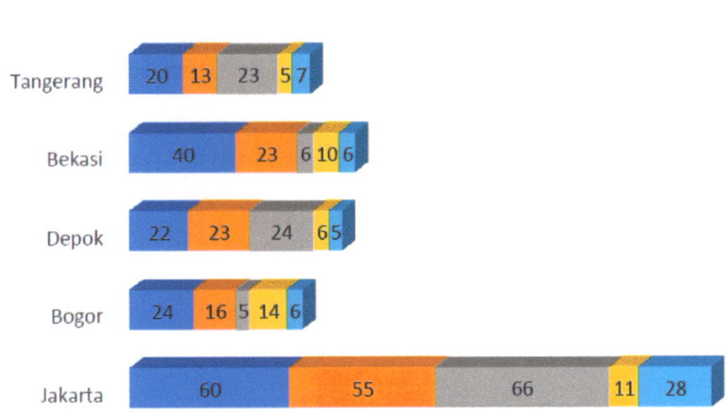

Figure 3. Level of awareness. Source: own elaboration.

Table 3. Fear and interest levels among respondents. Source: own elaboration.

Area	Participants	% of Total Participants	Very Fear (1)	Poor Fear (2)	Not Sure (3)	Interest (4)	Very Interesting (5)
Jakarta	220	0.44	65	40	59	38	18
Bogor	65	0.13	22	15	13	15	7
Depok	80	0.16	13	27	13	10	5
Bekasi	85	0.17	28	27	16	10	4
Tangerang	50	0.1	42	16	18	8	4

The results show that many respondents have serious reservations about using CCUS technology (see Figure 4). This may imply that respondents are aware of the difficulties associated with CCUS technology and that further public education and communication about the advantages of CCUS technologies are required to allay fears about their use. However, research also shows that people are at least somewhat curious about CCUS technologies.

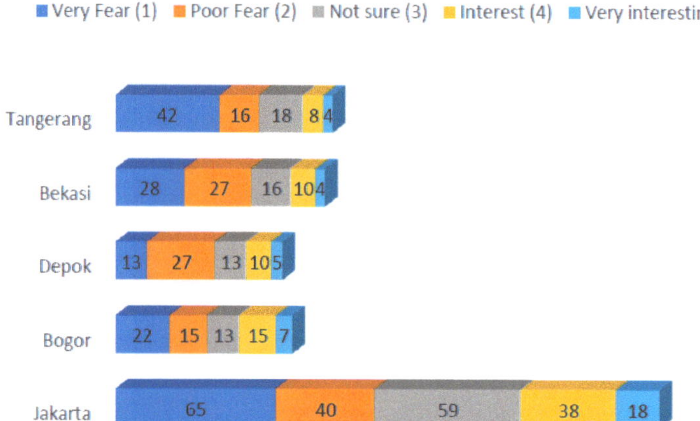

Figure 4. Fear and interest levels among respondents. Source: own elaboration.

3.4. Descriptive Statistics and Independent Sample T-Test Results for Demographic Factors and Social Acceptance of CCUS

The next analysis examined the relationship between demographic factors and social acceptance of carbon capture, utilisation, and storage (CCUS) technology. The analysis involved descriptive statistics for demographic factors and social acceptance scores and independent sample *t*-tests to determine whether there were significant differences in social acceptance scores between different demographic groups. Table 4 provides detailed information on the mean social acceptance scores, standard deviations, *t*-values, *p*-values, and test results for each demographic factor examined.

Table 4. Descriptive statistics and results of the independent sample *t*-test for demographic factors and social acceptance of CCUS. Source: own elaboration.

Demographic Factor	Mean (Males)	Standard Deviation (Males)	Mean (Females)	Standard Deviation (Females)	Score Difference	*t*-Value (Score)	*p*-Value (Score)	Result
Gender	3.97	0.65	3.98	0.64	−0.008	−0.36	0.719	No
Income	4.11	1.07	4.35	1.11	0.261	1.48	0.14	No
Age	4.02	0.81	4.03	0.83	−0.01	−0.39	0.697	No
Educational Status	4.01	0.95	4.05	1.01	0.007	0.34	0.734	No
Residence	3.98	0.69	4.04	0.88	0.053	2.06	0.04	Yes
Working Status	4.01	0.91	4.03	0.94	−0.015	−0.69	0.491	No

The analysis results indicate that there were no significant differences in the social acceptance scores between men and women ($t = -0.36$, $p = 0.719$), income groups ($t = 1.48$, $p = 0.14$), age groups ($t = -0.39$, $p = 0.697$), educational statuses ($t = 0.34$, $p = 0.734$), and working statuses ($t = -0.69$, $p = 0.491$). However, there was a significant difference in social acceptance scores between residents of different areas ($t = 2.06$, $p = 0.04$), and those living in urban areas showed greater social acceptance of CCUS technology than those living in rural areas.

3.5. Exploring the Acceptance of CCUS Technologies

To determine the acceptance of CCUS technologies, the researchers employed a multivariate analysis method using structural equation modelling (SEM) to examine direct and indirect determinants. The researchers proposed a model of acceptance to analyse public acceptance (see Table 5).

Table 5. Total, direct, and indirect standardised effects. Source: own elaboration.

Variable	Direct Effect	Indirect Effect	Total Effect
Attitudes	0.13	0.13 *	0.26 *
Beliefs	0.08	0.10 *	0.18 *
Perceived benefits/costs	0.18		0.18
Trust	0.13	0.13 *	0.26 *
Emotions	0.11		0.11
Social norms	0.05	0.07 *	0.12 *
Cognitive dissonance	-	0.02 *	0.02 *

* significant purpose.

The above table provides an overview of the direct, indirect, and overall effects of several variables on the social acceptance of carbon capture, use, and storage (CCUS) technology. Information was collected by interviewing 500 people and asking them about their thoughts and feelings towards CCUS.

According to the findings, an individual's feelings about CCUS had a direct effect of 0.13 on social acceptance. This indicates that those with more favourable attitudes towards this technology were more inclined to embrace it. Furthermore, attitudes also had a substantial indirect influence of 0.13 through other factors, which brought the total effect of attitudes to 0.26. This suggests that attitudes play a significant role in determining the degree to which CCUS is accepted in society, directly and indirectly, due to other factors.

Furthermore, beliefs about CCUS had a direct effect on social acceptance equal to 0.08 and a substantial indirect effect equal to 0.10, resulting in a total effect equal to 0.18. This suggests that individuals' opinions regarding the possible benefits and risks connected with CCUS are major determining factors in their willingness to accept the technology.

Other variables that have significant total impacts on societal acceptance of CCUS include perceptions of advantages and costs (total effect = 0.18), trust in energy corporations (total effect = 0.26), social norms (total effect = 0.12), and cognitive dissonance (total effect = 0.02). According to the findings, perceptions of the benefits and costs associated with CCUS, trust in the institutions and individuals involved in CCUS, their conformity to social norms, and their ability to reconcile conflicting attitudes and beliefs can play a role in determining whether or not they will accept technology.

4. Discussion

Research on the public assessment of CCS and CCUS is still very limited in developing countries such as Indonesia. However, it is interesting and important for governments and industries to understand public attitudes towards this technology, given its potential economic and environmental impacts if it is applied successfully [32–35]. Furthermore, to date, no studies have focused on targeting public acceptance in developing countries regarding the application of CCUS technology. Therefore, this study aims to fill this gap by analysing the social acceptance of CCUS technology and its relationship with various demographic factors. The results of this study provide information on the factors that influence public acceptance of CCUS technology in Indonesia, which could be used as a basis for policy making and industry practises [34,36–38]. Furthermore, the findings could also contribute to the literature on public attitudes towards emerging technologies in developing countries.

The research shows the importance of comprehending the distinct carbon capture and utilisation (CCU) and carbon capture and storage (CCS) technologies in the process of formulating and executing carbon capture, utilisation, and storage (CCUS) initiatives. In contemporary times, carbon capture and use (CCU) methods have garnered significant interest in mitigating greenhouse gas emissions through the capture and repurposing of carbon dioxide (CO_2) for alternative applications. Carbon dioxide (CO_2) has the potential to serve as a precursor for the synthesis of various chemical compounds, fuels, and building materials. The implementation of CCU technologies holds promise in mitigating emissions and generating new economic prospects and value chains. However, CCU technologies encounter various obstacles, such as exorbitant expenses and technological constraints.

Using carbon capture and storage (CCS) technologies encompasses the capture of CO_2 emissions from various sources, such as industrial processes and power plants, and their storage in geological formations, such as deep saline aquifers or depleted oil and gas reservoirs. Carbon capture and storage (CCS) technology exhibits the capacity to substantially mitigate greenhouse gas emissions originating from sizeable stationary sources, including power plants. This technology has already been implemented in numerous large-scale initiatives across the globe. Notwithstanding its potential benefits, CCS encounters various obstacles, such as exorbitant expenses, technological limitations, and societal approval.

To achieve the successful development and implementation of carbon capture, use, and storage (CCUS) programmes, a comprehensive and collaborative approach is necessary. This approach should consider the distinctive benefits and obstacles associated with each CCU and CCS technology. Collaboration among policymakers, industry stakeholders, and the general public is imperative to advance the growth and implementation of these technologies as a feasible approach to addressing the issue of climate change. Furthermore, it is imperative to continue to carry out research and development efforts to tackle the technological and economic obstacles linked to carbon capture and utilisation (CCU) and carbon capture and storage (CCS) and to guarantee their sustained feasibility.

The findings of this report indicate that, based on the survey results, only a small number of the respondents are aware of CCUS technology. This is followed by fear among the public about implementing this technology. This suggests that public knowledge of the implementation of CCUS technology is still weak, even though it could provide positive environmental benefits. These results are very similar to previous reports by Ericson et al. [39] or Porse et al. [40], which state that public awareness and knowledge of CCUS technology are limited.

CCUS technology, which stands for carbon capture, use, and storage, is a technology that aims to capture carbon dioxide emissions from various sources, such as power plants and industrial processes, and then store or use them for other purposes, such as improved oil recovery or chemical production [41–43]. Despite its potential benefits in reducing greenhouse gas emissions, public awareness and acceptance of this technology have been limited. The low level of awareness among the public regarding CCUS technology suggests a need for greater public education and engagement to increase awareness and understanding of the technology. This could involve government, academic institution, and industry stakeholder outreach campaigns to provide information and increase public participation in developing and implementing CCUS technology.

Additionally, addressing public fears and concerns about the implementation of CCUS technology is crucial to increasing public acceptance of the technology. This could involve transparent communication about the risks and benefits of the technology, as well as efforts to address any potential negative impacts on the environment and public health.

The research findings indicate that most of the respondents who participated in this study have a negative attitude towards CCUS technology. This is supported by the respondents' low awareness and high level of fear about implementing this technology. These results contradict previous research by Fernandez [16], in which respondents had a supportive attitude towards the implementation of CCUS because it was believed to provide environmental benefits. This difference may be due to the low level of knowledge about this technology. CCUS is perceived as dangerous and lacking innovation and not providing economic benefits to society, making it likely that rejection occurs in developing regions.

Increasing public knowledge and understanding of CCUS technology is important in addressing this issue. Efforts to increase public awareness of the potential benefits of this technology could include providing information on how CCUS can help reduce greenhouse gas emissions and mitigate climate change, as well as highlighting its economic benefits. Additionally, it is important to address concerns about the potential risks and negative impacts of CCUS on the environment and public health and to provide transparent information to increase public trust in technology.

In addition, it is important to engage the public in the development and implementation of CCUS technology. This could involve public participation in decision-making processes, such as consultations and public hearings, to ensure that the concerns and interests of the public are taken into account [23,41,44,45]. This would help build trust and confidence in this technology among the public and foster a more positive attitude towards its implementation.

Furthermore, another finding of this study is the attitude of the respondents towards CCUS technology. The analysis showed that the low acceptance of CCUS technology is due to the lack of knowledge about the technology, which is consistent with previous research by Oltra et al. [11]. Public concern is also a determining factor in the acceptance of technology, where when the public perceives that CCUS technology will provide better environmental safety, they tend to choose it as an alternative technology that provides better environmental protection [1,46,47]. Furthermore, differences in the sociodemographic characteristics of the respondents, who come from developing countries, also contribute to the rejection and low awareness of CCUS technology [14]. At the national level, there is a tendency for CCUS opponents to report less interest in the technology, as well as a higher level of fear associated with CCS relative to supporters. Opponents perceive CCS as more dangerous and less beneficial to the economy than supporters of the technology. They tend to have lower levels of trust in energy companies.

To address these concerns, providing accurate and transparent information about the benefits and risks of CCUS technology to the public is important. This can be achieved through public education and awareness campaigns and the participation of local communities in the development and implementation of CCUS projects. Involving local communities in decision making and addressing their concerns can help build trust and acceptance of this technology and ensure that their interests and concerns are considered. Furthermore, it is important to involve stakeholders from different sectors, including industry, government, and civil society, in discussions about CCUS technology to ensure that diverse perspectives are represented and to promote dialogue and collaboration.

An intriguing finding of this research is that the perception of risk and environmental damage, the perception of economic benefits for the region, and previous pro-technology attitudes are the most important individual-level predictors of CCUS adoption at the local level. This study's model does not entirely explain the acceptance of CCUS, but it aids in understanding its components. Other variables that were not taken into account in the study may explain CCUS acceptance. Furthermore, attitudes and opinions may be unstable and susceptible to contextual factors.

In general, our data indicate the need for (I) developing a positive CCUS project vision that is aligned with the community's values, expectations, and aspirations; (II) increasing local involvement in the project; (III) providing local benefits; (IV) integrating sustainability with existing community structures; and (V) ensuring effective communication and participation.

5. Conclusions

According to the results, people in developing nations such as Indonesia know little about CCUS technology. Because of the public's lack of understanding, CCUS technology is often met with scepticism and hostility, which threatens the effectiveness of CCUS initiatives. Therefore, outreach programmes by the government, academic institutions, and industry stakeholders are needed to educate the public and improve people's knowledge and understanding of technology.

In order to increase the public acceptance of CCUS technology, it is essential to address public anxieties and concerns about its implementation. This could include making people aware of the potential hazards and benefits of the technology and working to mitigate any adverse effects it may have on the environment or people's health. Further, public participation in CCUS research and implementation is essential for gaining the public's trust and confidence in the technology and encouraging a more optimistic outlook on its eventual deployment.

This study suggests that public awareness and understanding of CCUS technology is low in developing countries, such as Indonesia. To improve public acceptance of CCUS, it is essential to address public concerns and anxieties about the technology while providing accurate and complete information on its potential benefits and limitations. The study identified three concrete CCUS technologies: carbon capture, carbon storage, and carbon utilisation, each with advantages and challenges. Carbon capture technology can be energy-intensive and costly, while carbon storage technology may be limited by the availability of suitable storage sites and the risk of carbon leakage. Carbon utilisation technology may require significant infrastructure investments and could compete with other uses of captured carbon dioxide.

To enhance the public's acceptance of these CCUS technologies, outreach programmes by the government, academic institutions, and industry stakeholders are needed to educate the public and improve people's knowledge and understanding of the technology. Moreover, public participation in research and implementation is essential to gain public trust and confidence in the technology.

Finally, the public must receive complete and accurate information on the advantages and disadvantages of CCUS technology. Educating the public and receiving their input on the design and implementation of CCUS programmes are two ways to achieve this goal. Discussions about the benefits of CCUS technology should include representatives from various sectors, not just the private sector, government, and the nonprofit sector. These actions have the potential to raise public awareness of CCUS technology, which, in turn, could help reduce greenhouse gas emissions and slow the rate of climate change.

Despite the strengths of this study, several limitations need to be acknowledged. First, the sample size was relatively small and consisted mainly of participants from one geographic region, which may limit the generalisability of the findings to other populations. Second, the study relied on self-reported data, which may be subject to response and social desirability biases.

Several areas of future research could be based on the findings of this study. First, larger-scale studies with more diverse populations and geographic regions could help confirm the results' generalisability. Second, future research could investigate the possible mediating or moderating effects of other variables that may influence the relationship between X and Y. Third, longitudinal studies could help establish the causal relationship and add more variables that can determine human acceptance of new technology.

Author Contributions: Conceptualisation, S.E. and C.S.; methodology, C.S.; software, J.O. and C.S.; validation, C.S.; formal analysis, S.E. and J.O.; investigation, S.E.; resources, C.S. and J.O.; data curation, S.E.; writing—original draft preparation, C.S.; writing—review and editing, C.S. and J.O.; supervision, C.S. and J.O.; project administration, S.E. All authors have read and agreed to the published version of the manuscript.

Funding: This research received no external funding.

Institutional Review Board Statement: The study was conducted in accordance with the Declaration of Helsinki. An ethical evaluation was approved by the Ethics Secretariat of the JEP UKM Department (reference number: UKM/PPI/111/JEP-2023-113).

Informed Consent Statement: Informed consent was obtained from all subjects involved in the study.

Data Availability Statement: Data are contained within the article.

Conflicts of Interest: The authors declare no conflict of interest.

References

1.   Liu, B.; Xu, Y.; Yang, Y.; Lu, S. How Public Cognition Influences Public Acceptance of CCUS in China: Based on the ABC (Affect, Behavior, and Cognition) Model of Attitudes. *Energy Policy* **2021**, *156*, 112390. [CrossRef]

2. Nielsen, J.A.E.; Stavrianakis, K.; Morrison, Z. Community Acceptance and Social Impacts of Carbon Capture, Utilization and Storage Projects: A Systematic Meta-Narrative Literature Review. *PLoS ONE* **2022**, *17*, e0272409. [CrossRef] [PubMed]
3. Lu, S.; Fang, M.; Wang, Q.; Huang, L.; Sun, W.; Zhang, Y.; Li, Q.; Zhang, X. Public Acceptance Investigation for 2 Million Tons/Year Flue Gas CO_2 Capture, Transportation and Oil Displacement Storage Project. *Int. J. Greenh. Gas Control* **2021**, *111*, 103442. [CrossRef]
4. Wojakowski, D.; Langhelle, O.; Assadi, M.; Nagy, S. Public Acceptance of CCS/CCUS Technology in Onshore Areas in NW Poland. *Balt. Carbon Forum* **2022**, *1*, 16. [CrossRef]
5. Mulyasari, F.; Harahap, A.K.; Rio, A.O.; Sule, R.; Kadir, W.G.A. Potentials of the Public Engagement Strategy for Public Acceptance and Social License to Operate: Case Study of Carbon Capture, Utilisation, and Storage Gundih Pilot Project in Indonesia. *Int. J. Greenh. Gas Control* **2021**, *108*, 103312. [CrossRef]
6. Xu, S.; Dai, S. CCUS As a Second-Best Choice for China's Carbon Neutrality: An Institutional Analysis. *Clim. Policy* **2021**, *21*, 927–938. [CrossRef]
7. Li, Q.; Liu, G.; Leamon, G.; Liu, L.C.; Cai, B.; Chen, Z.A. A National Survey of Public Awareness of the Environmental Impact and Management of CCUS Technology in China. *Energy Procedia* **2017**, *114*, 7237–7244. [CrossRef]
8. Koukouzas, N.; Christopoulou, M.; Giannakopoulou, P.P.; Rogkala, A.; Gianni, E.; Karkalis, C.; Pyrgaki, K.; Krassakis, P.; Koutsovitis, P.; Panagiotaras, D.; et al. Current CO_2 Capture and Storage Trends in Europe in a View of Social Knowledge and Acceptance. A Short Review. *Energies* **2022**, *15*, 5716. [CrossRef]
9. Liu, B.; Xu, Y.; Chen, Y.; Lu, S.; Zhao, D. How to Garner Public Support for Carbon Capture, Utilization and Storage? Comparing Narrative and Statistical Frames among Chinese Citizens. *Energy Res. Soc. Sci.* **2022**, *91*, 102738. [CrossRef]
10. Verliac, M.; Le Calvez, J. Microseismic Monitoring for Reliable CO_2 injection and Storage—Geophysical Modeling Challenges and Opportunities. *Lead. Edge* **2021**, *40*, 418. [CrossRef]
11. Oltra, C.; Dütschke, E.; Preuß, S.; Gonçalves, L.; Germán, S. What Influences Public Attitudes and Acceptance of CCUS Technologies on the National and Regional Level? Results from a Survey Study in France and Spain. *SSRN Electron. J.* **2022**. [CrossRef]
12. Mendoza, N.; Mathai, T.; Boren, B.; Roberts, J.; Niffenegger, J.; Sick, V.; Zimmermann, A.W.; Weber, J.; Schaidle, J. Adapting the Technology Performance Level Integrated Assessment Framework to Low-TRL Technologies Within the Carbon Capture, Utilization, and Storage Industry, Part I. *Front. Clim.* **2022**, *4*, 818786. [CrossRef]
13. Li, Q.; Liu, G.; Cai, B.; Leamon, G.; Liu, L.C.; Chen, Z.A.; Li, X. Public Awareness of the Environmental Impact and Management of Carbon Dioxide Capture, Utilization and Storage Technology: The Views of Educated People in China. *Clean Technol. Environ. Policy* **2017**, *19*, 2041–2056. [CrossRef]
14. Liu, G.; Cai, B.; Li, Q.; Leamon, G.; Cao, L.; Zhou, Y. Perceptions and acceptance of CCUS environmental risk assessment, by experts and government officers, in China. *Int. J. Big Data Min. Glob. Warm.* **2019**, *1*, 1950010. [CrossRef]
15. Li, S.; Zhang, Y.; Lin, Q.; Li, Y.; Liang, X. Topography Modelling for Potentially Leaked CO_2 Diffusion and Its Application in Human Health Risk Assessment for Carbon Capture, Utilisation, and Storage Engineering in China. *Int. J. Greenh. Gas Control* **2022**, *119*, 103714. [CrossRef]
16. Fernández-Canteli Álvarez, P.; García Crespo, J.; Martínez Orío, R.; Mediato Arribas, J.F.; Ramos, A.; Berrezueta, E. Techno-Economic Evaluation of Regional CCUS Implementation: The STRATEGY CCUS Project in the Ebro Basin (Spain). *Greenh. Gases Sci. Technol.* **2022**, *13*, 197–215. [CrossRef]
17. Martínez-Gordón, R.; Sánchez-Diéguez, M.; Fattahi, A.; Morales-España, G.; Sijm, J.; Faaij, A. Modelling a Highly Decarbonised North Sea Energy System in 2050: A Multinational Approach. *Adv. Appl. Energy* **2022**, *5*, 100080. [CrossRef]
18. Prajna Indrasuta, M.B.; Hutagalung, A.S.; Putra, S.G.Z.; Saga, R.; Rizky, A.A. Integration of Clustering System and Joint Venture Business Model for CCUS Deployment. *Indones. J. Energy* **2023**, *6*, 30–47. [CrossRef]
19. Perreault, P.; Kummamuru, N.B.; Gonzalez Quiroga, A.; Lenaerts, S. CO_2 Capture Initiatives: Are Governments, Society, Industry and the Financial Sector Ready? *Curr. Opin. Chem. Eng.* **2022**, *38*, 100874. [CrossRef]
20. Olfe-Kräutlein, B.; Armstrong, K.; Mutchek, M.; Cremonese, L.; Sick, V. Why Terminology Matters for Successful Rollout of Carbon Dioxide Utilization Technologies. *Front. Clim.* **2022**, *4*, 830660. [CrossRef]
21. Frank-Collins, J.; Greenberg, S.; Sung, J.; Hawkins, J.; Gupta, N.; Haagsma, A. A Case Study Strategic CCUS Communications across a 20-State Region of the United States. In Proceedings of the 16th Greenhouse Gas Control Technologies Conference, Lyon, France, 23–27 October 2022. [CrossRef]
22. Usman, U.; Saputra, D.; Firdaus, N. Source Sink Matching for Field Scale CCUS CO_2-EOR Application in Indonesia. *Sci. Contrib. Oil Gas* **2021**, *44*, 97–106. [CrossRef]
23. Jiutian, Z.; Zhiyong, W.; Jia-Ning, K.; Xiangjing, S.; Dong, X. Several Key Issues for CCUS Development in China Targeting Carbon Neutrality. *Carbon Neutrality* **2022**, *1*, 17. [CrossRef]
24. Rakhiemah, A.N.; Xu, Y. Economic Viability of Full-Chain CCUS-EOR in Indonesia. *Resour. Conserv. Recycl.* **2022**, *179*, 106069. [CrossRef]
25. Marbun, B.T.H.; Santoso, D.; Kadir, W.G.A.; Wibowo, A.; Suardana, P.; Prabowo, H.; Susilo, D.; Sasongko, D.; Sinaga, S.Z.; Purbantanu, B.A.; et al. Improvement of Borehole and Casing Assessment of CO_2-EOR/CCUS Injection and Production Well Candidates in Sukowati Field, Indonesia in a Well-Based Scale. *Energy Rep.* **2021**, *7*, 1598–1615. [CrossRef]

26. Jonek-Kowalska, I. Towards the Reduction of CO2 Emissions. Paths of pro-Ecological Transformation of Energy Mixes in European Countries with an above-Average Share of Coal in Energy Consumption. *Resour. Policy* **2022**, *77*, 102701. [CrossRef]

27. Brodny, J.; Tutak, M. Assessing the Energy and Climate Sustainability of European Union Member States: An MCDM-Based Approach. *Smart Cities* **2023**, *6*, 339–367. [CrossRef]

28. Kuzior, A.; Vyshnevskyi, O.; Trushkina, N. Assessment of the Impact of Digitalization on Greenhouse Gas Emissions on the Example of EU Member States. *Prod. Eng. Arch.* **2022**, *28*, 407–419. [CrossRef]

29. Kuzior, A.; Postrzednik-Lotko, K.A.; Postrzednik, S. Limiting of Carbon Dioxide Emissions through Rational Management of Pro-Ecological Activities in the Context of CSR Assumptions. *Energies* **2022**, *15*, 1825. [CrossRef]

30. Tota, V.; Viganò, F.; Gatti, M. Application of CCUS to the WtE Sector. In Proceedings of the 15th Greenhouse Gas Control Technologies Conference, Online, 15–18 March 2021. [CrossRef]

31. Lui, L.C.; Leamon, G.; Li, Q.; Cai, B. Developments towards Environmental Regulation of CCUS Projects in China. *Energy Procedia* **2014**, *63*, 6903–6911. [CrossRef]

32. Brauer, D.C.; Larrick, D. Public Communication and Collaboration for Carbon Capture, Utilization, and Storage Technology: Acceptance, Education, and Outreach. In Proceedings of the 31st Annual International Pittsburgh Coal Conference: Coal—Energy, Environment and Sustainable Development, PCC 2014, Pittsburgh, PA, USA, 6–9 October 2014.

33. Shogenov, K.; Shogenova, A. New CO2 and Hydrogen Storage Site Marketing: How to Make Your Storage Site Unique and Attractive? *Balt. Carbon Forum* **2022**, *1*, 2–3. [CrossRef]

34. Torres, J.; Bogdanov, I.; Boisson, M. A Modeling Workflow for Geological Carbon Storage Integrated with Coupled Flow and Geomechanics Simulations. In Proceedings of the ECMOR 2020—17th European Conference on the Mathematics of Oil Recovery, Online Event, 14–17 September 2020. [CrossRef]

35. Da Coelho, S.; Almeida, A.; Lima, R.; Câmara, G.A.; Rocha, P.; Andrade, J.C. Quick Screening for CO$_2$-EOR as CCUS Technology: A Case Study in Recôncavo Basin. In *SPE International Conference and Exhibition on Health, Safety, Environment, and Sustainability*; OnePetro: Manama, Bahrain, 2020. [CrossRef]

36. Romanak, K.; Fridahl, M.; Dixon, T. Attitudes on Carbon Capture and Storage (Ccs) as a Mitigation Technology within the Unfccc. *Energies* **2021**, *14*, 629. [CrossRef]

37. Koning, M.; Zikovic, V.; Van der Valk, K.; Pawar, R.; Williams, J.; Opedal, N.; Dudu, A. Development of a Screening Framework for Re-Use of Existing Wells for CCUS Projects Considering Regulatory, Experimental and Technical Aspects. In *Middle East Oil, Gas and Geosciences Show*; OnePetro: Manama, Bahrain, 2023. [CrossRef]

38. Corrales, M.; Mantilla Salas, S.; Tasianas, A.; Hoteit, H.; Afifi, A. The Potential for Underground CO$_2$ Disposal Near Riyadh. In *International Petroleum Technology Conference*; OnePetro: Manama, Bahrain, 2022. [CrossRef]

39. Erickson, D.D.; Haghighi, H.; Phillips, C. The Importance of CO$_2$ Composition Specification in the CCUS Chain. In Proceedings of the Annual Offshore Technology Conference, Houston, TX, USA, 2–5 May 2022. [CrossRef]

40. Porse, S.L.; Wade, S.; Hovorka, S.D. Can We Treat CO$_2$ Well Blowouts like Routine Plumbing Problems? A Study of the Incidence, Impact, and Perception of Loss of Well Control. *Energy Procedia* **2014**, *63*, 7149–7161. [CrossRef]

41. Wesche, J.; Germán, S.; Gonçalves, L.; Jödicke, I.; López-Asensio, S.; Prades, A.; Preuß, S.; Algado, C.O.; Dütschke, E. CCUS or No CCUS? Societal Support for Policy Frameworks and Stakeholder Perceptions in France, Spain, and Poland. *Greenh. Gases Sci. Technol.* **2023**, *13*, 48–66. [CrossRef]

42. Duguid, A.; Hawkins, J.; Keister, L. CO$_2$ Pipeline Risk Assessment and Comparison for the Midcontinent United States. *Int. J. Greenh. Gas Control* **2022**, *116*, 103636. [CrossRef]

43. Wadea, S.; Catherb, M.; Cummingc, L.; Dalyd, D.; Garrette, G.; Greenbergf, S.; Myhreg, R.; Stoneg, M.; Tollefsonh, L. Digital Communications: Status and Potential Applications for CCUS Public Outreach. *Energy Procedia* **2014**, *63*, 7070–7086. [CrossRef]

44. Zhang, H.; Li, D.; Gu, X.; Chen, N. Three Decades of Topic Evolution, Hot Spot Mining and Prospect in CCUS Studies Based on CitNetExplorer. *Chin. J. Popul. Resour. Environ.* **2022**, *20*, 91–104. [CrossRef]

45. Chen, S.; Liu, J.; Zhang, Q.; Teng, F.; McLellan, B.C. A Critical Review on Deployment Planning and Risk Analysis of Carbon Capture, Utilization, and Storage (CCUS) toward Carbon Neutrality. *Renew. Sustain. Energy Rev.* **2022**, *167*, 112537. [CrossRef]

46. Han, J.; Li, J.; Tang, X.; Wang, L.; Yang, X.; Ge, Z.; Yuan, F. Coal-Fired Power Plant CCUS Project Comprehensive Benefit Evaluation and Forecasting Model Study. *J. Clean. Prod.* **2023**, *385*, 135657. [CrossRef]

47. Greig, C.; Uden, S. The Value of CCUS in Transitions to Net-Zero Emissions. *Electr. J.* **2021**, *34*, 107004. [CrossRef]

Article

Biomass Based N/O Codoped Porous Carbons with Abundant Ultramicropores for Highly Selective CO_2 Adsorption

Congxiu Guo [1], Ya Sun [1], Hongyan Ren [2], Bing Wang [1,*], Xili Tong [3,*], Xuhui Wang [1], Yu Niu [1] and Jiao Wu [1,*]

[1] School of Electric Power, Civil Engineering and Architecture, Shanxi University, Taiyuan 030006, China; guocongxiu@sxu.edu.cn (C.G.); 17835342913@163.com (Y.S.); wangxuhui@sxu.edu.cn (X.W.); niuyu@sxu.edu.cn (Y.N.)

[2] Jinzhong College of Information, Taigu 030800, China; renhongyan@jzci.edu.cn

[3] State Key Laboratory of Coal Conversion, Institute of Coal Chemistry, Chinese Academy of Sciences, Taiyuan 030001, China

* Correspondence: wangbing@sxu.edu.cn (B.W.); tongxili@sxicc.ac.cn (X.T.); jiaow@sxu.edu.cn (J.W.)

Abstract: In this work, N/O codoped porous carbons (NOPCs) were derived from corn silk accompanied by Na_2CO_3 activation. The porous structures and surface chemical features of as-prepared carbon materials were tailored by adjusting the Na_2CO_3 mass ratio. After activation, the optimized sample (NOPC1) with abundant ultramicropores and pyrrolic N displays an enhanced CO_2 adsorption capacity of 3.15 mmol g^{-1} and 1.95 mmol g^{-1} at 273 K and 298 K at 1 bar, respectively. Moreover, this sample also exhibited high IAST selectivity (16.9) and Henry's law selectivity (15.6) for CO_2/N_2 at 298 K as well as moderate heat adsorption. Significantly, the joint effect between ultramicropore structure and pyrrolic N content was found to govern the CO_2 adsorption performance of NOPCs samples.

Keywords: biomass; corn silk; ultramicropore; nitrogen and oxygen co-doping; CO_2 adsorption

Citation: Guo, C.; Sun, Y.; Ren, H.; Wang, B.; Tong, X.; Wang, X.; Niu, Y.; Wu, J. Biomass Based N/O Codoped Porous Carbons with Abundant Ultramicropores for Highly Selective CO_2 Adsorption. *Energies* **2023**, *16*, 5222. https://doi.org/10.3390/en16135222

Academic Editors: Haris Ishaq and Liwei Zhang

Received: 14 June 2023
Revised: 4 July 2023
Accepted: 5 July 2023
Published: 7 July 2023

1. Introduction

Parallel to the massive consumption of fossil fuels, the continuous emissions of various air pollutants have led to global warming and resulted in a series of serious ecological issues. To alleviate these environmental problems, various technologies have been developed to inhibit CO_2 from releasing into the atmosphere, such as chemical absorption [1], physical absorption [2], membrane separation [3], cryogenic distillation [4] and others [5]. Among these technologies, physical absorption with the merits of a cost-effective process, including low equipment requirements, low power consumption and efficient regeneration [6,7], is considered one of the most competitive techniques to curb CO_2 emissions. To date, various porous materials, such as zeolite [8], porous organic polymers [9], mesoporous silica [10], graphene [11], metal-organic frameworks (MOFs) [12,13], covalent organic frameworks (COFs) [14] and porous carbons [15], have been developed as physical adsorbents for CO_2 capture. In this regard, porous carbons have been attracting greater attention due to their large specific surface area, adjustable porosity and surface functionality [15,16]. However, the low polarity of pristine porous carbons commonly causes poor adsorption selectivity of CO_2 over N_2 and low adsorption heat [6]. Activation is considered to be the optimal approach to improve the physicochemical properties of carbon materials, including physical activation and chemical activation. During physical activation, the porosity of carbon materials significantly increase through high temperature carbonization or partial gasification of a precursor. By contrast, during chemical activation, the activator can create a multi-scale porous structure for carbon materials via chemical etching under high temperatures [17,18]. Various activators have been widely used in the literature, such as KOH, NaOH, H_3PO_4, Na_2CO_3, K_2CO_3, $ZnCl_2$ and others [18,19]. Among these activators, KOH is the most frequently used as a chemical agent to modify various carbon materials to enhance their CO_2 adsorption capacity. For instance, sucrose-derived porous carbons were prepared by adding different ratios of KOH and urea, and displayed a remarkable CO_2 uptake of 8.19 mmol g^{-1} at 273 K and 1 bar [19]. Liu et al. [20] synthesized a category of porous carbon monoliths via a novel in-situ KOH activation methodology. The reported carbon monoliths had a high ultramicropore volume of 0.3 m^3 g^{-1} and therefore demonstrated remarkable static CO_2 uptake (5.0 mmol g^{-1}). Similarly, the N, S, and O co-doped porous carbons were fabricated via KOH activation of polybenzoxazine and exhibited a high CO_2 adsorption capacity of 7.04 mmol g^{-1} and CO_2/N_2 selectivity of 27 at 273 K [21]. In fact, KOH activation can induce large amounts of micropores and small mesopores in carbon frameworks [22]. However, KOH activation can result in potentially serious environmental pollution and heavy corrosion damage to the experimental apparatus. In contrast, carbonate can simultaneously carry out chemical and physical activation of the carbon framework via the release of CO_2 at high temperatures [17]. For example, through a Na_2CO_3 assisted method, a 3D holey N-doped graphene with a large specific surface area of 1173 m^2 g^{-1} was synthesized via fast pyrolysis of alanine [23]. Caglayan et al. demonstrated that the CO_2 mass uptake of carbon material activated by Na_2CO_3 was eight times high than that of the air oxidized and nitric acid oxidized counterparts [24].

Heteroatom doping, especially strongly electronegative atoms (such as N, O and P), can effectively improve the surface physical-chemical properties of carbon materials [25,26] and further enhance CO_2 adsorption performance. Substantially, incorporation of nitrogen into the carbon skeleton can provide abundant basic sites that may enhance the Lewis acid-base interaction between functional groups on the surface and acidic CO_2 molecules [19,22,27]. Furthermore, introduction of another heteroatom can increase the surface polarities of N-doped carbon materials [19], which is also conducive to CO_2 adsorption. For instance, N and P co-doped porous carbon was fabricated through pyrolysis of a highly-cross-linked triazine polymer and demonstrated an enhanced CO_2 adsorption capacity of 1.52 mmol g^{-1} at 1 bar [28]. Similarly, N and O co-doped carbon nanotubes synthesized via coal pyrolysis activation achieved a remarkable CO_2 uptake capacity of 3.77 mmol g^{-1} at 298 K and 1 bar [29]. In addition, micropores, especially ultramicropores (d < 1 nm), present a vital contribution to the adsorption capacity and adsorption selectivity of CO_2, while mesopores and macropores can reduce CO_2 transport resistance and accelerate CO_2 diffusion, respectively [29–31]. Therefore, rational design of carbon materials with a hierarchical porous structure of combined micropores, mesopores and macropores can optimize the adsorption performance of CO_2.

Biomass-derived porous carbons have received a great deal of attention for their physical adsorption of CO_2 considering their low cost, high accessibility, environmental sustainability and high chemical stability [32,33]. More significantly, heteroatom self-doping stemmed from the organic elements in plants that can be formed through chemical activation processes, and the multi-scale pores originated from biological structures with interconnected channels that can effectively increase the rates of CO_2 diffusion [33]. In a recent study, heteroatom-doped porous carbons derived from cornstalk [15], bamboo [34], tobacco stems [35], lotus stalks [36], hazelnut shells [37] and Entada rheedii shells [38] have been prepared and employed as the adsorbents for CO_2 adsorption, demonstrating the decisive effect of ultramicropore and heteroatom doping on CO_2 adsorption capacity. Corn silk, as a cheap and readily available biomass raw material, containing abundant N and O elements [39], is a potential candidate for preparing porous adsorbents toward CO_2 capture, and the adsorption performance of carbon materials based on corn silk has not yet been investigated. Herein, N/O codoped porous carbons were synthesized from corn silk, while pore-structure and surface chemical features were adjusted through Na_2CO_3 activation at high temperatures. The optimized sample, with abundant ultramicropores and pyrrolic N, displays an enhanced CO_2 adsorption capacity, in comparison to its counterparts. This work is expected to pave an accessible pathway for synthesizing a cost-effective alternative that enhances CO_2 capture.

2. Materials and Methods

2.1. Materials

The original corn silk was obtained from a corn field in Daixian, Shanxi province, China. The chemical reagents, such as sulfuric acid (H_2SO_4), hydrochloric acid (HCl), and sodium carbonate (Na_2CO_3) were purchased from Sinopharm Chemical Reagent Co., Ltd., Shanghai, China and used without further purification.

2.2. Fabrication of N/O Codoped Porous Carbons (NOPCs)

The synthesizing process of NOPCs is shown in Figure 1. Typically, corn silks were dispersed in 0.5 M H_2SO_4 solution and maintained at 80 °C for 5 h. The products were treated with 1 M HCl, and subsequently washed with deionized H_2O until neutral. Finally, the silks were dried at 100 °C for 12 h to obtain acid-activated corn silks (a-CS). Next, the a-CS was thoroughly blended with a Na_2CO_3 solution (the weight ratio value of Na_2CO_3 to a-CS is 0:1/1:1/3:1/9:1) and the obtained mixture was impregnated and stirred for 12 h, followed by separation and airing at 100 °C for 24 h. Afterwards, 1 g of the mixture powder was placed in a tube furnace under an N_2 atmosphere and activated at 800 °C for 3 h. Subsequently, the products were soaked in 1 M HCl for 24 h and washed with deionized water until a pH value of seven was reached. Finally, a N/O codoped porous carbon was obtained after drying overnight at 100 °C, which was named NOPCx, where x represented the Na_2CO_3/a-CS mass ratio.

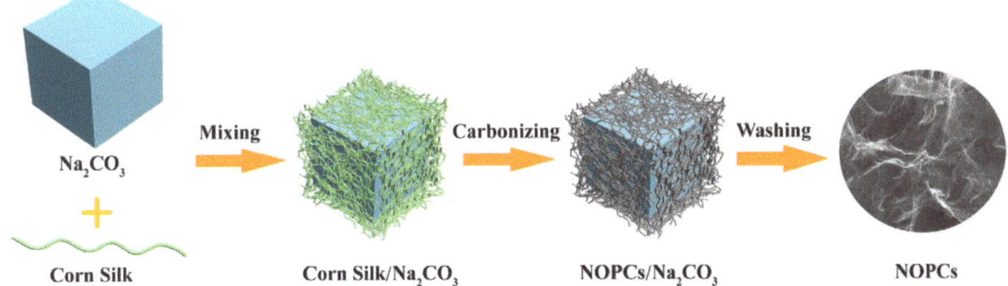

Figure 1. Schematic illustration of fabrication of NOPCs.

2.3. Characterization

The surface morphology and microstructure of NOPCs were observed via scanning electron microscopy (SEM, Thermo Fisher Quattro S, Thermo Fisher Scientific, Waltham, MA, USA) and transmission electron microscopy (TEM, Thermo Scientific Talos F200S, Thermo Fisher Scientific, USA). The crystallinity of samples was characterized via X-ray diffractometer (XRD, Brucker D8 Advance, Brucker, Germany) with Cu-Ka radiation (λ = 1.5418 Å). The graphitization degree of NOPC samples was investigated using a Raman spectrometer (Raman, Thermofisher Dxr2xi, Thermo Fisher Scientific, USA) with an excitation wavelength of 514 nm. The elemental composition of samples was analyzed using an elemental analyzer (EA, Elementar Vario EL cube, Elementar, Germany). The surface functional groups and elemental contents of NOPCs were determined using a Fourier transformed infrared spectrometer (FTIR, Thermo Scientific Nicolet iS50, Thermo Fisher Scientific, USA) and X-ray photoelectron spectroscopy (XPS, Thermo Fisher Scientific K-alpha, Thermo Fisher Scientific, USA) with a monochromatic Al Ka X-ray source (1486.6 eV photons). Shirley background subtraction was used for the deconvolution of XPS spectra. The textural properties of NOPCs were measured at 77 K using a gas sorption analyzer (Micromeritics ASAP 2460, Micromeritics, Norcross, GA, USA). The specific surface area and pore size distribution were determined using the BET method and non-local density functional theory (NLDFT).

2.4. CO$_2$ Adsorption Experiment

The CO$_2$ and N$_2$ adsorption isotherms were measured on a Belsorp Max II instrument under pressures of 0–100 kPa. Prior to the adsorption measurement, the samples were degassed at 120 °C for 12 h to remove impurities from the pores.

3. Results

3.1. Physico-Chemical Properties of NOPCs

The morphological features of the NOPCs are shown in Figure 2. The NOPC0 sample directly carbonized from a-CS shows a relatively dense block structure and smooth surface without apparent pores (Figure 2a,b). Compared to NOPC0, the NOPC1 displays a wrinkled flake surface (Figure 2c,d), implying that the new surface with pore structures in the sample can be produced through Na$_2$CO$_3$ activation. With the increase of Na$_2$CO$_3$ dosage, a large number of pores, even macropores, are observed on the surface (Figure 2e,f). It is worth noting that, NOPC9 exhibits a sponge-like overall structure with multiscale pore holes present (Figure 2g,h), indicating that micropores can be transformed into mesopores and macropores in the cases of higher dosage of Na$_2$CO$_3$, which is mainly caused by the escape of a large amount of gas molecules, such as CO or CO$_2$, during the activation process. In addition, there are no obvious macropores on the flake surface in NOPC1 (Figure 2i), suggesting micropores and mesopores dominated carbon materials. The corresponding HR-TEM image (Figure 2i inset image) shows a heterogeneous worm-like structure with irregular lattice fringes, manifesting the characteristics of amorphous carbon.

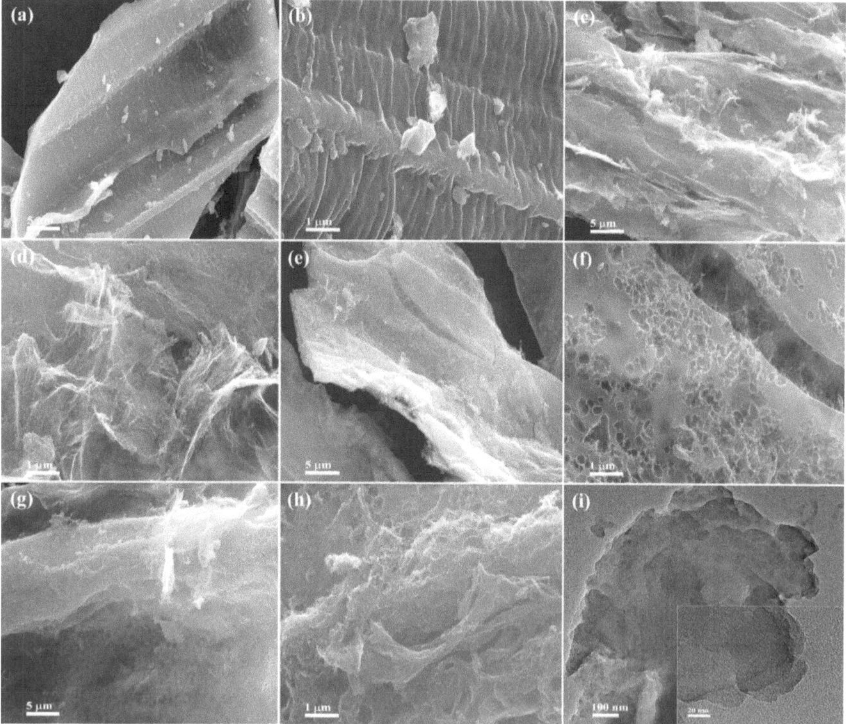

Figure 2. SEM images of (**a,b**) NOPC0, (**c,d**) NOPC1, (**e,f**) NOPC3, (**g,h**) NOPC9. TEM image of (i) NOPC1 and HR-TEM image of NOPC1 in inset.

The XRD spectra of all samples (Figure 3a) displays two broad peaks at around 24° and 43°, which belong to (002) and (100) plane reflections of carbon, respectively [15,27].

Notably, compared to the unactivated NOPC0, the diffraction peaks of other samples become wider, signifying more defects and disorderly structures. The graphitization degree of all samples was further analyzed via Raman spectra as shown in Figure 3b. Two prominent broad peaks located at approximately 1344 cm^{-1} (D band) and 1575 cm^{-1} (G band) are clearly observed, which is ascribed to amorphous/defective carbon and graphitic carbon, respectively [27,40]. The ratios of D band to G band (I_D/I_G) for NOPC1, NOPC3 and NOPC9 are higher than NOPC0, indicating that more defects were introduced into carbon network by the activation treatment of Na_2CO_3 at high temperatures. In fact, this enhanced degree of disorder can offer more active sites for improving CO_2 adsorption capacity [28,40].

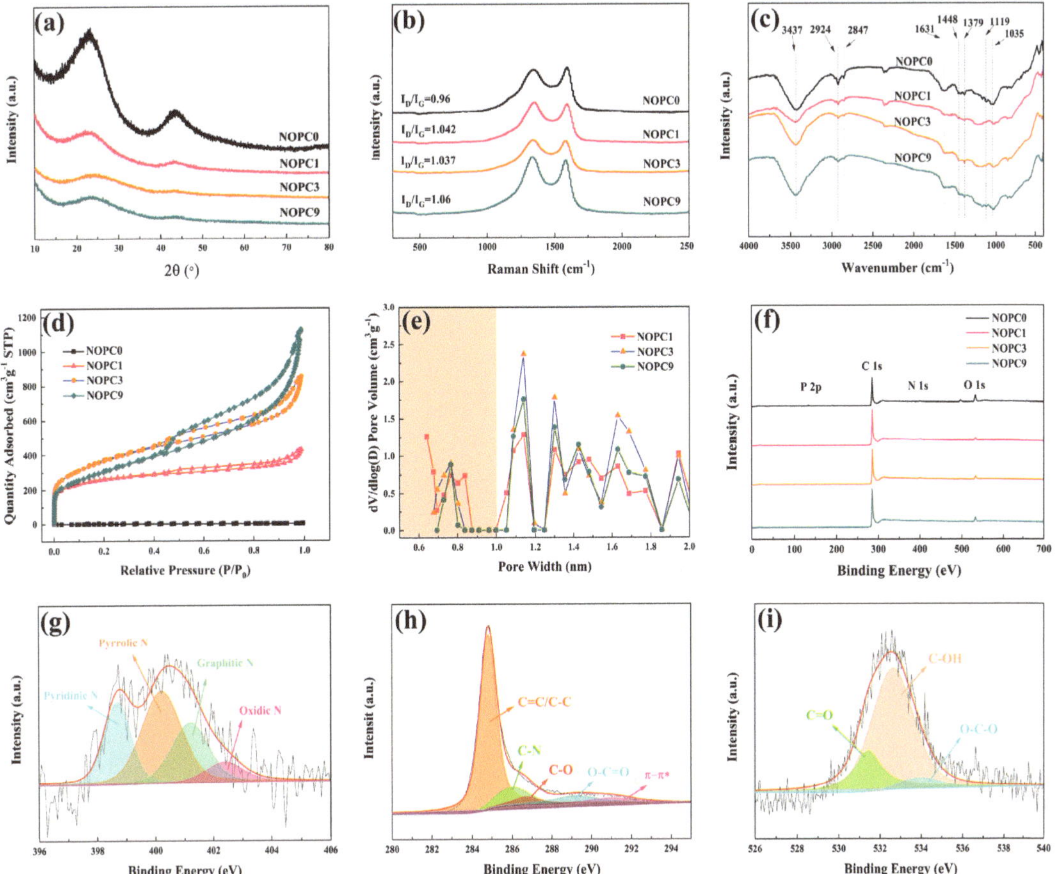

Figure 3. (**a**) XRD patterns, (**b**) Raman spectra, (**c**) FT−IR spectra and (**d**) N_2 adsorption-desorption isotherms of NOPCs. (**e**) Pore size distributions of NOPC1, NOPC3 and NOPC9. (**f**) XPS survey scan for all samples. (**g**) N 1s spectrum, (**h**) C 1s and (**i**) O 1s spectrum of NOPC1.

The surface features of all samples were further analyzed by FT−IR, as illustrated in Figure 3c. The broad band at 3437 cm^{-1} is assigned to O−H and/or N−H stretching vibrations [15,41,42], while two weak peaks at 2924 cm^{-1} and 2847 cm^{-1} are ascribed to the C−H stretching vibrations [40,43]. The moderately intense peak centered at 1631 cm^{-1} belongs to C=C/C−N/C−O [11,15]. Two weak peaks at 1448 and 1119 cm^{-1} are attributed to pyridine-like groups and C−N, respectively [15,41,43]. In addition, the peaks located at 1379 and 1035 cm^{-1} are related to O−H [11,43] and C−O bonds [42], respectively.

The results of the FT−IR analysis clearly confirm that N and O elements are successfully introduced into the carbon skeleton structure through the activation of Na_2CO_3.

The textural properties of NOPCs are measured via N_2 adsorption-desorption at 77 K, as shown in Figure 3d. The NOPC0 with a non-porous structure exhibits a low BET surface area ($0.47 \ m^2 \ g^{-1}$) and a negligible adsorption capacity. Different from NOPC0, other samples prominently increase in their N_2 adsorption capacity at $P/P_0 < 0.05$ and show an obvious hysteresis loop at $P/P_0 > 0.4$, implying the characteristics of microporous and mesoporous structures [22,28]. In addition, a sharp increase of N_2 uptakes in the high pressures ($0.9 < P/P_0 < 1.0$), suggesting the existence of a small number of macropores [15], testified in Figures 2 and S1. For further analysis, the textural properties of all samples are listed in Table 1. After chemical activation, the specific areas of NOPC1, NOPC3 and NOPC9 are significantly higher than that of NOPC0, verifying the indispensable role of Na_2CO_3 in forming porous structures in carbon materials. Notably, the NOPC9 has lower surface area and pore volumes than NOPC3, which is most likely due to the pore collapse phenomenon caused by higher etching with more dosages of Na_2CO_3 at high temperatures. As summarized in Table 1, the pore size distributions obtained via the NLDFT method show that the cumulative micropore volume of NOPC1 is $0.3668 \ cm^3 \ g^{-1}$, higher than NOPC3 ($0.3618 \ cm^3 \ g^{-1}$) and NOPC9 ($0.2899 \ cm^3 \ g^{-1}$). In addition, the micropore diameters (<1 nm) of NOPC1, NOPC3 and NOPC9 are mainly concentrated at 0.8 nm, and the NOPC1 possesses the highest cumulative ultramicropore volume. More impressively, as reflected in Figure 3e and Table 1, the cumulative ultramicropore volume (<0.7 nm) for NOPC1 is much larger than that of other samples, which is beneficial for enhancing CO_2 adsorption due to the overlapping interaction of the van der Waals' force [20,27,29,32].

Table 1. Textural properties of all samples.

Sample	S_{BET} $(m^2 \ g^{-1})$	V_t [a] $(cm^3 \ g^{-1})$	V_m [b] $(cm^3 \ g^{-1})$	V_1 [c] $(cm^3 \ g^{-1})$	$V_{0.7}$ [d] $(cm^3 \ g^{-1})$
NOPC0	0.47	8.9×10^{-4}	-	-	-
NOPC1	850.16	0.82	0.3668	0.1795	0.1194
NOPC3	1269.79	1.63	0.3618	0.1169	0.0762
NOPC9	1096.95	1.42	0.2899	0.0885	0.0604

[a]: Total pore volume (Vt) obtained at P/P_0 of 0.99. [b,c,d]: Cumulative micropore volume of V_m, (d < 2 nm), V_1, (d < 1 nm) and $V_{0.7}$ (d < 0.7 nm) determined by NLDFT model.

The elemental contents of NOPCs samples were analyzed via EA as listed in Table S1. With the increase of the mass ratio of Na_2CO_3, the oxygen contents of activated samples display a continuously increasing tendency, while nitrogen contents increase at first, and then decrease. Interestingly, the nitrogen content of NOPC9 is lower than NOPC3, which may be due to the release of nitrogen species by reacting with excessive Na_2CO_3. In addition, all samples have higher levels of oxygen than nitrogen, revealing that oxygen-containing functional groups are more stable at high temperatures. Considering that all samples retain a considerable amount of N and O, it can be speculated that N and O elements are successfully induced into the carbon backbone. The surface elemental compositions of NOPCs were further determined via XPS. The C, O, N and trace P elements were detected on the surface of all samples, as shown in Figure 3f and Table S2, and the elemental contents of nitrogen and oxygen calculated by XPS results exhibit the same variation pattern as the EA analysis. The deconvolution of N 1s spectra (Figure 3g) for all samples yielded four peaks at binding energies of 398.7, 400.2, 401.2 and 402.4 eV, which are attributed to the pyridinic N, pyrrolic N, graphitic N and oxidized N, respectively [44,45]. The relative contents of different N species and O species for all samples are summarized in Table 2. With the increase in the mass ratio of Na_2CO_3, the contents of pyridinic N and oxidized N increase significantly, while that of pyrrolic N and graphitic N decrease. In fact, pyrrolic N and pyridinic N play a dominant role in enhancing the CO_2 adsorption performance via Lewis acid-base interactions and hydrogen bonding between surface functional groups and CO_2 molecules [15,19,22,27,41,42]. Therefore, the NOPC1 sample with the highest relative contents of pyrrolic N can be employed as a potential adsorbent for CO_2 adsorption. In addition, a certain number of oxygen groups in porous carbons can also boost CO_2 adsorption capacity [34,35]. Therefore, the characteristics of oxygen-containing groups are assessed by the high-resolution C 1s and O 1s spectra, as described in Figure 3h,i. The high-resolution C 1s spectroscopy shows five major peaks, corresponding to C=C/C—C (284.8 eV), C—N (286.0 eV), C—O (286.7 eV), O—C=O (289.2 eV) and π—π^* bonds (291.0 eV), respectively [46,47]. Likewise, the O 1s spectra can be fitted into three peaks located at 531.4 eV, 532.6 eV and 533.9 eV, corresponding to C=O, C—OH and C—O—C, respectively [44,48]. These XPS results confirm the existence of abundant oxygen-containing functional groups on the surface of NOPC samples, which is beneficial for further promoting CO_2 adsorption [16,29,43]. More significantly, the NOPC1 sample has the highest content of C—OH among NOPCs, as listed in Table 2, suggesting a stronger affinity for CO_2 through the hydrogen bonding interactions [21].

Table 2. The contents of different N species and O species for NOPCs samples.

Sample	N Species (%)				O Species (%)		
	Pyridinic N (%)	Pyrrolic N (%)	Graphitic N (%)	Oxidized N (%)	C=O (%)	C—OH (%)	C—O—C (%)
NOPC0	27.26	41.34	22.07	9.34	14.97	64.77	20.26
NOPC1	25.55	41.21	24.98	8.27	15.24	79.53	5.23
NOPC3	29.37	38.01	19.54	13.08	20.3	75.38	4.32
NOPC9	32.21	34.31	18.49	15	16.18	75.92	7.89

3.2. CO_2 Adsorption Performance of NOPCs

The CO_2 adsorption capacity of NOPC1, NOPC3 and NOPC9 were measured at 273 K and 298 K under 1 bar, respectively. As displayed in Figure 4a,b, the CO_2 adsorption uptakes of all samples significantly decrease as the temperature is increased from 273 K to 298 K, verifying the exothermic adsorption characteristic in thermodynamics. It is noteworthy that NOPC1, with the lowest specific surface area, exhibits the highest CO_2 uptakes (3.15 and 1.95 mmol g^{-1} at 273 K and 298 K, respectively, Table 3). To demonstrate these results, the relationships between CO_2 uptake and the physicochemical characteristics are investigated and shown in Figure 4c and Figure S2. It can be seen that the CO_2 uptake is positively correlated with the V_1 (d < 1 nm), $V_{0.7}$ (d < 0.7 nm) and content of pyrrolic N, however, it is rarely related to BET surface area, nitrogen content and micropore volume. In fact, when the slit pore (below 0.7 nm) is smaller than twice the kinetic diameter of CO_2 molecules (0.33 nm), the interaction between CO_2 and adsorbent surfaces can be enhanced through the van der Waals force [20]. Additionally,, as listed in Table 2, the higher contents of pyrrolic N and C—OH in NOPC1 can strengthen the affinity for CO_2 molecules, thereby effectively improving its CO_2 adsorption capacity. Numerous studies have claimed that ultramicropores (d < 0.7 nm) play the decisive role in enhancing CO_2 adsorption compared to other factors [32,35,36,49]. More comparisons of the CO_2 adsorption capacity of NOPC1 and other activated carbon materials are listed in Table 4. The NOPC1 sample in our work exhibits competitive CO_2 adsorption performance compared to other adsorbents in previous studies.

Table 3. CO_2 and N_2 uptakes, and CO_2/N_2 selectivity of NOPC1, NOPC3 and NOPC9.

Sample	CO_2 Uptake (mmol g^{-1})		N_2 Uptake (mmol g^{-1})	CO_2/N_2 Selectivity (298 K)	
	273 K	298 K	298 K	IAST (1 bar)	Henry's Law
NOPC1	3.15	1.95	0.196	16.9	15.6
NOPC3	2.63	1.55	0.223	8.8	10.2
NOPC9	2.17	1.27	0.145	12.2	16.0

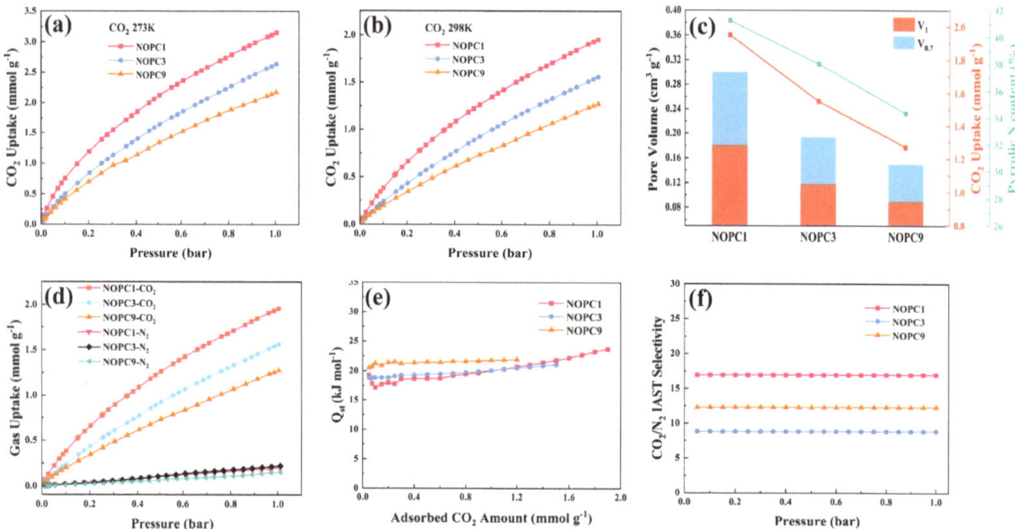

Figure 4. CO_2 adsorption isotherms of NOPC1, NOPC3 and NOPC9 at (**a**) 273 K and (**b**) 298 K. (**c**) The relationship of CO_2 uptake with ultramicropore volume and pyrrolic N content. (**d**) Isosteric heat of CO_2 adsorption for NOPC1, NOPC3 and NOPC9. (**e**) Isotherms of CO_2 and N_2 of NOPC1, NOPC3 and NOPC9 at 298 K. (**f**) CO_2/N_2 IAST selectivity of NOPC1, NOPC3 and NOPC9 at 298 K.

Table 4. CO_2 adsorption performance of NOPC1 compared with other works.

Sample	Activation	S_{BET} (m^2 g^{-1})	CO_2 Uptake (mmol g^{-1}) (298 K, 1 Bar)	IAST Selectivity	Ref.
A-TDP-12	KOH, 700 °C	1332	1.52	6	[28]
C0800	800 °C	431	2.4	33.7	[31]
C2800	hydroquinonesulfonic acid potassium salt, 800 °C	-	1.9	12.5	[31]
AC-TBG	KOH, 800 °C	971	3.2	-	[33]
AC-UK	KOH, 650 °C	532	2.63	-	[34]
SRC-3K-500	KOH, 500 °C	743	2.78	6.7	[50]
Z0800	Zn(NO$_3$)$_2$, 800 °C	569	2.5	25.0	[51]
1:2/900/2	H$_3$PO$_4$, 800 °C	773	1.88	-	[52]
A-TCLP-700-1	KOH, 700 °C	1353	2.39	19	[53]
NOPC1	Na$_2$CO$_3$, 500 °C	850.16	1.95	16.9	This work

Isosteric heat adsorption (Q_{st}) is a significant thermodynamic parameter to evaluate the affinity of adsorbents for CO_2. To determine the value of Q_{st}, the pure adsorption isotherms of CO_2 on NOPCs are first fitted using the single-site Langmuir model [54,55]:

$$q = \frac{q_{sat}bp}{1 + bp} \tag{1}$$

where q is the adsorption capacity (mmol g^{-1}), q_{sat} presents saturation adsorption capacity (mmol g^{-1}), b is Langmuir constant (kPa^{-1}), and p is the pressure (kPa). Next, Q_{st} can be calculated based on the fitted CO_2 adsorption data at 273 and 298 K via the Clausius-Clapeyron equation [27]:

$$Q_{st} = -RT^2 \left(\frac{\partial lnP}{\partial T} \right) \tag{2}$$

where R, P and T represent the gas constant (8.314 J mol K^{-1}), pressure (kPa) and temperature (K), respectively. From Figure 4d, it is calculated that the Q_{st} values of all samples are in the range of 18.7–23.6 kJ mol^{-1}, implying the physisorption mechanism of the adsorption process [15]. Notably, the Q_{st} values of all adsorbents exhibit a slight increase with the improved adsorption capacity, which may be attributed to the enhanced adsorbate-adsorbate interactions between CO_2 molecules in the ultramicropore [22,27]. In substance, CO_2 molecules preferentially occupy the active sites during the adsorption process, and as the amount of CO_2 increased, the ultramicropore (d < 0.7 nm) played the dominant role [27,56]. The higher Q_{st} for NOPC9 at low CO_2 loading is most likely due to its higher content of oxygen-containing functional groups (Tables S1 and S2), which provide more active adsorption sites. Furthermore, at high surface coverage, the value of Q_{st} for NOPC1 is higher than those for NOPC3 and NOPC9, verifying the decisive effect of ultramicropores (d < 0.7 nm) enhancing the affinity of adsorbents for CO_2.

Moreover, the CO_2/N_2 selectivity is a significant parameter to determine the practical applications of adsorbents. Figure 4e shows the CO_2 and N_2 adsorption uptakes of NOPC1, NOPC3 and NOPC9 at 298 K. The CO_2 adsorption uptakes are significantly higher than the N_2 adsorption uptakes over the entire pressure range at 298 K, demonstrating the higher adsorption selectivity of as-prepared samples for CO_2, which is in favor of separating CO_2 from combustion flue gas. To further evaluate the performance of adsorbents, the CO_2/N_2 adsorption selectivity can be calculated by the ideal adsorption solution theory (IAST) using CO_2 and N_2 adsorption data fitted with a single-site Langmuir model (Equation (1)). The main components of flue gas are assumed to be N_2 (85%) and CO_2 (15%). The IAST selectivity (S) can be defined as [55,57]:

$$S = \frac{x_1/y_1}{x_2/y_2} \tag{3}$$

where x and y represent the mole fraction of different components in the adsorbed phase mixture and the gas phase mixture, respectively. Next, the IAST selectivity of CO_2/N_2 for NOPC1, NOPC3 and NOPC9 at 298 K was calculated according to the IAST method and presented in Figure 4f and Table 3. It is found that the IAST selectivity of NOPC1 is much higher than those of NOPC3 and NOPC9. In addition, the CO_2/N_2 selectivity is also estimated using Henry's law (Supporting information) and the corresponding results are summarized in Table 3. The CO_2/N_2 selectivity determined by two theoretical models turn in a consistent performance. Significantly, the NOPC1 sample with the highest selectivity may benefit from its higher ratio of ultramicropores (Figure 3e and Table 1), since the deep well potential in the ultramicropore wall can thoroughly soak CO_2 molecules and increase the adsorption potential energy, thereby boosting the CO_2 adsorption selectivity [27,49]. On the other hand, the CO_2/N_2 selectivity of NOPC9 is significantly higher than that of NOPC3, which is most likely due to the fact that abundant oxygen-contained functional groups on the surface of NOPC9 enhance interactions with CO_2 molecules.

4. Conclusions

In summary, NOPCs were fabricated via direct carbonization of corn silk involving Na_2CO_3 activators. The obtained NOPC1 sample has a large specific surface, high ultra-micropore volume (d < 0.7 nm and d < 1 nm) and high content of pyrrolic N. Therefore, NOPC1 provides 3.15 and 1.95 mmol g^{-1} of the CO_2 uptake at 273 K and 298 K under 1 bar, respectively. Additionally, the CO_2/N_2 selectivity determined by the IAST model and Henry's law is 16.9 and 15.6 at 298 K, respectively. It has been demonstrated that ultramicropore and pyrrolic N exert greater influence on the CO_2 adsorption capacity and CO_2/N_2 selectivity. Notably, the CO_2 adsorption capacity of NOPCs needs to be further improved and the essential factor (ultramicropore and pyrrolic N) for enhancing the CO_2 adsorption performance cannot be independently studied in this work. However, the current research provides valuable information on synthesis of N/O codoped carbon adsorbents derived from biomass for selective CO_2 adsorption.

Supplementary Materials: The following supporting information can be downloaded at: https://www.mdpi.com/article/10.3390/en16135222/s1, Figure S1: Pore size distribution curves of NOPC1, NOPC3 and NOPC9. Figure S2: The linear fitting of CO_2 uptake (298 K, 1 bar) against (a) BET surface area, (b) Nitrogen content, (c) Pyrrolic N content, (d) V_m, (e) V_1 and (f) $V_{0.7}$. Figure S3: The CO_2/N_2 selectivity by Henry's law on NOPC1 (a), NOPC3 (b) and NOPC9 (c) at 298 K. Table S1: The elemental composition of NOPCs analyzed by EA. Table S2: The surface elemental composition of NOPCs calculated by XPS results.

Author Contributions: C.G.: conceptualization, methodology, resources, data analysis, formal analysis, funding acquisition, writing—original draft. Y.S.: data analysis. H.R.: resources, supervision. B.W.: formal analysis, funding acquisition, supervision. X.T.: formal analysis, supervision. X.W.: funding acquisition, supervision. Y.N.: funding acquisition, supervision. J.W.: formal analysis, supervision. All authors have read and agreed to the published version of the manuscript.

Funding: This work was funded by Science and Technology Innovation Project of Shanxi Colleges and Universities, China (2021L014 and 2021L006) and the Applied Basic Research Project of Shanxi province, China (20210302124126 and 202103021223031).

Data Availability Statement: Not applicable.

Conflicts of Interest: The authors declare no conflict of interest.

References

1. Jahromi, F.B.; Elhambakhsh, A.; Keshavarz, P.; Panahi, F. Insight into the application of amino acid-functionalized MIL-101 (Cr) micro fluids for high-efficiency CO_2 absorption: Effect of amine number and surface area. *Fuel* **2023**, *334*, 126603. [CrossRef]
2. Xin, K.; Annaland, M.V.S. Diffusivities and solubilities of carbon dioxide in deep eutectic solvents. *Sep. Purif. Technol.* **2023**, *307*, 122779. [CrossRef]
3. Pazani, F.; Shariatifar, M.; Maleh, M.S.; Alebrahim, T.; Lin, H.Q. Challenge and promise of mixed matrix hollow fiber composite membranes for CO_2 separations. *Sep. Purif. Technol.* **2023**, *308*, 122876. [CrossRef]
4. Park, K.H.; Lee, J.W.; Lim, Y.; Seo, Y. Life cycle cost analysis of CO_2 compression processes coupled with a cryogenic distillation unit for purifying high-CO_2 natural gas. *J. CO2 Util.* **2022**, *60*, 102002. [CrossRef]
5. Wang, L.; Yao, Y.; Tran, T.; Lira, P.; Sternberg, S.P.E.; Davis, R.; Sun, Z.; Lai, Q.H.; Toan, S.; Luo, J.M.; et al. Mesoporous MgO enriched in Lewis base sites as effective catalysts for efficient CO_2 capture. *J. Environ. Manag.* **2023**, *332*, 117398. [CrossRef]
6. Cui, H.M.; Xu, J.G.; Shi, J.S.; Yan, N.F.; Zhang, C.; You, S.Y. Oxamic acid potassium salt as a novel and bifunctional activator for the preparation of N-doped carbonaceous CO_2 adsorbents. *J. CO2 Util.* **2022**, *64*, 102198. [CrossRef]
7. Ma, C.D.; Bai, J.L.; Demir, M.; Yu, Q.Y.; Hu, X.; Jiang, W.H.; Wang, L.L. Polyacrylonitrile-derived nitrogen enriched porous carbon fiber with high CO_2 capture performance. *Sep. Purif. Technol.* **2022**, *303*, 122299. [CrossRef]
8. Cavallo, M.; Dosa, M.; Porcaro, N.G.; Bonino, F.; Piumetti, M.; Crocella, V. Shaped natural and synthetic zeolites for CO_2 capture in a wide temperature range. *J. CO2 Util.* **2023**, *67*, 102335. [CrossRef]
9. Song, K.S.; Fritz, P.W.; Coskun, A. Porous organic polymers for CO_2 capture, separation and conversion. *Chem. Soc. Rev.* **2022**, *51*, 9831. [CrossRef]
10. Al-Absi, A.A.; Mohamedali, M.; Domin, A.; Benneker, A.M.; Mahinpey, N. Development of in situ polymerized amines into mesoporous silica for direct air CO_2 capture. *Chem. Eng. J.* **2022**, *447*, 137465. [CrossRef]
11. Sun, H.F.; Zhang, Q.G.; Hagio, T.; Ryoichi, I.; Kong, L.; Li, L. Facile synthesis of amine-functionalized three-dimensional graphene composites for CO_2 capture. *Surf. Interfaces* **2022**, *33*, 102256. [CrossRef]
12. Zhu, W.F.; Wang, L.Z.; Cao, H.H.; Guo, R.L.; Wang, C.F. Introducing defect-engineering 2D layered MOF nanosheets into Pebax matrix for CO_2/CH_4 separation. *J. Membr. Sci.* **2023**, *669*, 121305. [CrossRef]
13. Ding, M.L.; Robinson, W.F.; Jiang, H.L.; Yaghi, O.M. Carbon capture and conversion using metal-organic frameworks and MOF-based materials. *Chem. Soc. Rev.* **2019**, *48*, 2783–2828. [CrossRef]
14. Yang, L.X.; Yang, H.; Wu, H.; Zhang, L.L.; Ma, H.Z.; Liu, Y.T.; Wu, Y.Z.; Ren, Y.X.; Wu, X.Y.; Jiang, Z.Y. COF membranes with uniform and exchangeable facilitated transport carriers for efficient carbon capture. *J. Mater. Chem. A* **2021**, *9*, 12636. [CrossRef]
15. Yuan, X.F.; Xiao, J.F.; Yılmaz, M.; Zhang, T.C.; Yuan, S.J. N, P Co-doped porous biochar derived from cornstalk for high performance CO_2 adsorption and electrochemical energy storage. *Sep. Purif. Technol.* **2022**, *299*, 121719. [CrossRef]
16. Lu, T.Y.; Ma, C.D.; Demir, M.; Yu, Q.Y.; Aghamohammadi, P.; Wang, L.L.; Hu, X. One-pot synthesis of potassium benzoate-derived porous carbon for CO_2 capture and supercapacitor application. *Sep. Purif. Technol.* **2022**, *301*, 122053. [CrossRef]
17. Petrovic, B.; Gorbounov, M.; Soltani, S.M. Influence of surface modification on selective CO_2 adsorption: A technical review on mechanisms and methods. *Microporous Mesoporous Mater.* **2021**, *312*, 110751. [CrossRef]
18. Hakami, O. Urea-doped hierarchical porous carbons derived from sucrose precursor for highly efficient CO_2 adsorption and separation. *Surf. Interfaces* **2023**, *37*, 102668. [CrossRef]
19. Shen, Y.F. Preparation of renewable porous carbons for CO_2 capture—A review. *Fuel Process. Technol.* **2022**, *236*, 107437. [CrossRef]

20. Liu, B.G.; Shi, R.; Ma, X.C.; Chen, R.F.; Zhou, K.; Xu, X.; Sheng, P.; Zeng, Z.; Li, L.Q. High yield nitrogen-doped carbon monolith with rich ultramicropores prepared by in-situ activation for high performance of selective CO_2 capture. *Carbon* **2021**, *181*, 270–279. [CrossRef]

21. Guo, Z.; Lu, X.; Xin, Z. N, S, O co-doped porous carbons derived from bio-based polybenzoxazine for efficient CO_2 capture. *Colloid Surf. A* **2022**, *646*, 128845. [CrossRef]

22. Zhou, Y.B.; Tan, P.; He, Z.Q.; Zhang, C.; Fang, Q.Y.; Chen, G. CO_2 adsorption performance of nitrogen-doped porous carbon derived from licorice residue by hydrothermal treatment. *Fuel* **2022**, *311*, 122507. [CrossRef]

23. Cui, H.J.; Jiao, M.G.; Chen, Y.N.; Guo, Y.B.; Yang, L.P.; Xie, Z.J.; Zhou, Z.; Guo, S.J. Molten-salt-assisted synthesis of 3D holey N-doped graphene as bifunctional electrocatalysts for rechargeable Zn-Air batteries. *Small Methods* **2018**, *2*, 1800144. [CrossRef]

24. Caglayan, B.S.; Aksoylu, A.E. CO_2 adsorption on chemically modified activated carbon. *J. Hazard. Mater.* **2013**, *252–253*, 19–28. [CrossRef]

25. Li, C.F.; Zhao, J.W.; Xie, L.J.; Wang, Y.; Tang, H.B.; Zheng, L.R.; Li, G.R. N coupling with S-coordinated Ru nanoclusters for highly efficient hydrogen evolution in alkaline media. *J. Mater. Chem. A* **2021**, *9*, 12659–12669. [CrossRef]

26. Wang, Y.C.; Shao, Y.; Wang, H.; Yuan, J.Y. Advanced heteroatom-doped porous carbon membranes assisted by poly(ionic liquid) design and engineering. *Acc. Mater. Res.* **2020**, *1*, 16–29. [CrossRef]

27. Chen, C.; Zhang, Y.K.; Li, Q.H.; Wang, Y.L.; Ma, J. Guanidine-embedded poly (ionic liquid) as a versatile precursor for self-templated synthesis of nitrogen-doped carbons: Tailoring the microstructure for enhanced CO_2 capture. *Fuel* **2022**, *329*, 125357. [CrossRef]

28. Wang, Y.; Xiao, J.F.; Wang, H.Z.; Zhang, T.C.; Yuan, S.J. Binary doping of nitrogen and phosphorus into porous carbon: A novel di-functional material for enhancing CO_2 capture and super-capacitance. *J. Mater. Sci. Technol.* **2022**, *99*, 73–81. [CrossRef]

29. Yuan, J.C.; Wang, Y.; Tang, M.F.; Hao, X.D.; Liu, J.; Zhang, G.J.; Zhang, Y.F. Preparation of N, O co-doped carbon nanotubes and activated carbon composites with hierarchical porous structure for CO_2 adsorption by coal pyrolysis. *Fuel* **2023**, *333*, 126465. [CrossRef]

30. Rehman, A.; Nazir, G.; Rhee, K.Y.; Park, S.J. A rational design of cellulose-based heteroatom-doped porous carbons: Promising contenders for CO_2 adsorption and separation. *Chem. Eng. J.* **2021**, *420*, 130421. [CrossRef]

31. Shi, J.S.; Cui, H.M.; Xu, J.G.; Yan, N.F.; Zhang, C.; You, S.Y. Synthesis of nitrogen and sulfur co-doped carbons with chemical blowing method for CO_2 adsorption. *Fuel* **2021**, *305*, 121505. [CrossRef]

32. Ma, X.C.; Yang, Y.H.; Wu, Q.D.; Liu, B.G.; Li, D.P.; Chen, R.F.; Wang, C.H.; Li, H.L.; Zeng, Z.; Li, L.Q. Underlying mechanism of CO_2 uptake onto biomass-based porous carbons: Do adsorbents capture CO_2 chiefly through narrow micropores? *Fuel* **2020**, *282*, 118727. [CrossRef]

33. Prasankumar, T.; Salpekar, D.; Bhattacharyya, S.; Manoharan, K.; Yadav, R.M.; Mata, M.A.C.; Miller, K.A.; Vajtai, R.; Jose, S.; Roy, S.; et al. Biomass derived hierarchical porous carbon for supercapacitor application and dilute stream CO_2 capture. *Carbon* **2022**, *199*, 249–257. [CrossRef]

34. Dilokekunakul, W.; Teerachawanwong, P.; Klomkliang, N.; Supasitmongkol, S.; Chaemchuen, S. Effects of nitrogen and oxygen functional groups and pore width of activated carbon on carbon dioxide capture: Temperature dependence. *Chem. Eng. J.* **2020**, *389*, 124413. [CrossRef]

35. Ma, X.C.; Su, C.Q.; Liu, B.G.; Wu, Q.D.; Zhou, K.; Zeng, Z.; Li, L.Q. Heteroatom-doped porous carbons exhibit superior CO_2 capture and CO_2/N_2 selectivity: Understanding the contribution of functional groups and pore structure. *Sep. Purif. Technol.* **2021**, *259*, 118065. [CrossRef]

36. Li, Q.; Lu, T.Y.; Wang, L.L.; Pang, R.X.; Shao, J.W.; Liu, L.L.; Hu, X. Biomass based N-doped porous carbons as efficient CO_2 adsorbents and high-performance supercapacitor electrodes. *Sep. Purif. Technol.* **2021**, *275*, 119204. [CrossRef]

37. Ma, C.D.; Lu, T.Y.; Shao, J.W.; Huang, J.M.; Hu, X.; Wang, L.L. Biomass derived nitrogen and sulfur co-doped porous carbons for efficient CO_2 adsorption. *Sep. Purif. Technol.* **2022**, *281*, 119899. [CrossRef]

38. Mallesh, D.; Anbarasan, J.; Kumar, P.M.; Upendar, K.; Chandrashekar, P.; Rao, B.V.S.K.; Lingaiah, N. Synthesis, characterization of carbon adsorbents derived from waste biomass and its application to CO_2 capture. *Appl. Surf. Sci.* **2020**, *530*, 147226. [CrossRef]

39. Zhang, Y.L.; Zhao, R.; Li, Y.Q.; Zhu, X.X.; Zhang, B.; Lang, X.Y.; Zhao, L.J.; Jin, B.; Zhu, Y.F.; Jiang, Q. Potassium-ion batteries with novel N, O enriched corn silk-derived carbon as anode exhibiting excellent rate performance. *J. Power Sources* **2021**, *481*, 228644. [CrossRef]

40. Ruan, W.; Wang, Y.; Liu, C.R.; Xu, D.W.; Hu, P.; Ye, Y.Y.; Wang, D.C.; Liu, Y.Q.; Zheng, Z.F.; Wang, D. One-step fabrication of N-doped activated carbon by NH_3 activation coupled with air oxidation for supercapacitor and CO_2 capture applications. *J. Ana. Appl. Pyrol.* **2022**, *168*, 105710. [CrossRef]

41. Xiao, J.F.; Yuan, X.F.; Zhang, T.C.; Ouyang, L.K.; Yuan, S.J. Nitrogen-doped porous carbon for excellent CO_2 capture: A novel method for preparation and performance evaluation. *Sep. Purif. Technol.* **2022**, *298*, 121602. [CrossRef]

42. Jin, Z.H.; Jiang, X.; Dai, Z.D.; Xie, L.L.; Wang, W.; Shen, L. Continuous synthesis of nanodroplet-templated, N-doped microporous carbon spheres in microfluidic system for CO_2 capture. *ACS Appl. Mater. Interfaces* **2020**, *12*, 52571–52580. [CrossRef]

43. Wu, D.W.; Liu, J.; Yang, Y.J.; Zheng, Y. Nitrogen/oxygen co-doped porous carbon derived from biomass for low-pressure CO_2 capture. *Ind. Eng. Chem. Res.* **2020**, *59*, 14055–14063. [CrossRef]

44. Xin, S.S.; Huo, S.Y.; Zhang, C.L.; Ma, X.M.; Liu, W.J.; Xin, Y.J.; Gao, M.C. Coupling nitrogen/oxygen self-doped biomass porous carbon cathode catalyst with $CuFeO_2$/biochar particle catalyst for the heterogeneous visible-light driven photo-electro-Fenton degradation of tetracycline. *Appl. Catal. B* **2022**, *305*, 121024. [CrossRef]

45. Guo, J.R.; Xu, X.T.; Hill, J.P.; Wang, L.P.; Dang, J.J.; Kang, Y.Q.; Li, Y.L.; Guan, W.S.; Yamauchi, Y. Graphene-carbon 2D heterostructures with hierarchically-porous P, N-doped layered architecture for capacitive deionization. *Chem. Sci.* **2021**, *12*, 10334–10340. [CrossRef] [PubMed]

46. Khandelwal, M.; Tran, C.V.; In, J.B. Nitrogen and phosphorous co-doped laser-induced graphene: A high-performance electrode material for supercapacitor applications. *Appl. Surf. Sci.* **2022**, *576*, 151714. [CrossRef]

47. Ren, G.Y.; Li, Y.N.; Chen, Q.S.; Qian, Y.; Zheng, J.G.; Zhu, Y.A.; Teng, C. Sepia-derived N, P co-doped porous carbon spheres as oxygen reduction reaction electrocatalyst and supercapacitor. *ACS Sustain. Chem. Eng.* **2018**, *6*, 16032–16038. [CrossRef]

48. Zhu, C.W.; Yan, J.J. Fabricating of N/O-codoping porous carbon interpenetrating networks for high energy aqueous supercapacitor. *J. Energy Storage* **2022**, *52*, 105047. [CrossRef]

49. Wang, Y.H.; Wang, M.H.; Wang, Z.W.; Wang, S.M.; Fu, J.W. Tunable-quaternary (N, S, O, P)-doped porous carbon microspheres with ultramicropores for CO_2 capture. *Appl. Surf. Sci.* **2020**, *507*, 145130. [CrossRef]

50. Li, D.N.; Yang, J.Z.; Zhao, Y.; Yuan, H.R.; Chen, Y. Ultra-highly porous carbon from wasted soybean residue with tailored porosity and doped structure as renewable multi-purpose absorbent for efficient CO_2, toluene and water vapor capture. *J. Clean. Prod.* **2022**, *337*, 130283. [CrossRef]

51. Cui, H.M.; Xu, J.G.; Shi, J.S.; Yan, N.F.; Liu, Y.W.; Zhang, S.W. Zinc nitrate as an activation agent for the synthesis of nitrogen-doped porous carbon and its application in CO_2 adsorption. *Energy Fuels* **2020**, *34*, 6069–6076. [CrossRef]

52. Botomé, M.L.; Poletto, P.; Junges, J.; Perondi, D.; Dettmer, A.; Godinho, M. Preparation and characterization of a metal-rich activated carbon from CCA-treated wood for CO_2 capture. *Chem. Eng. J.* **2017**, *321*, 614–621. [CrossRef]

53. Wang, Y.; Xiao, J.F.; Wang, H.Z.; Zhang, T.C.; Yuan, S.J. N-doped porous carbon derived from solvent-free synthesis of cross-linked triazine polymers for simultaneously achieving CO_2 capture and supercapacitors. *Chem. Eur. J.* **2021**, *27*, 7908–7914. [CrossRef] [PubMed]

54. Fu, Z.Y.; Jia, J.Z.; Li, J.; Liu, C.K. Transforming waste expanded polystyrene foam into hyper-crosslinked polymers for carbon dioxide capture and separation. *Chem. Eng. J.* **2017**, *323*, 557–564. [CrossRef]

55. Shao, L.S.; Sang, Y.F.; Liu, N.; Liu, J.; Zhan, P.; Huang, J.H.; Chen, J.N. Selectable microporous carbons derived from poplar wood by three preparation routes for CO_2 capture. *ACS Omega* **2020**, *5*, 17450–17462. [CrossRef]

56. Wang, Z.; Goyal, N.; Liu, L.Y.; Tsang, D.C.W.; Shang, J.; Liu, W.J.; Li, G. N-doped porous carbon derived from polypyrrole for CO_2 capture from humid flue gases. *Chem. Eng. J.* **2020**, *396*, 125376. [CrossRef]

57. Krishna, R.; Baten, J.M.V. Elucidation of selectivity reversals for binary mixture adsorption in microporous adsorbents. *ACS Omega* **2020**, *5*, 9031–9040. [CrossRef]

Article

Influence of PPD and Mass Scaling Parameter on the Goodness of Fit of Dry Ice Compaction Curve Obtained in Numerical Simulations Utilizing Smoothed Particle Method (SPH) for Improving the Energy Efficiency of Dry Ice Compaction Process

Jan Górecki [1,*], Maciej Berdychowski [1], Elżbieta Gawrońska [2] and Krzysztof Wałęsa [1]

[1] Faculty of Mechanical Engineering, Institute of Machine Design, Poznan University of Technology, 60-965 Poznań, Poland

[2] Faculty of Mechanical Engineering and Computer Science, Czestochowa University of Technology, Dabrowskiego 69, 42-201 Czestochowa, Poland

* Correspondence: jan.gorecki@put.poznan.pl

Abstract: The urgent need to reduce industrial electricity consumption due to diminishing fossil fuels and environmental concerns drives the pursuit of energy-efficient production processes. This study addresses this challenge by investigating the Smoothed Particle Method (SPH) for simulating dry ice compaction, an intricate process poorly addressed by conventional methods. The Finite Element Method (FEM) and SPH have been dealt with by researchers, yet a gap persists regarding SPH mesh parameters' influence on the empirical curve fit. This research systematically explores Particle Packing Density (PPD) and Mass Scaling (MS) effects on the agreement between simulation and experimental outputs. The Sum of Squared Errors (SSE) method was used for this assessment. By comparing the obtained FEM and SPH results under diverse PPD and MS settings, this study sheds light on the SPH method's potential in optimizing the dry ice compaction process's efficiency. The SSE based analyses showed that the goodness of fit did not vary considerably for PDD values of 4 and up. In the case of MS, a better fit was obtained for its lower values. In turn, for the ultimate compression force F_C, an empirical curve fit was obtained for PDD values of 4 and up. That said, the value of MS had no significant bearing on the ultimate compression force F_C. The insights gleaned from this research can largely improve the existing sustainability practices and process design in various energy-conscious industries.

Keywords: Smoothed Particle Method; SPH; FEM; mass scaling; dry ice; compaction

Citation: Górecki, J.; Berdychowski, M.; Gawrońska, E.; Wałęsa, K. Influence of PPD and Mass Scaling Parameter on the Goodness of Fit of Dry Ice Compaction Curve Obtained in Numerical Simulations Utilizing Smoothed Particle Method (SPH) for Improving the Energy Efficiency of Dry Ice Compaction Process. *Energies* **2023**, *16*, 7194. https://doi.org/10.3390/en16207194

Academic Editors: Nikolaos Koukouzas, Manoj Khandelwal, Michael Zhengmeng Hou, Liehui Zhang, Cheng Cao, Yulong Zhao and Hejuan Liu

Received: 21 August 2023
Revised: 26 September 2023
Accepted: 20 October 2023
Published: 22 October 2023

1. Introduction

Reductions in electricity consumption in production are becoming an issue of growing importance and also an engineering challenge throughout the present-day world [1]. The underlying causes include limited availability of fossil fuels, which are the primary source of electrical energy [2]. In addition, lowered consumption of fossil fuels very often reduces greenhouse gas emissions with a desired effect on the product carbon footprint [3,4].

This explains the growing interest of researchers in studies to improve the energy efficiency of the existing production processes, as confirmed by our literature review [5,6].

Processes that involve mechanical processing of materials have been successfully studied experimentally [7] with numerical simulations as a complement [8,9]. The Discrete Element Method [10] may be considered an option of choice for studying loose materials processing, for example, the mixing materials during a transport process [11,12] or compacting [13]. However, when dealing with densification (compaction) and extrusion processes, densities tended to exceed 0.8 relative density; Harthong et al., 2009, pointed to the Finite Element Method (FEM) method as the preferred choice, as it gives better fit to the experimental data (FEM) [14].

Both methods are used for numerical model discretization [15]. Jagota et al., 2013, showed that the solid is made up of finite elements connected at nodes in the FEM method [16]. Thus, the elements create a consistent mesh representing the shape of the solid under analysis. The loading of a non-rigid element changes the distances between the nodes, resulting in deformations of this element. The element type and size are taken depending on the simulation type and the discretized model. However, in compression simulations, in which the elements are not allowed to penetrate each other, it is possible to exceed the null hypothesis. This is hard to achieve in the case of processes characterised by high deformations [17]. As a result, there are a number of adaptive finite element method (AFEM) and freemesh solutions reported in the literature, which allow relative displacement of the elements making up the finite element model [18].

The application of the DEM method for the simulation of compaction processes was described by Bahrami et al. in 2022 [19]. The discretization process involves particle size description and reactions between the particles, including friction. However, compared to FEM analysis, a different contact model is used in this case, allowing free relative movement of the model elements [20]. While widely used, this method offers worse goodness of fit values than the FEM method, which, in the case of density, fall in the range of 0.8 to 1 relative density.

The above-described methods are an option of choice in the simulation of densification and extrusion processes using single-cavity dies. This is not the case for multiple-cavity dies, where the final density of the material is clearly above 0.8 relative density, and the material is split in the process into the respective extrusion cavities. The results of these simulations would not allow obtaining the desired goodness of fit to the experimental data. Libersky et al., 1993, proposed an extended application of Smoothed Particle Hydrodynamics (SPH) to processes involving high solid body deformations [21].

Our earlier FEM studies included determinations of the Drucker–Prager/Cap (DPC), Mohr–Columb, and modified Cam–Clay model parameters [22,23]. The simulation results were compared with the experimental data, showing the highest accuracy of representation given by the DPC method. The above-mentioned models may be used in simulation programs, such as Abaqus, developed by Dassults Systemes in both FEM- and SPH-based analyses.

The authors used, to a good effect, the SPH method in a simulation study on dry ice extrusion through single-cavity dies [24]. However, as the literature review revealed [25], the reported findings must be supplemented with the information on the effect of SPH mesh division parameters on the goodness of fit to the empirical curve.

The numerical study results presented in this article fill the gap in knowledge, as identified by the literature review, on the influence of SPH discretization parameters on the goodness of fit to the experimental data presented in the previously published articles [22]. The following part of this article compares the FEM and SPH results obtained with different number of particles to be generated per isoparametric direction (PPD) and Mass Scaling (MS) values.

Goodness of fit was assessed using an SSE value, a parameter described in detail in the literature [22].

2. Materials and Methods

2.1. Materials

2.1.1. Powder

Powdery dry ice was obtained through the process of expanding liquid carbon dioxide at a specific temperature of 255 K, obtained by keeping it in a tightly sealed container under 20 bar pressure. For liquid to solid transition, the liquid carbon dioxide was rapidly expanded to atmospheric pressure in a process leading to its crystallization. In the powdered form, dry ice had a density of 0.55 g/cm^3, as reported in [22].

In the context of empirical testing, the loose dry ice material was placed and secured in a DRICE 30 container manufactured by MELFROM of Monasterolo di Savigliano, Italy, as shown in Figure 1. The purpose was to reduce the material sublimation rate. In addition, the container was used for cooling the test stand components.

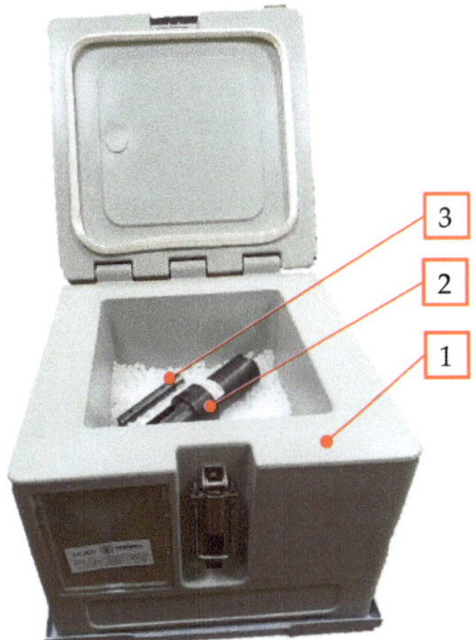

Figure 1. Insulated container used to store dry ice powder and condition the experimental set-up components: 1—insulated container; 2—dry ice; and 3—experimental set-up [26].

The cooling aspect of the experiment was particularly critical, as it entailed reducing the temperature of the test stand components to a level that closely approximated the temperature of the tested material. The aim was to minimize the influence of any external factors or variations that could introduce bias or errors into the experiment.

Summing up, the loose dry ice sample was meticulously prepared by subjecting liquid carbon dioxide to controlled temperature and pressure conditions, thereby inducing crystallization. The sample obtained in this way was then stored in the DRICE 30 container to reduce sublimation effects and simultaneously facilitate temperature control during the experiment. This created a controlled and consistent testing environment, enhancing the accuracy and reliability of the empirical observations and results.

2.1.2. Compaction

In the experimental part of this research, cylindrical pellets were compressed by an extrusion ram. The authors employed a custom-designed test apparatus, tailored for compressing dry ice specimens. An MTS Insight 50 kN universal testing machine manufactured by MTS Systems Corporation based in Eden Prairie, MN, USA, was used. For a detailed description of the experimental setup, the reader is referred to the previously published article [23].

The primary focus of these experiments was to measure the force F_Z applied to the 20 mm diameter ram. The measurements were executed with high precision, using a 50 kN MTS Insight strain gauge sensor of 0.01 accuracy class.

About 30 g portion of dry ice powder was subjected to compression in order to achieve a predefined specimen strain value. As a result, specimens with density values that approached the limit of approximately 1650 kg/m^3 were obtained.

2.2. Numerical Model

All the numerical calculations of this study were performed using the software Abaqus 2020 (Dassault Systèmes, Vélizy-Villacoublay, France). The geometry of the same numerical model that was described in [22,23] was used to simulate compaction. It consisted of a 30 mm diameter stamp and a corresponding sleeve. This allowed for a comparison with the previously obtained values in [23]. The numerical model shown in Figure 2 is made up of 4 elements: the material being compressed (Figure 2, Label 3), the forming sleeve (Figure 2, Label 1), sleeve bottom closing disc (Figure 2, Label 4), and the upper disc acting as a compacting stamp (Figure 2, Label 2). The material being compacted was the only element modelled as a deformable body. It was illustrated as a cylinder with diameter $D_C = 30$ mm and height $h_C = 39.95$ mm. The other modelled elements, which act as a sleeve (Figure 2, Label 1), in which the compaction process takes place, were modelled as discrete rigid part surface object, which is assumed to be rigid and is used in contact analyses to model bodies that cannot deform. The stamp (Figure 2, Label 1) was modelled in the same way as the cylinder (Figure 2, Label 2). In the model, it is represented by a flat disc. The same geometry was also used to represent the sleeve base (Figure 2, Label 4), the place where compaction takes place.

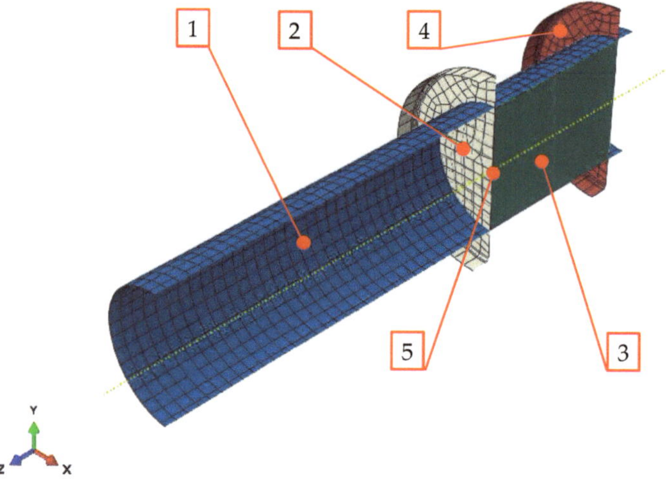

Figure 2. Numerical model: 1—sleeve; 2—stamp; 3—compacted dry ice; 4—bottom closing disc; and 5—compaction force measuring point [22].

The model was then discretized with a finite element grid of 0.0028 m approximate global size. C3D8R Hex-type elements were used. In the next step, for all the elements making up the compressed materials, the convert to particles option was selected in the finite element properties (Figure 3). This conversion took place in the first step of the calculation.

In order to represent the actual lab compaction process, the compacting stamp was allowed to move linearly at a speed of 5 mm/s along the z-axis, in line with the sleeve/cylinder axis. The simulation time was 3 s, corresponding to 15 mm stamp travel distance. All the remaining components, i.e., the cylinder and the lower disc representing the sleeve end, had all degrees of freedom fixed. A measuring point was set on the stamp for logging and reading of the calculated reaction force values during the compaction process and to measure the displacement. It was positioned in the axis of symmetry on the inner surface of the stamp. Its readout values were directly compared with the obtained experimental data.

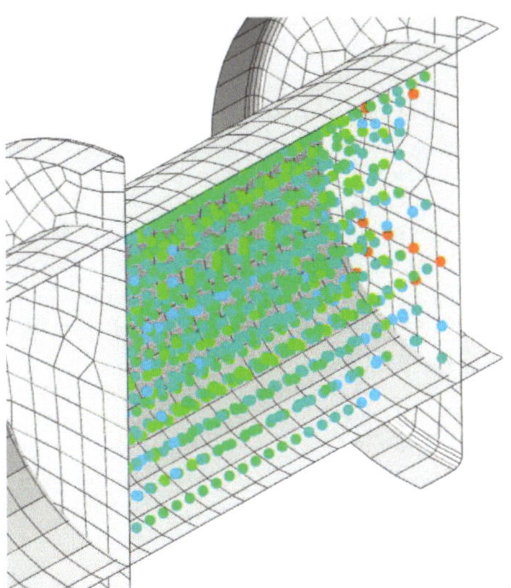

Figure 3. Numerical model after conversion into particles after the first step of calculation (particles visualization).

General Contact was the contact type used in the model. The value of the friction coefficient $\mu = 0.1$ was set in the contact properties. In the next step, the compacted material was assigned the parameters determined from the experimental studies presented by Berdychowski et al. in 2022 [22]. As mentioned, the described results indicated that as the degree of compaction of the test material increases, the properties of the material change accordingly. In order to obtain the best fit between the simulation and experimental data, Abaqus Subroutines VUSDFLD was applied to map these changes as the criterion for determining the change in material properties, in which PEEQ values were defined (Table 1). This means that the material property input values that Abaqus will take for a given calculation step will depend on the PEEQ (equivalent plastic strain) values determined in the previous calculation step.

Table 1. DPC model parameters.

Material Cohesion (MPa)	Angle of Friction (deg)	Cap Eccentricity (Pa)	Init Yld Surf Pos (-)	Yield Stress (MPa)	Vol Plas Strain (-)	Young's Modulus (MPa)	Poisson's Ratio (-)
1.07	25.92	0.68	0.02	1.24	0	136.94	0.023
1.75	21.46	0.76	0.02	1.57	0.048	194.18	0.059
2.22	20.51	0.78	0.02	1.92	0.095	251.42	0.102
2.53	20.59	0.79	0.02	3.20	0.139	308.66	0.150
2.73	21.05	0.80	0.02	4.50	0.182	365.9	0.200
2.85	21.59	0.81	0.02	6.54	0.223	423.14	0.249
2.93	22.08	0.83	0.02	7.92	0.262	480.38	0.295
3.00	22.43	0.84	0.02	10.25	0.300	537.62	0.335
3.07	22.61	0.87	0.02	13.1	0.336	594.86	0.370
3.14	22.64	0.90	0.02	16.57	0.371	652.1	0.399
3.23	22.60	0.93	0.02	21.51	0.405	709.34	0.422
3.32	22.63	0.98	0.02	27.22	0.438	766.58	0.441
3.40	22.88	1.03	0.02	32.06	0.470	823.82	0.456

The SPH (Smoothed Particle Hydrodynamics) method was used for the calculations. It is a numerical method, a representative of the meshfree family of methods. For these methods, nodes and elements are not defined as they are in a finite element analysis. To represent a given body, this method needs just a set of points. In the SPH method, these nodes are commonly referred to as particles or pseudo-particles. In principle, however, the method is not based on discrete particles that collide during compression or exhibit cohesion-like behaviour in tension, as the term "particles" might suggest. Rather, it is more a discretization method of continuum partial differential equations. In this respect, the SPH method resembles the FEM method quite closely [27].

For moderate deformations, the SPH method is generally less accurate than Lagrangian Finite Element analyses. The same holds true for large deformations analyses. Here, coupled Euler–Lagrange analyses yield slightly better results, but this is at a cost of significantly greater computing power, as compared to the SPH methods. Hence, they are a good alternative when computing large deformations [27].

In the SPH method, the object at the initial moment is divided into subareas that are subsequently replaced by material particles. These particles represent the respective physical parameters, such as the position vector r_i or the mass m_i. In this method, the particles interact, yet they are not discrete points and feature some degree of diffusion. This diffusion is modelled by a certain function (smoothing kernel) applied to the particles:

$$W\left(\left|r_i - r_j\right|, h\right) \tag{1}$$

where $\left|r_i - r_j\right|$ is the relative distance between the particles, and h is the smoothing parameter, defining the range of interaction between the particles.

In the SPH method, the changes in the physical values describing the particle state result from aggregate effects of the particles that surround it. For example, for a set of N j-particles that surround an i-particle, we can define the density ρ_i at the point r_i as follows [28]:

$$\rho_i = \sum_{j=1}^{N} m_j W\left(\left|r_i - r_j\right|, h\right) \tag{2}$$

and speed distribution may be represented by the following equation:

$$\frac{dv_i}{dt} = -\sum_{j=1}^{N} m_j \left(\frac{p_i}{\rho_i^2} + \frac{p_j}{\rho_j^2}\right) \nabla W\left(\left|r_i - r_j\right|, h\right) \tag{3}$$

In our model, we decided to use the SPH method to test its suitability for dry ice compaction process analysis. For this purpose, the material being compacted was defined in particles. It was assumed that conversion of finite elements to particles would take place at the beginning of the simulation (Threshold = 0). In addition, it was also possible to define the number of particles to be generated. This is performed with the PPD parameter, which can take integer values of 1–7. Thus, for a cubic finite element with PPD = 1, one (1) particle per element will be generated, with PPD = 2, there will be eight (8) particles, and, with PPD = 3, there will be twenty seven particles (27), and so on. Hence, the PPD value gives the number of generated particles when raised to the power of 3, as it is represented in Figure 4. In this study, calculations were carried out for all the seven PPD value cases. Thus, it was possible to relate the initial particle distance $\left|r_i - r_j\right|$ to PDD using the following equation:

$$\left|r_i - r_j\right| = f(PDD) \tag{4}$$

Considering the non-linear effect of the change in $f(PDD)$, it was decided to carry out numerical tests to determine and compare the effect of PDD on the accuracy of FEM representation in order to determine the effective value of this parameter.

In addition, the mass scaling (MS) method with values of 0.0001 and 0.00001 was used to reduce the required computation time. This followed from the results analysis, which indicated the need to reduce the mass scaling (MS) value to avoid computational (numerical) errors.

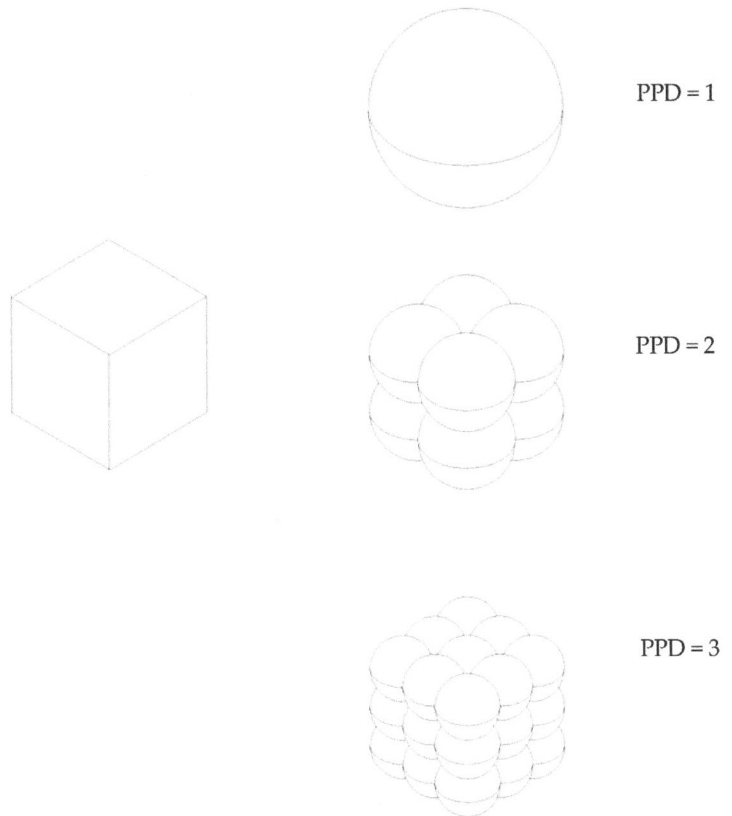

Figure 4. Graphical interpretation of PPD values.

3. Results and Discussion

This section presents the results of the experimental and simulation studies carried out to verify the following hypotheses:

1. The greater the PPD value, the better the fit obtained between the simulation and experimental curves.
2. An increase in PPD reduces the difference between the calculated and experimental ultimate force values.
3. The greater the MS value, the better the fit obtained between the simulation and experimental curves.
4. An increase in MS reduces the difference between the calculated and experimental ultimate force values.

The results of FEM, SPH with variable PPD simulations, and experimental data are presented as an F_C vs. displacement s curve in Figure 5. The experimental and FEM results are also shown in Figure 6. Note, however, that the MS value has changed for the SPH model from 1×10^{-4} in Figure 5 to 1×10^{-5} in Figure 6.

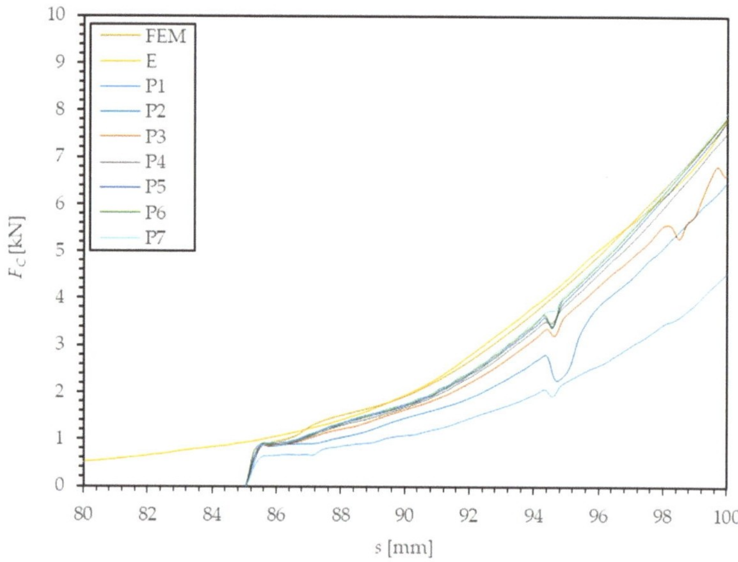

Figure 5. Compression force F_C vs. stamp displacement. E—empirical test results; FEM—results obtained using FEM discretisation; Pn—SPH simulation data with different PPD values; and MS = 1×10^{-4}.

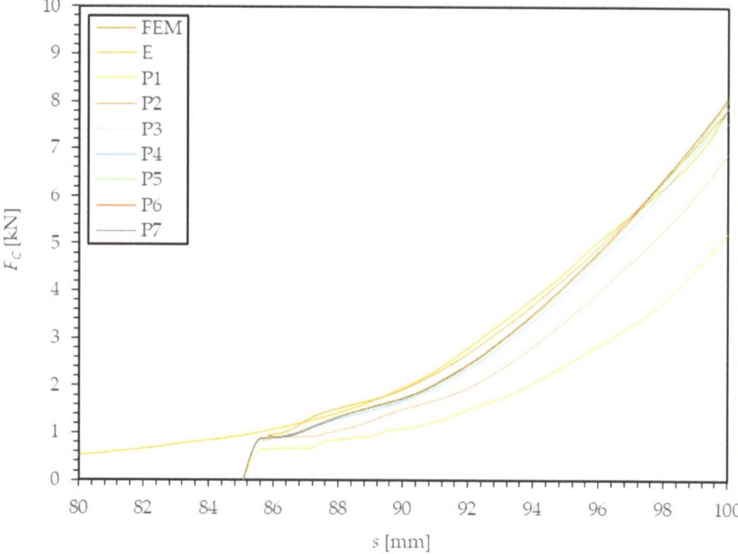

Figure 6. Compression force F_C vs. stamp displacement. E—empirical test results; FEM—results obtained using FEM discretisation; Pn—SPH simulation data with different PPD values; and MS = 1×10^{-5}.

The goodness of fit between these curves and the experimental curve was assessed using the SSE (sum of squared errors) value. SSE was calculated with the following equation, described in previous studies [22]:

$$\text{SSE} = \sum_{s=86}^{100} \left(F_s^S - F_s^E \right)^2. \tag{5}$$

where

F^S—value of F_C obtained in numerical simulation;

F^E—experimental F_C value used as reference.

The calculated SSE values for the curves in Figure 5 are given in Table 2 and shown in Figure 7. SSE values in 1 mm intervals and SSE totals can be compared. The simulation results for MS = 1×10^{-5} are given in Table 3 and shown in Figure 8.

Table 2. SSE values obtained in SPH simulation with a variable value of P and MS = 1×10^{-4}.

	FEM	P1	P2	P3	P4	P5	P6	P7
<86; 87)	0.18	2.30	0.92	0.45	0.37	0.33	0.31	0.25
<87; 88)	0.05	3.82	1.60	0.43	0.23	0.17	0.14	0.11
<88; 89)	0.02	4.43	2.39	0.57	0.21	0.11	0.09	0.06
<89; 90)	0.02	7.02	3.36	0.84	0.53	0.35	0.29	0.21
<90; 91)	0.11	10.39	4.54	1.54	1.03	0.82	0.69	0.61
<91; 92)	0.22	16.47	7.62	2.68	1.61	1.27	1.04	0.91
<92; 93)	0.45	25.04	13.35	4.34	2.67	2.13	1.80	1.47
<93; 94)	0.49	34.09	17.54	5.00	2.73	2.08	1.70	1.33
<94; 95)	0.43	45.03	21.23	6.16	3.39	3.25	3.02	1.36
<95; 96)	0.52	62.68	45.20	6.56	3.19	2.23	1.58	1.11
<96; 97)	0.40	74.89	25.44	6.50	2.76	1.64	1.07	0.62
<97; 98)	0.12	83.02	21.56	5.37	1.46	0.44	0.19	0.07
<98; 99)	0.16	92.89	20.18	9.42	0.48	0.06	0.11	0.31
<99; 100)	0.15	113.79	22.58	7.96	0.41	0.06	0.16	0.41
Total	3.34	575.86	207.52	57.81	21.07	14.95	12.20	8.81

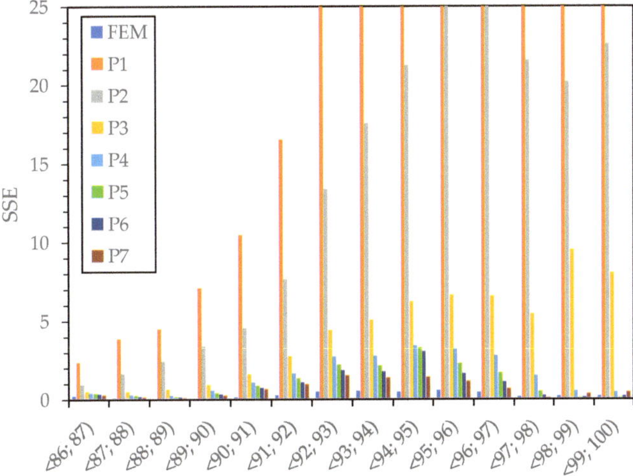

Figure 7. SSE values depending on the set P value of SPH mesh for MS 1×10^{-4} in the respective intervals of the compaction process.

Table 3. SSE values obtained in SPH simulation with a variable value of P and MS = 1×10^{-5}.

	FEM	P1	P2	P3	P4	P5	P6	P7
<86; 87)	0.18	2.12	0.58	0.41	0.36	0.32	0.29	0.27
<87; 88)	0.05	2.99	1.29	0.40	0.23	0.17	0.14	0.13
<88; 89)	0.02	3.83	1.66	0.52	0.20	0.09	0.06	0.05
<89; 90)	0.02	5.79	2.00	0.72	0.50	0.31	0.23	0.19
<90; 91)	0.11	9.11	2.88	1.26	0.95	0.73	0.65	0.57
<91; 92)	0.22	13.67	5.22	2.13	1.44	1.08	0.95	0.83
<92; 93)	0.45	20.22	8.56	3.38	2.33	1.76	1.50	1.31
<93; 94)	0.49	27.53	9.99	3.42	2.22	1.57	1.33	1.12
<94; 95)	0.43	34.25	10.82	3.38	2.11	1.39	1.06	0.84
<95; 96)	0.52	42.96	11.92	3.70	2.03	1.20	0.80	0.60
<96; 97)	0.40	52.21	12.01	3.46	1.55	0.77	0.48	0.31
<97; 98)	0.12	56.07	9.80	2.14	0.42	0.09	0.07	0.12
<98; 99)	0.16	57.63	8.05	0.64	0.07	0.33	0.58	0.81
<99; 100)	0.15	61.08	7.43	0.42	0.12	0.52	0.90	1.20
Total	3.34	389.46	92.21	25.98	14.53	10.33	9.04	8.35

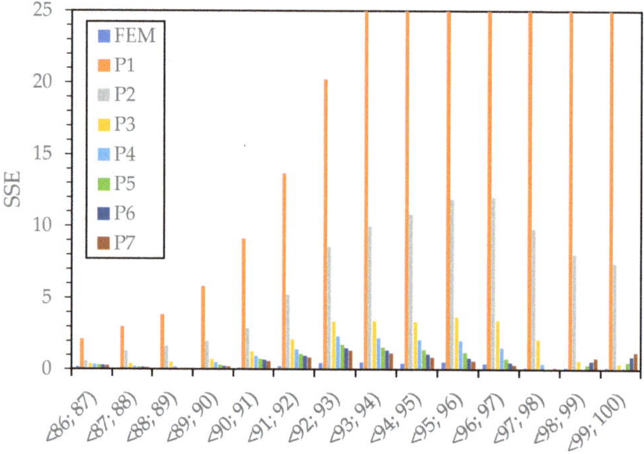

Figure 8. SSE values depending on the set PPD value of the SPH mesh for MS 1×10^{-5} in different compaction spans.

The above results support the first study hypothesis, i.e., the increasing the PPD value improves the goodness of fit between the simulation and the experimental reference curves. It is worthwhile to note the over 2.5 times higher value of SSE for PPD = 7 compared to the FEM simulation value.

The second study hypothesis was that increasing the set PPD value decreased the differences between the simulated SPH and the reference ultimate forces. The results given in Table 3 support this hypothesis. It was observed that there were no significant differences between the simulated ultimate values for PPD set at 5, 6 and 7, respectively, with MS = 1×10^{-4}. This lack of difference was observed also for simulations with MS = 1×10^{-5}, in this case, for PPD values of 4 and up.

To verify the third hypothesis, we should compare the charts in Figures 7 and 8 and the data given in Tables 2 and 3. The effect of the studied parameter on the SSE value for PPD of 1 to 6 became apparent at lower MS values. With PPD = 7, the differences in SEE values were not significant.

The fourth hypothesis was that a change in MS would reduce the difference between the simulated and experimental ultimate force values. Comparing the simulation results given in Table 4 for different PPD and MS values, we see that the force values were, right from the beginning, closer to the reference than in the simulation, with MS = 1×10^{-4}, for

MS = 1×10^{-5}. Simulations were performed also for MS = 1×10^{-3}, yet the ultimate forces were so different from the reference values that the results were left out of consideration. In addition, as mentioned in the previous paragraph, no significant effect on the ultimate force representation was observed for PPD values of 4 and up.

Table 4. Ultimate forces in kN taken from the respective simulation curves and from the experimental reference curve.

MS	E	FEM	P1	P2	P3	P4	P5	P6	P7
1×10^{-4}			4.3	6.2	6.9	7.7	7.9	8.0	8.0
1×10^{-5}	7.9	8.1	5.4	7.0	7.7	8.0	8.2	8.1	8.1

4. Conclusions

The study results demonstrated the applicability of the SPH method for the numerical simulation of loose dry ice compaction process. It is worth noting, however, that the PPD value had a significant effect on the degree of fit, with better fit to the reference curve obtained even with the highest available PPD value of 7.

Based on the literature review, the SPH method should be considered suitable to simulate the extrusion process using single-cavity dies. Hence, the information presented in this article may be useful in the selection of PPD and MS values in future simulations. This will be particularly relevant to research efforts to optimise the shapes of tools such as multiple-cavity dies.

The results presented in this article showed that PDD values should not be lower than 4 in future investigations. When a high quality of change in the compression force during the process is required, it is suggested to set the MS value at 1×10^{-5}. However, in simulations limited to the ultimate force analysis, the authors suggest to set the MS value at 1×10^{-4} in order to reduce the computation time. The time of calculation should be taken into account for both parameters, as it may be strongly influenced by their values in studies using evolutionary algorithms.

The authors believe that this study of the effect of simulation parameters on the fit between the calculated and experimental reference may be used in future efforts to reduce the ultimate force values in extrusion processes of dry ice and similar loose materials. As the final outcome, these studies will allow reductions in the energy consumption of these production processes.

Author Contributions: Conceptualization, J.G.; methodology, M.B. and J.G.; validation, M.B.; formal analysis, J.G., M.B., and E.G.; investigation, J.G. and M.B.; data curation, M.B. and J.G.; writing—original draft preparation, J.G. and M.B.; writing—review and editing, J.G.; visualization, J.G., M.B., and K.W.; supervision, J.G.; project administration, J.G.; funding acquisition, J.G. All authors have read and agreed to the published version of the manuscript.

Funding: This research is a part of the project: "Developing an innovative method using the evolutionary technique to design a shaping dies used in the extrusion process of crystallized CO_2 to reduced consumption of electricity and raw material", number: "LIDER/3/0006/L-11/19/NCBR/2020" financed by National Centre for Research and Development in Poland, https://www.gov.pl/web/ncbr (accessed on 15 December 2021).

Data Availability Statement: Not applicable.

Conflicts of Interest: The authors declare no conflict of interest.

References

1. Gierz, Ł.; Warguła, Ł.; Kukla, M.; Koszela, K.; Zawiachel, T. Computer Aided Modeling of Wood Chips Transport by Means of a Belt Conveyor with Use of Discrete Element Method. *Appl. Sci.* **2020**, *10*, 9091. [CrossRef]
2. Omer, O.; Alireza, K. *Alternative Energy in Power Electronics: Chapter 2 Energy in Power Electronics*; Butterworth-Heinemann: Oxford, UK, 2015; pp. 81–154. [CrossRef]

3. Tahmasebi, M.M.; Banihashemi, S.; Hassanabadi, M.S. Assesment of the Variation Impacts og Windwos on Energy Consumption and Carbon Footprint. *Procedua Eng.* **2011**, *21*, 820–828. [CrossRef]

4. Lohri, C.; Rajabu, H.; Sweeney, D.; Zurbrügg, C. Char fuel production in developing countries—A review of urban biowaste carbonization. *Renew. Sustain. Energy Rev.* **2016**, *59*, 1514–1530. [CrossRef]

5. Gawrońska, E.; Dyja, R. A Numerical Study of Geometry's Impact on the Thermal and Mechanical Properties of Periodic Surface Structures. *Appl. Sci.* **2021**, *14*, 427. [CrossRef] [PubMed]

6. Wilczyński, D.; Talaśka, K.; Wojtokwiak, D.; Wałęsa, K.; Wojciechowski, S. Selection of the Electric Drive for the Wood Waste Compacting Unit. *Energies* **2022**, *15*, 7488. [CrossRef]

7. Ishiguro, M.; Dan, K.; Nakasora, K.; Morino, Y.; Yoshii, Y.; Chaki, T. Increase of Snow Compaction Density by Repeated Articial Snow Consolidation Formation. *J. Inst. Ind. Appl. Eng.* **2020**, *8*, 104. [CrossRef]

8. Baiul, K.; Vashchenko, S.; Khudyakov, O.; Bembenek, M.; Zinchenko, A.; Krot, P.V.; Solodka, N. A Novel Approach to the Screw Feeder Design to Improve the Reliability of Briquetting Process in the Roller Press. *Eksploat. Niezawodn.—Maint. Reliab.* **2023**, *25*, 167967. [CrossRef]

9. Warguła, Ł.; Wojtkowiak, D.; Kukla, M.; Talaśka, K. Modelling the process of splitting wood and chipless cutting Pinus sylvestris L. wood in terms of designing the geometry of the tools and the driving force of the machine. *Eur. J. Wood Wood Prod.* **2022**, *81*, 223–237. [CrossRef]

10. Gierz, Ł.; Kolankowska, E.; Markowski, P.; Koszela, K. Measurements and Analysis of the Physical Properties of Cereal Seeds Depending on Their Moisture Content to Improve the Accuracy of DEM Simulation. *Appl. Sci.* **2022**, *12*, 549. [CrossRef]

11. Lin, L.; Zeheng, G.; Weixin, X.; Yunfeng, T.; Xinghua, F.; Dapeng, T. Mixing mass transfer mechanism and dynamic control of gas-liquid-solid multiphase flow based on VOF-DEM coupling. *Energy* **2023**, *272*, 127015. [CrossRef]

12. Ge, M.; Chen, J.; Zhao, L.; Zheng, G. Mixing Transport Mechanism of Three-Phase Particle Flow Based on CFD-DEM Coupling. *Processes* **2023**, *11*, 1619. [CrossRef]

13. Giannis, K.; Schilde, C.; Finke, J.H.; Kwade, A. Modeling of High-Density Compaction of Pharmaceutical Tablets Using Multi-Contact Discrete Element Method. *Pharmaceutics* **2021**, *13*, 2194. [CrossRef]

14. Harthong, B.; Jérier, J.-F.; Dorémus, P.; Imbault, D.; Donzé, F.V. Modeling of high-density compaction of granular materials by the Discrete Element Method. *Int. J. Solids Struct.* **2009**, *46*, 3357–3364. [CrossRef]

15. Wu, Z.; Yu, F.; Zhang, P.; Liu, X. Micro-mechanism study on rock breaking behavior under water jet impact using coupled SPH-FEM/DEM method with Voronoi grains. *Eng. Anal. Bound. Elem.* **2019**, *108*, 472–483. [CrossRef]

16. Jagota, V.; Sethi, A.P.; Kumar, K. Finite Element Method: An Overview. *Walailak J. Sci. Technol.* **2013**, *10*, 1–8. Available online: https://wjst.wu.ac.th/index.php/wjst/article/view/499 (accessed on 6 July 2023).

17. Wang, W.; Wu, Y.; Wu, H.; Yang, C.; Feng, Q. Numerical analysis of dynamic compaction using FEM-SPH coupling method. *Soil Dyn. Earthq. Eng.* **2021**, *140*, 106420. [CrossRef]

18. Yagawa, G.; Furukawa, T. Recent development of free mesh method. *Int. J. Numer. Meth. Eng.* **2000**, *47*, 1419–1443.

19. Bahrami, M.; Naderi-Boldaji, M.; Ghanbarian, D.; Keller, T. Simulation of soil stress under plate sinkage loading: A comparison of finite element and discrete element methods. *Soil Tillage Res.* **2022**, *223*, 105463. [CrossRef]

20. Gierz, Ł.; Kruszelnicka, W.; Robakowska, M.; Przybył, K.; Koszela, K.; Marciniak, A.; Zwiachel, T. Optimization of the Sowing Unit of a Piezoelectrical Sensor Chamber with the Use of Grain Motion Modeling by Means of the Discrete Element Method. Case Study: Rape Seed. *Appl. Sci.* **2022**, *12*, 1594. [CrossRef]

21. Libersky, L.; Perschek, A.; Carney, T.; Hipp, C.; Allahdadi, F. High Strain Lagrangian Hydrodynamics: A Three-Dimensional SPH Code for Dynamic Material Response. *J. Comput. Phys.* **1993**, *109*, 67–75. [CrossRef]

22. Berdychowski, M.; Górecki, J.; Biszczanik, A.; Wałęsa, K. Numerical Simulation of Dry Ice Compaction Process: Comparison of Drucker-Prager/Cap and Cam Clay Models with Experimental Results. *Materials* **2022**, *15*, 5771. [CrossRef] [PubMed]

23. Berdychowski, M.; Górecki, J.; Wałęsa, K. Numerical Simulation of Dry Ice Compaction Process: Comparison of the Mohr-Coulomb Model with the Experimental Results. *Materials* **2022**, *15*, 7932. [CrossRef] [PubMed]

24. Wałęsa, K.; Górecki, J.; Berdychowski, M.; Biszczanik, A.; Wojtkowiak, D. Modelling of the Process of Extrusion of Dry Ice through a Single-Hole Die Using the Smoothed Particle Hydrodynamics (SPH) Method. *Materials* **2022**, *15*, 8242. [CrossRef] [PubMed]

25. Jankowiak, T.; Łodygowski, T. Using of Smoothed Particle Hydrodynamics (SPH) method for concrete application. *Bull. Pol. Acad. Sci. Tech. Sci.* **2013**, *61*, 111–121. [CrossRef]

26. Górecki, J.; Wiktor, Ł. Influence of Die Land Length on the Maximum Extrusion Force and Dry Ice Pellets Density in Ram Extrusion Process. *Materials* **2023**, *16*, 4281. [CrossRef]

27. *Abaqus Documentation*; Dassault Systemes: Paris, France, 2017.

28. Jach, K. *Computer Modeling of Dynamic Interactions of Bodies Using the Free Point Method*; Wydawnictwo Naukowe PWN: Warsaw, Poland, 2001. (In Polish)

Article

Effect of Geochemical Reactivity on ScCO$_2$–Brine–Rock Capillary Displacement: Implications for Carbon Geostorage

Felipe Cruz *, Son Dang, Mark Curtis and Chandra Rai

Mewbourne School of Petroleum and Geological Engineering, University of Oklahoma, Norman, OK 73019-1003, USA; dangthaison@ou.edu (S.D.); mcurtis@ou.edu (M.C.); crai@ou.edu (C.R.)
* Correspondence: felipecruz@ou.edu

Abstract: The displacement efficiency of supercritical CO$_2$ (scCO$_2$) injection in the storage zone and its primary trapping mechanism in the confining zone are strongly tied to the capillary phenomenon. Previous studies have indicated that the capillary phenomenon can be affected by geochemical reactivity induced by scCO$_2$ dissolution in formation brine. To quantify such changes, thin disk samples representing a sandstone storage reservoir, siltstone confining zone, and mudstone confining zone were treated under a scCO$_2$-enriched brine static condition for 21 days at 65 °C and 20.7 MPa. Geochemical alterations were assessed at the surface level using scanning electron microscopy coupled with energy-dispersive X-ray spectroscopy and X-ray fluorescence. Before and after treatment, the wettability of the scCO$_2$–brine–rock systems was determined using the captive-bubble method at fluid-equilibrated conditions. Pore size distributions of the bulk rocks were obtained with mercury injection capillary pressure, nuclear magnetic resonance, and isothermal nitrogen adsorption. The results indicate the dissolution of calcite at the surface, while other potentially reactive minerals (e.g., clays, feldspars, and dolomite) remain preserved. Despite alteration of the surface mineralogy, the measured contact angles in the scCO$_2$–brine–rocks systems do not change significantly. Contact angle values of 42 ± 2° for sandstone and 36 ± 2° for clay-rich siltstone/calcite-rich mudstone were determined before and after treatment. The rocks studied here maintained their water-wettability at elevated conditions and after geochemical reactivity. It is also observed that surface alteration by geochemical effects did not impact the pore size distributions or porosities of the thin disk samples after treatment. These results provide insights into understanding the impact of short-term geochemical reactions on the scCO$_2$–brine capillary displacement in the storage zone and the risks associated with scCO$_2$ breakthrough in confining zones.

Keywords: carbon geostorage; geochemical reactivity; supercritical CO$_2$; wettability; capillary pressure

Citation: Cruz, F.; Dang, S.; Curtis, M.; Rai, C. Effect of Geochemical Reactivity on ScCO$_2$–Brine–Rock Capillary Displacement: Implications for Carbon Geostorage. *Energies* **2023**, *16*, 7333. https://doi.org/10.3390/en16217333

Academic Editor: Fernando Rubiera González

Received: 29 September 2023
Revised: 18 October 2023
Accepted: 27 October 2023
Published: 29 October 2023

1. Introduction

Carbon geostorage (CGS) is considered a primary solution to limit the global average temperature according to Paris Agreement and achieve net-zero emissions [1]. In CGS, gaseous CO$_2$ is captured at large-scale point sources and injected in porous geological formations at depths beyond 800 m to ensure a supercritical state (scCO$_2$), hence maximizing storage resources [2]. Sandstones are frequently chosen as host rocks for CGS because of their high injectivity and storage capacity [3,4], whereas siltstones [5–8] and mudstones [9,10], characterized by low permeability, are well-suited as confining zones. In addition to the macroscopic fluid displacement within the host rock, which is a function of its heterogeneity, the efficiency of the storage process is intricately linked to microscopic displacement factors such as interfacial tension (IFT), wettability, and pore size distribution [11–13]. During and after brine displacement, scCO$_2$ is primarily trapped by capillary forces acting on the confining zone (structural trapping) and in the storage zone (residual trapping) [14]. Secondary immobilization of scCO$_2$ occurs by long-term geochemical interactions with formation fluids and rock, including dissolution in fluid (solubility trapping) and formation of stable mineral species (mineral trapping) [15]. Understanding the outcomes of these chemical reactions to the scCO$_2$ microscopic displacement and trapping mechanisms is a challenge, which requires exposure experiments to be conducted at different conditions representing time-spatial distribution within the storage and confining zones [16,17].

The influencing factors on the scCO$_2$ microscopic displacement and its primary trapping mechanisms can be simplified by the capillary phenomenon [17,18]. The net pressure required to displace a non-wetting phase (e.g., scCO$_2$) through a pore throat saturated by a wetting phase (e.g., brine) is given by the Laplace equation [19,20] (Equation (1)):

$$P_c = P_{scCO_2} - P_{H_2O} = \frac{2\gamma\cos\theta}{r},\tag{1}$$

where P_c is the capillary pressure, γ is the scCO$_2$–brine interfacial tension, θ is the contact angle, and r is the pore throat radius. As described in Equation (1), the capillary pressure is directly proportional to the product of the tension formed between the fluid phases and the cosine of the contact angle, and inversely proportional to the pore size. If the scCO$_2$ pressure exceeds the breakthrough pressure in the confining zone, upward fluid migration takes place, and the maximum scCO$_2$ height that can be stored in the caprock (h) is given by Equation (2):

$$h = \frac{2\gamma\cos\theta}{\Delta\rho g r},\tag{2}$$

where $\Delta\rho$ is the difference in fluid densities between scCO$_2$ and brine, g is the gravity constant, and r is the connected pore throat radius during breakthrough [21]. Equation (2) shows that in addition to the capillary properties, knowledge of fluid densities in scCO$_2$–brine systems is essential to determine the capillary seal capability [22].

Regarding the interfacial tension, several authors have shown that the IFT of CO$_2$–H$_2$O systems is primarily controlled by the density difference of the two fluid phases [23–26]. It is generally agreed that the IFT follows a bilinear trend with $\Delta\rho$. For gaseous CO$_2$, the IFT decreases steeply and linearly with decreasing $\Delta\rho$ (alternatively, increasing pressure) due to an increase in gas solubility [21]. Above the scCO$_2$ condition ($\Delta\rho\sim$600 kg/m^3, ~7.3 MPa, ~31 °C), the IFT decreases mildly, reaching a pseudo plateau at ~26 $\pm$ 2 mN/m [23,27]. The pseudo-plateau IFT is independent of temperature [23], increases linearly with salt molality [23,28,29], and is more pronounced in the presence of bivalent cations [23,28,29].

Numerous authors have described the wettability of scCO$_2$–brine–rock systems [30–74]. During CGS, water-wettability over scCO$_2$ is often preferred because it limits upward vertical migration and increases residual trapping in the reservoir [75–77], as well as increasing structural trapping, thus reducing the risk of scCO$_2$ breakthrough in the confining zone [78,79]. Figure 1 shows a literature compilation of scCO$_2$–brine–rock contact angles as a function of pressure for sandstone [66], limestone [66], dolomite [55], fine-grained caprocks [49], and source rock [58], measured at similar temperatures (~50 °C). A water-wet behavior with respect to scCO$_2$ is observed for most rocks. An exception is dolomite, which has lower hydrophilicity and shows strong negative pressure dependence with cos θ [55]. Aside from pressure, other parameters such as temperature and salinity exhibit less clear influence on θ. Studies conducted on quartz show increasing θ with increasing temperatures [40,44,51,66]; sandstones exhibit no clear trend [62,66]; whereas calcite [56] and carbonate [32] show decreasing θ trends with increasing temperature. Regarding salinity, some papers report quartz and calcite surfaces becoming slightly less water-wet in high salinity [38,56,66]. However, studies conducted on real rocks show either no trend in θ with salinity for a silica/dolomite-rich shale caprock [54], or a slight decrease with salinity for sandstone [62]. Inconsistencies in θ values are mostly attributed to differences in surface cleaning methods [44,71], rock heterogeneity [31], and surface roughness [57,62].

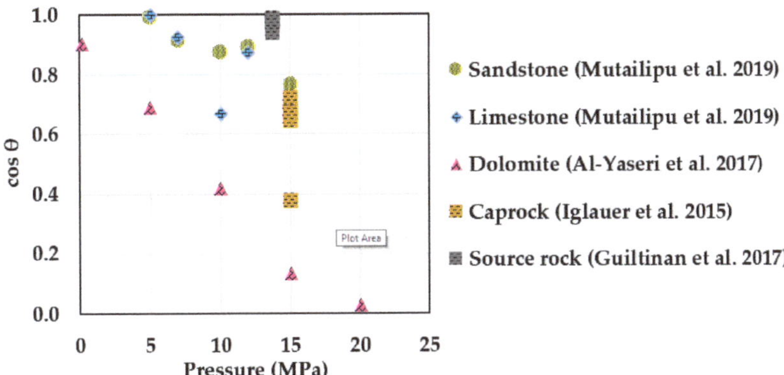

Figure 1. ScCO$_2$–brine–rock wettability as function of pressure at similar temperature (~50 °C) and salinity conditions. Most rocks exhibit water-wettability with respect to scCO$_2$ even at elevated pressures. An exception is dolomite, showing strong negative pressure dependence on θ (lower hydrophilicity). Note: sandstone [66], limestone [66], and source rock [58] contact angle values were conducted in static measurements. Caprock [49] represents receding contact angles, while dolomite [55] is an average of retreating and advancing values.

The interaction between scCO$_2$ and brine in the subsurface can trigger geochemical reactions with the rock and potentially impact the microscopic displacement and trapping mechanisms during CGS. These geochemical reactions are generally classified as short- and long-term reactions [80,81]. At the early stage, the dissolution of CO$_2$ in brine occurs, generating carbonic acid, which further ionizes into carbonate and bicarbonate ions, ultimately reducing the pH down to ~3 [80,82,83]:

$$CO_2 \text{ (gas)} + H_2O \rightleftharpoons H_2CO_3 \tag{3}$$

$$H_2CO_3 \rightleftharpoons H^+ + HCO_3^- \tag{4}$$

$$HCO_3^- \rightleftharpoons H^+ + CO_3^- \tag{5}$$

The increase in H$^+$ species in the brine accelerates carbonate dissolution given their fast reactivity in acidic environments [84]. Among carbonates, calcite exhibits the largest dissolution rate, being ~2–3 and ~3–4 orders of magnitude faster than dolomite and siderite, respectively [85–87]. During calcite dissolution, the reactant is consumed quickly, and the reaction ceases at an equilibrium pH of 4.5–5.0 [80,83]:

$$CaCO_3 \text{ (calcite)} + H^+ \rightleftharpoons Ca^{2+} + HCO_3^- \tag{6}$$

Alternatively, aluminosilicates such as feldspars, micas, and clays can yield more cations in the brine and buffer the pH up to ~8.0. However, their reaction rates are small and can take up to thousands of years to occur [80,83]. A secondary alteration of feldspars can occur when the concentration of calcium ions deriving from the dissolution of carbonates plus originally occurring in the brine is considerable [80,88]. An example is the dissolution of albite consuming Ca^{2+} and leading to the precipitation of calcite and kaolinite [80]:

$$2NaAlSi_3O_8 \text{ (albite)} + CO_2 + 2H_2O + Ca^{2+} > 4SiO_2 \text{ (chalcedony)} + CaCO_3 \text{ (calcite)} + Al_2Si_2O_5(OH)_4 \text{ (kaolinite)} + 2Na^+ \tag{7}$$

Recent studies have evaluated the role of geochemical alterations induced by scCO$_2$–brine interactions in the capillary properties of the storage zone [89–91] and confining zone [60,69,73,92–98]. Due to their high reactivity, limestones are generally the storage rock of focus, given that the reactivity of sandstones (mainly quartz) is extremely slow and pH-independent [16,83]. Dynamic scCO$_2$-enriched brine injection tests on limestone cores show strong calcite dissolution and enlargement of pore throats in the inlet of the core, followed by slight mineral precipitation in the far region [89,90]. Additionally, an increase in the limestone scCO$_2$-wettability was observed after scCO$_2$-enriched brine treatment [90,91].

With respect to confining zones, most scCO$_2$ treatment studies were conducted on crushed samples, observing significant geochemical alteration in the form of dissolution and/or precipitation of preferentially carbonate minerals [92,94,95,98,99]. Mouzakis, et al. [92] reported an increase in porosity and connectivity in all pore length scales of a carbonate-rich shale. Sanguinito, et al. [94], Goodman, et al. [95] performed experiments on the same Utica Shale (clay/calcite-rich) but varying reactor fluid, from dry scCO$_2$ to scCO$_2$-enriched deionized water (DIW), respectively. They reported increased reactivity when water was introduced into the system. Some studies indicate that dry scCO$_2$ can also affect the hydration of clays by possibly releasing interstitial water [60,94]. Regarding wettability alteration, Qin, et al. [60] described an increase in water contact angles ($57 \pm 2°$ to $69 \pm 2°$) on quartz/clay-rich shales after exposure to dry scCO$_2$ for 12 days. However, their θ measurements were conducted at ambient conditions and do not reflect the equilibrium between scCO$_2$–brine in the subsurface [76]. Conversely, Gholami, et al. [69] conducted equilibrated scCO$_2$–DIW θ measurements on quartz/clay-rich shales exposed to scCO$_2$ for 6 months. They reported an average increase from $40 \pm 5°$ to $49 \pm 5°$, along with surface dissolution of quartz and precipitation of kaolinite. To the best of our knowledge, changes in scCO$_2$–brine–rock wettability by geochemical reactivity of carbonates present in confining zones and the extent of reactivity in the pore size distributions of intact rock samples have not been addressed.

The principal aim of this study is to evaluate the impact of geochemical reactivity on the scCO$_2$–brine–rock wettability and pore size distribution (PSD) of quartz, clay, and carbonate-rich intact rocks, both before and after treatment with scCO$_2$. The specific objectives are as follows: (i) assess geochemical alterations on thin disk samples after exposing to scCO$_2$-enriched brine environment at 20.7 MPa and 65 °C for 21 days, (ii) monitor changes in fluid-equilibrated scCO$_2$–brine–rock contact angles, with a particular emphasis on the impact of surface reactivity on the wettability of the scCO$_2$–brine–rock system under elevated pressure conditions (~20 MPa), and (iii) determine alterations in PSD using three different methods: mercury injection capillary pressure (MICP), nuclear magnetic resonance (NMR), and isothermal nitrogen (N$_2$) adsorption, and analyze the resulting data to assess the effects of geochemical reactions on the PSD of the thin disk samples. This study aims to provide insights into how geochemical reactions occurring within an experimental timeframe may influence the wettability and pore structure of rocks relevant to carbon geostorage (CGS).

2. Materials and Methods

2.1. Samples Description

For this study, we selected potential storage and confining zone rock samples varying in mineralogy and petrophysical properties (Table 1). The core samples were obtained from drilling sites where horizontal plugs (~2.5 cm diameter and length) were extracted in the "as-received" state. The plugs were cleaned using Soxhlet extraction with 80/20 toluene/methanol solution to remove hydrocarbons, water, and salt, and dried at 100 °C until weight stabilized. Total porosity was measured using the protocol described in our previous work [100]. Mineralogy determined using transmission Fourier transform infrared spectroscopy (FTIR) [101,102] and microstructural examination with scanning electron microscopy (SEM) indicates that sample S1 is a quartz-rich sandstone, S2 is a quartz/clay-rich siltstone, and S3 is a finer-grained carbonate/clay-rich mudstone. Total organic carbon (TOC) measured with the LECO® method on the mudstone (S3) shows low TOC (~2.0%). To prepare the sample for our measurements, four adjacent thin disks (~0.6 cm length) were cut from each plug. Vertical heterogeneity from the same plug was minimized by conducting several X-ray fluorescence (XRF) measurements on the surface of each disk and confirming low elemental variation. To minimize surface roughness, the measurement surface of each disk was subjected to fine polishing up to 2400 grit sandpaper followed by ion-milling (Fischione 1060 SEM, Export, USA). Then, MICP data were taken on one "untreated" disk from each sample (Table 1). The other disks were used in other pre-treatment measurements such as (1) surface characterization using SEM coupled with energy dispersive X-ray spectroscopy (EDS) (detailed imaging procedure can be found in Curtis, et al. [103]), (2) fluid-equilibrated scCO$_2$–brine–rock captive-bubble contact angle (θ) measurements, (3) NMR T$_2$ porosity measurements, and (4) isothermal N$_2$ adsorption. After initial measurements, the samples were exposed to static scCO$_2$-enriched brine. The same measurements were conducted after exposure to assess changes in the samples due to $scCO_2$-enriched brine exposure.

Table 1. Mineralogy, porosity, TOC, and MICP measurements on the samples used in this study.

Sample	Lithology	Quartz + Feldspars (wt.%)	Clays (wt.%)	Carbonates (wt.%)	Porosity (%)	TOC (wt.%)	Mean Pore Radius (nm)
S1	Sandstone	80	14	6	9.3	-	440
S2	Siltstone	47	40	13	7.1	-	7.7
S3	Mudstone	14	31	55	5.9	2.0	3.0

Note: mean pore radius determined with MICP measurement.

2.2. scCO$_2$-Enriched Brine Treatment

According to Gaus, et al. [80], most caprock reactivity induced by $scCO_2$ will occur close to the boundary of the reservoir. For that reason, the samples used in this study were treated in a $scCO_2$-enriched brine condition in order to maintain maximum brine saturation and avoid undesirable effects such as drying out by CO_2 [104,105]. The brine was synthetically formulated using a 2.5 wt.% KCl solution (DIW). For the $scCO_2$-enriched brine treatments, each sample was exposed in individual reactors (Parr Instrument, Moline, IL, USA) that can operate in pressure and temperature conditions up to 58 MPa and 350 °C, respectively. The reactors were cleaned between each test with toluene, isopropanol, and acetone to avoid cross-contamination. To determine the experimental conditions, a desirable injection depth of 2000 m was selected [106]. Assuming a hydrostatic pressure gradient for brine of ~1.05 $\frac{MPa}{100m}$ and temperature gradient of ~24 $\frac{°C}{km}$, experimental conditions of ~20.7 MPa and ~65 °C were determined. The exposure time selected was 21 days, which, according to previous studies, is sufficient to observe geochemical reactions such as carbonate dissolution/precipitation [92,94,95,98,99] and clay reactivity [60,94]. Note that given the mineral reaction rates, quartz dissolution [69] and secondary feldspar alteration [80] are not expected to be observed since they could take up to 180 days and 19 months to experimentally occur, respectively. After selecting the exposure conditions, the samples were placed in the reactors inside an oven and immersed in brine at a ~10:1 brine/rock volume ratio. Then, $scCO_2$ was slowly introduced into the system using a piston pump (Teledyne ISCO, Lincoln, Dearborn, MI, USA) at a ~0.4 MPa/h pressure gradient to avoid microfracturing of samples. Pressure and temperature were monitored during the process. After exposure, the system was cooled and depressurized at the same gradient as injection and the samples were sequentially used in further measurements.

2.3. scCO$_2$–Brine–Rock Contact Angle Measurement

Several configurations have been used to measure the contact angle of $scCO_2$–brine–rock systems in subsurface conditions [64]. The captive-bubble technique was chosen (Figure 2) in order to keep the brine sample saturated during the measurement and avoid possible drying out if surrounded by a $scCO_2$ environment. The measurements were conducted using a drop shape analyzer (DSA100, Krüss GmbH, Hamburgh, Germany) coupled with an HPHT unit (Eurotechnica, Bargteheide, Germany) with a maximum working pressure and cell temperature of 69 MPa and 200 °C, respectively. Prior to the measurement, a brine-saturated sample was mounted on a custom-made, chemically inert holder (PEEK) that slides through a known outside diameter needle (~1.587 mm) that orients with the center of the sample. This orientation step allows the $scCO_2$ bubble to contact the ion-milled surface of the sample, reducing the roughness impact on θ and minimizing lateral heterogeneity. Then, roughly 30 mL of brine was introduced into the system until it covered the top surface of the sample. The unit was heated slowly up to 65 °C. ScCO$_2$ was injected from the top (gas cap), allowing diffusion into the brine and maintaining a constant pressure of 20.7 MPa, which was also monitored. After the fluid equilibration had taken place (~2 h), one $scCO_2$ bubble was carefully extruded from the needle using a separate dosing unit. Image calibration was performed, and the software (ADVANCE, version 1.10.0) recorded the static $scCO_2$–brine–rock contact angle over time. The exact same procedure was conducted on the samples after $scCO_2$–brine–rock treatment.

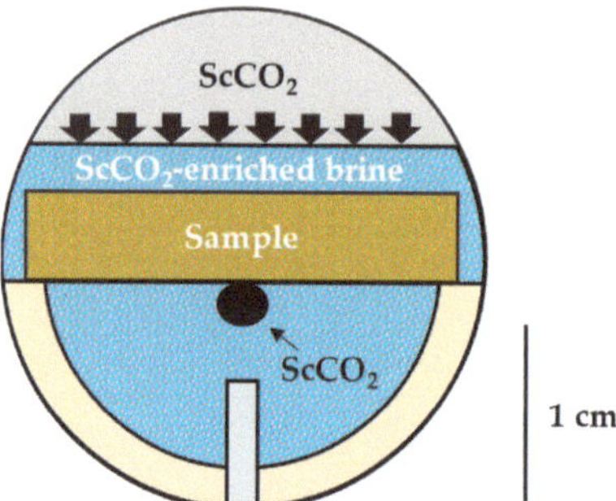

Figure 2. Schematic of the scCO$_2$–brine–rock captive-bubble measurement including approximate scale in centimeters. The brine surrounding the rock sample is enriched in scCO$_2$ diffusing downward from the "gas-cap" at desired pressure and temperature conditions. After equilibration, one scCO$_2$ bubble is carefully extruded from the needle (bottom), contacting the center of the ion-milled bottom surface of the sample.

2.4. Pore Size Distribution Measurements

Porosity and pore size distribution measurements were conducted before and after scCO$_2$ treatment. Pore throat size distributions (PTSDs) were obtained on clean and dried thin disk samples using MICP measurements (AutoPore IV, Micromeritics, Norcross, GA, USA). The AutoPore IV mercury porosimeter has a capability of 60,000 psi, which covers a nanopore throat radius of up to 1.5 nm [107]. For porosity, 12 MHz NMR T$_2$ measurements (GeoSpec II, Oxford Instruments, Abindgon, UK) were conducted on 2.5 wt.% KCl brine-saturated disk samples. In the NMR measurements, three disks were stacked in order to increase volume and enhance the signal-to-noise ratio (SNR). Echo spacing of 200 μs was selected for sample S1 (storage zone), while 100 μs was used in samples S2 and S3 (confining zone). Isothermal N$_2$ adsorption measurements (TriStar II, Micromeritics, Norcross, GA, USA) were conducted on the confining zones (S2 and S3) to assess changes in the nanopore structure. For the N$_2$ adsorption measurements, each sample was crushed, homogenized, and sieved into micrometer sizes (250–425 μm). N$_2$ adsorption pore size distributions were obtained with the density functional theory (DFT) using a slit-shaped model. Specific surface areas were determined with the Brunauer–Emmett–Teller (BET) method, fitting the adsorption data in the low relative pressure range (0.5–0.35).

3. Results

3.1. Geochemical Alteration

Figure 3 shows the backscattered electron (BSE) images of the three samples obtained over the same areas before and after scCO$_2$-enriched brine treatment. Figure 3a–c represent the microstructure before treatment of samples S1, S2, and S3, respectively. Images of the same areas for each sample after treatment are sequentially shown in Figure 3d–f. Before treatment, sample S1 is composed mainly of quartz grains and some clays (illite, chlorite, and kaolinite) filling the pore spaces. Sample S2 is covered by silicates, aluminosilicates, and minor carbonates. S3 is predominantly carbonate-rich, including clays, minor silicates, and pyrite (white). After treatment, no geochemical reactivity is observed on sample S1, except for salt precipitation. Conversely, sample S2 exhibits preferential dissolution of carbonate minerals (highlighted in red), while the main silicate and aluminosilicate framework remains preserved. Given the larger carbonate content of sample S3 (Table 1), it shows significant surface alteration by dissolution of carbonates after treatment. It is observed that other potentially reactive minerals such as clays and pyrite (white) are unaltered during this process.

Figure 4 shows BSE images on a strongly reacting area of the sample S3 before (Figure 4a) and after (Figure 4b) treatment. In Figure 4c, the spatial elemental mapping before treatment obtained with EDS shows predominantly calcite with minor amounts of silicate and dolomite. Figure 4d shows that, after treatment, calcite is the only carbonate mineral dissolved, whereas dolomite remains unaltered. XRF measurements performed before and after treatment at the surface of sample S3 indicate a reduction in the relative content of calcium oxide (CaO) from 55 wt.% to 14 wt.%, respectively.

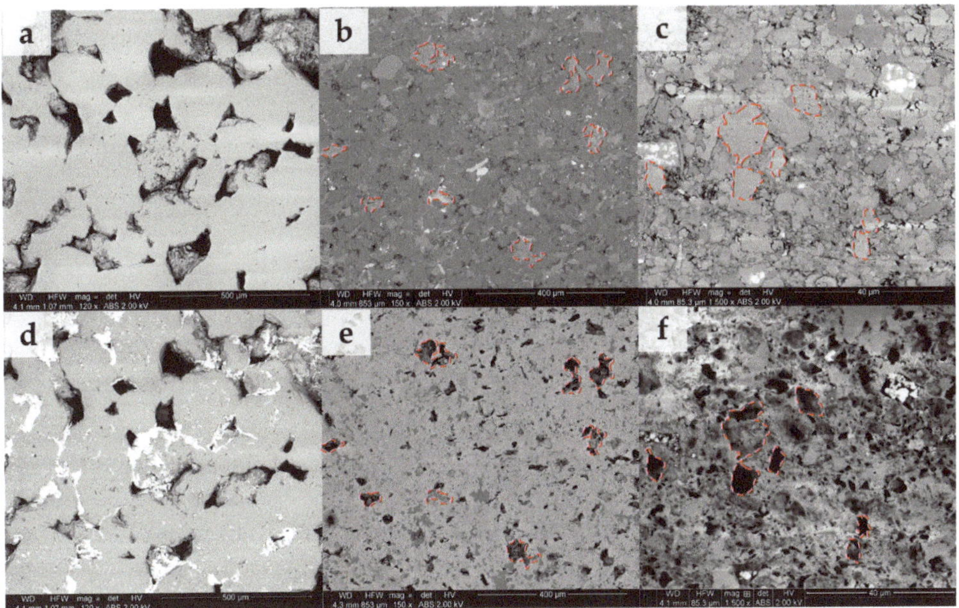

Figure 3. Backscattered images (BSEs) of the same sample area before and after scCO$_2$-enriched brine treatment. (**a**–**c**) represent areas before treatment for samples S1, S2, and S3, respectively. Their same areas after treatment are sequentially shown in (**d**–**f**). Sample S1 indicates no geochemical reactivity aside from salt precipitation. Sample S2 shows preferential dissolution of carbonates (highlighted in red), while the silicate/aluminosilicate framework remains preserved. Sample S3 reveals intense surface alteration by dissolution of carbonates, which was expected given their larger initial carbonate content. Other potentially reactive minerals such as clays and pyrite were little affected during the process.

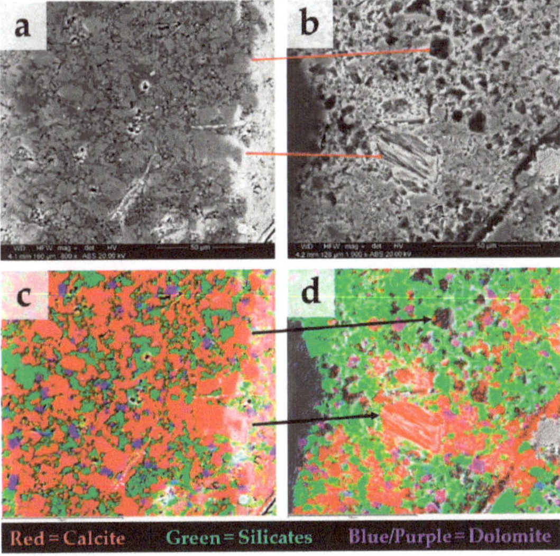

Figure 4. BSE images before (**a**) and after (**b**) treatment on the same area of sample S3 showing strong alteration due to carbonate dissolution. Note: in (**b**), the depicted image exhibits a slight displacement

relative to (**a**), while the arrows serve to locate the same calcite grains. (**c**) spatial elemental mapping with EDS before treatment, the area has predominantly calcite with minor amount of dolomite. (**d**) after treatment, calcite appears as the only mineral dissolved, while dolomite remains unaltered.

3.2. ScCO$_2$–Brine–Rock Wettability

Figure 5 shows the summary of the captive-scCO$_2$ bubble–brine–rock contact angle measurements conducted at the same experimental conditions (20.7 MPa, 65 °C) before and after scCO$_2$-enriched-brine treatment. Before treatment, contact angles of 42 $\pm$ 1°, 37 $\pm$ 1°, and 35 $\pm$ 2° are observed for samples S1, S2, and S3, respectively. These contact angle values are comparable to ones reported in the literature for sandstones, caprock, and source rocks at similar experimental conditions (Figure 1) [49,55,58,66]. It can be observed that these rocks maintain their water-wettability even at larger pressures up to ~20 MPa. After scCO$_2$-enriched brine treatment, negligible changes in contact angles are observed, despite the surface reactivity (primarily S3) shown in Figure 3. Trapping of surface bubbles can also be noted on sample S3, which develops during the fluid-equilibration phase and indicates that in-situ reactions could be taking place as the brine becomes saturated in scCO$_2$.

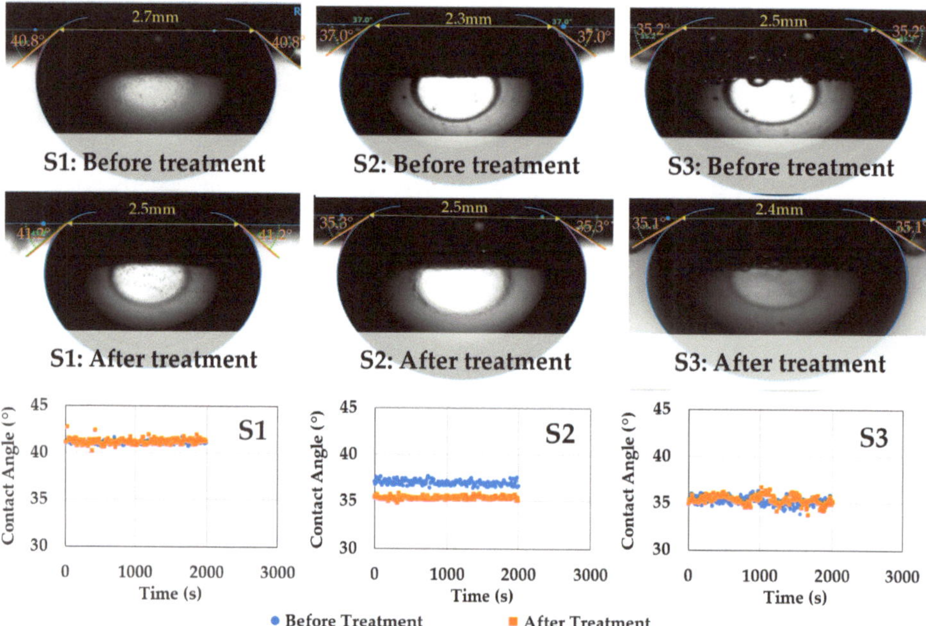

Figure 5. Summary of the scCO$_2$–brine–rock captive-bubble contact angle measurements before and after scCO$_2$ enriched brine treatment. The equilibrated contact angles remain unaltered despite surface alteration (primarily S3). The samples maintain their water-wettability even at elevated pressure (~20 MPa) and temperature (65 °C) conditions and after geochemical reactivity.

3.3. Pore Size Distribution

Pore size distribution alterations of the thin disk samples were assessed before and after scCO$_2$-enriched brine treatment with three different techniques. Figure 6 shows MICP incremental and cumulative pore throat size distributions (PTSDs) of the samples before and after treatment. Figure 6a–c represent the incremental curves for samples S1, S2, and S3, respectively. Their cumulative curves are sequentially shown in Figure 6d–f. The PTSD for sample S1 (storage zone) indicates initial intrusion in the 2–10 µm pore throat size range followed by a main peak between 200 and 600 nm. Conversely, the main peaks for the confining zone samples S2 and S3 are considerably lower, at around 6–10 nm and 2–4 nm, respectively. After treatment, little alteration is observed in the PTSD of samples S1 and

S2. Sample S3 indicates minor changes, aside from a slight increase in amplitude in the pore range of 100–1000 nm, followed by a drop in the main peak at 1–10 nm.

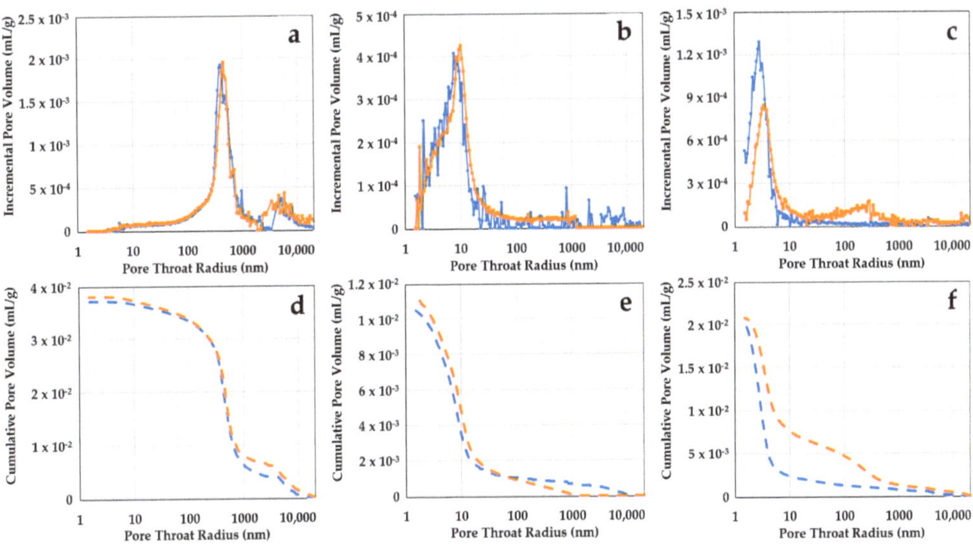

Figure 6. Pore throat size distributions (PTSDs) obtained with MICP before and after scCO$_2$-enriched brine treatment. The incremental curves for samples S1, S2, and S3 are shown in (**a**–**c**), respectively. Their cumulative curves are sequentially represented in (**d**–**f**). Little alteration is observed in the PTSD of the samples, aside from a small increase in amplitude in the 100–1000 nm range for sample S3 followed by a drop in the 1–10 nm range.

Figure 7 shows NMR T$_2$ porosity distributions of the three samples before and after scCO$_2$-enriched-brine treatment. Figure 7a–c represent the incremental curves for samples S1, S2, and S3, respectively. Their cumulative curves are sequentially shown in Figure 7d–f. Before treatment, sample S1 (storage zone) shows a main peak around 100 ms T$_2$ relaxation, while the main peaks for the confining samples S2 and S3 are in shorter T$_2$ times of ~1.1 ms and 0.8 ms, respectively. For the confining zone samples, the second small peak above 100 ms represents brine trapped in the surface irregularities of the samples. Nevertheless, the porosity distributions obtained with NMR indicate no significant alteration in the three samples after treatment. Sample S3 exhibits a small inflection in the spectrum after treatment around 1–20 ms, which does not impact the cumulative porosity within this range.

Figure 8 displays N$_2$ adsorption and desorption isotherms for confining zone samples S2 and S3 before (Figure 8a) and after (Figure 8b) scCO$_2$-enriched-brine treatment. These curves, along with hysteresis loop analysis, yield insights into pore structure and connectivity in tight porous media, such as shales [108–110]. Before treatment, both samples exhibited type IV isotherm characteristics with a distinctive hysteresis loop, indicating capillary condensation in meso/macro-pores [111]. There is no limiting uptake at high relative pressures (p/p^0), indicating an H3 hysteresis loop with slit-shaped pores forming plate-like structures [112]. A wider hysteresis loop for sample S3 is also observed, suggesting a more complex pore network. After treatment, isotherm shapes indicate minimal changes in adsorption behavior. PSDs were extracted using the density functional theory (DFT) with a slit-shaped model. Incremental pore volume curves of S2 (Figure 8c) and S3 (Figure 8d) show a main peak in the mesopore region (20–40 nm) and broader distribution for S3, with a larger fine-mesopore volume (2–20 nm) than S2. The increase in fine mesopores for S3 results in a larger BET surface area (Figure 8e). After treatment, no significant alterations in PSD or BET surface areas are observed. Sample S3 exhibits a slight decrease in mesopore volume followed by a minor increase in the fine-mesopore range, resulting in a slight decrease in surface area, but these differences are small and do not significantly impact cumulative pore volumes in the nanopore range.

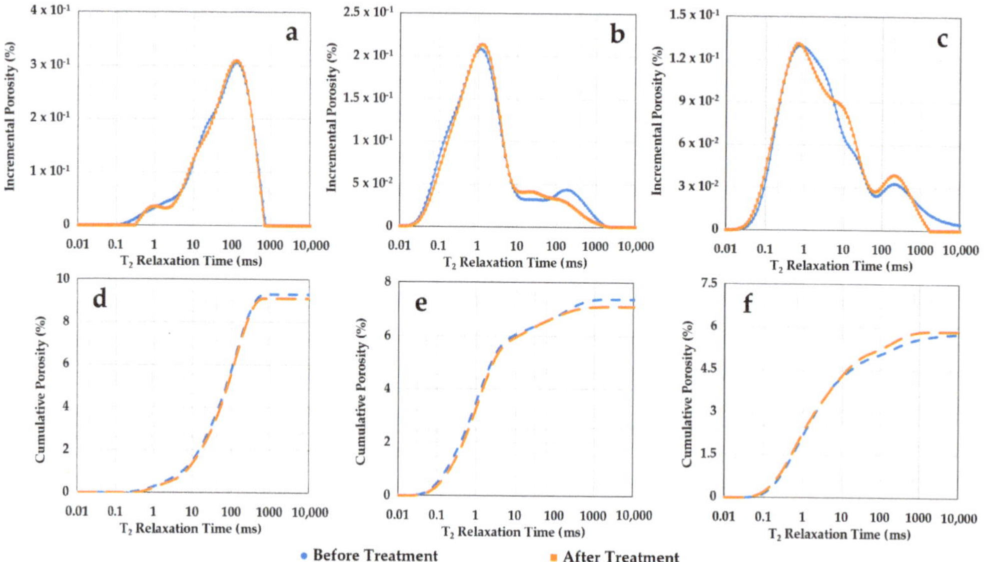

Figure 7. NMR porosities as function of T_2 relaxation time before and after scCO$_2$-enriched brine treatment. The incremental curves for samples S1, S2, and S3 are shown in (**a**–**c**), respectively. Their cumulative curves are sequentially represented in (**d**–**f**). Negligible alteration in the NMR porosity distributions is observed, aside from a small inflection around 1–20 ms for sample S3, which does not affect its cumulative porosity.

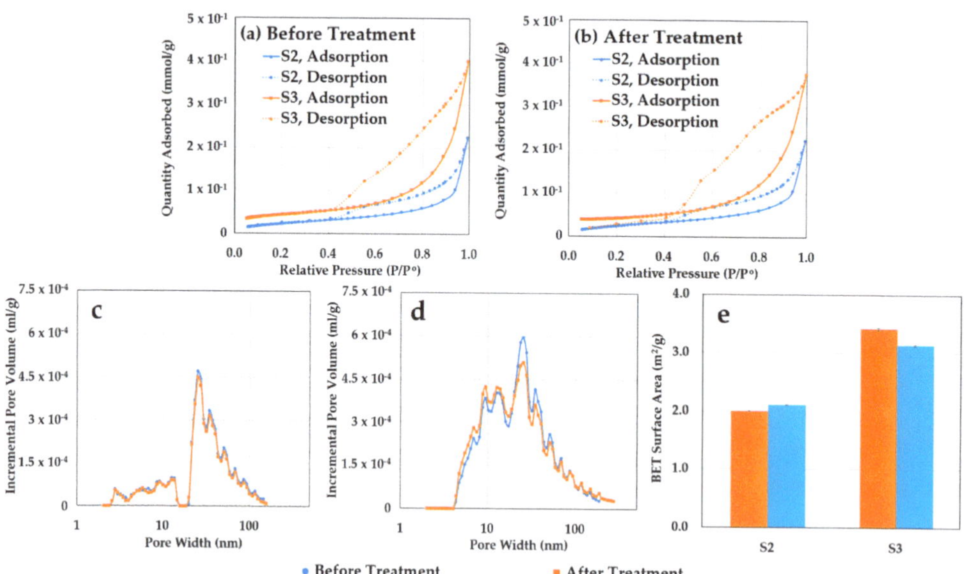

Figure 8. Isothermal N$_2$ adsorption results for the confining zone samples S2 and S3 before and after scCO$_2$-enriched brine treatment. (**a**) Before treatment, both samples exhibit type IV isotherms with

distinctive hysteresis loops and larger adsorption for sample S3. (**b**) After treatment, isotherm shapes remain stable, indicating minimal changes in adsorption behavior. (**c**) Incremental pore volume curves for S2 show a main peak in the mesopore region (20–40 nm) and constant PSD after treatment. (**d**) Incremental pore volume curves for S3 reveal a broader distribution, with a larger fine mesopore volume (2–20 nm) compared to S2, and slight alteration in PSD. (**e**) The increase in fine mesopores for S3 results in a larger BET surface area and slight alteration after treatment.

4. Discussion

Geochemical reactions arising from the dissolution of $scCO_2$ in brine have been experimentally proven to impact the microstructure of storage zones [89–91] and confining zones [60,69,73,92–98] even at short treatment times of typically few weeks. The impact on storage zones is commonly observed by dynamic injection flooding of $scCO_2$-enriched brine in limestone core samples. Significant dissolution (wormholes) has been reported on the inlet of the core plugs, followed by a small amount of precipitation in the far region. Although dynamic tests are important to evaluate geochemical effects in high flow regions (i.e., around injection wells), the extent of reactivity is prolonged because "fresh" reactant species (e.g., H^+) are continuously fed in open flow systems. On the other hand, geochemical alteration of confining zones is generally assessed by static $scCO_2$ treatments in closed systems conducted on crushed (micrometer size) samples. Significant geochemical reactivity has been reported by preferential dissolution (and precipitation) of carbonate minerals, which impact pore size distributions. However, the ultra-low transport properties of nanoporous confining zones imply concentration profiles of reactant species (e.g., H^+) across the samples [113,114]. Such concentration profiles favor surface reactivity and augment reactions on crushed samples due to their higher specific surface areas as compared to core plugs. In addition, the brine/rock volume ratio and brine composition selected in experiments can also impact reactivity [94,95,99]. When a large brine/rock volume ratio or calcium-depleted brines are used in closed systems, a higher Ca^{2+} uptake from calcite dissolution is needed before the reaction ceases (pH~4.5–5). Nonetheless, reactive transport studies show that geochemical reactions driven by diffusion in nanoporous confining zones are expected to impact properties on the scale of hundreds to thousands of years [80]. A strategy to overcome long experimental times is the combination of quantitative tools that can simultaneously resolve such alterations at the surface level up to the nanometer, micrometer, and millimeter depth of invasion.

Here, the impact of geochemical reactions is quantified by both surface characterization and three different PSD measurements conducted on thin disk samples. SEM/EDS images reveal surface dissolution of calcite grains after treatment, while other potentially reactive minerals remain unaltered (e.g., dolomite, clays, pyrite). Although surface reactivity occurred, little alteration is seen on the PSDs obtained with MICP, NMR, and N_2 adsorption after treatment. Minor changes were consistently observed in sample S3: MICP indicated a small amplitude increase in the 100–1000 nm range, followed by an amplitude drop in the main peak at 1–10 nm; the NMR T_2 spectrum exhibited a small inflection around 1–20 ms; and N_2 adsorption showed a slight increase in the main peak at 20–40 nm followed by small reduction at fine mesopores (1–20 nm). Combined with SEM/EDS data, it is understood that the small changes in PSD of sample S3 could be indicative of the surface dissolution of calcite. Potential precipitation below the surface that can cause a reduction in micropore volume is possible due to a local increase in the concentration of Ca^{2+} driven by surface dissolution (Equation (6)). Based on this study's experimental conditions, the geochemical reactivity on the confining zone samples preferentially occurs at the surface and does not impact pore size distributions or porosities at a core scale.

The alteration in surface mineralogy by calcite dissolution could change surface hydrophilicity and impact the wettability and hence the trapping mechanisms in the $scCO_2$–brine–rock system. However, the fluid-equilibrated $scCO_2$–brine–rock contact angles (θ) reported here are little influenced by the reactivity occurring at the surface. The rocks studied here maintain their water-wettability at high-pressure conditions (~20 MPa) and even after moderate or strong surface reactivity. Calcite dissolution in the forms of etching and pitting could locally increase the surface roughness (Figure 3f), which has been found to decrease θ in hydrophilic surfaces of pure minerals (e.g., quartz, calcite) [51,56]. In fine-grained heterogeneous surfaces such as confining zones, the large difference in scale between the static $scCO_2$ bubble (millimeter size) and grain size (micrometer to nanometer) accounts for small roughness changes due to dissolution that could impact the contact angle. It is also found that using NMR to assess wettability alteration due to dissolution in nanoporous media [115,116] is experimentally challenging because the geochemical reactivity shown here does not appear to influence the surface relaxivity of the bulk rock (Figure 8).

The microscopic displacement efficiency and primary trapping mechanisms during $scCO_2$ storage are tied to capillary properties including interfacial tension (IFT), wettability, and pore size. Miller, et al. [99] showed that the aqueous geochemistry of high-concentration analytes that could impact the IFT remains fairly constant during $scCO_2$ treatment. Therefore, changes in brine ionic concentration due to dissolution or precipitation of species are not expected to alter the $scCO_2$–brine–rock IFT. Regarding wettability, this work demonstrated that contact angles are barely affected by induced geochemical reactions in $scCO_2$–brine–rock systems. With the contact angles reported here and knowledge of the $scCO_2$–brine IFT at subsurface conditions, Hg–air capillary pressure measurements can be converted to subsurface $scCO_2$–brine conditions

$$P_{c,scCO_2:H_2O} = \frac{\gamma_{scCO_2:H_2O} \times \cos\theta_{scCO_2:H_2O}}{\gamma_{Hg:air} \times \cos\theta_{Hg:air}} \times P_{c,Hg:air} \approx \left| \frac{\left(\frac{27\ mN}{m}\right) \times \cos(35°)}{\left(\frac{485\ mN}{m}\right) \times \cos(130°)} \right| \times P_{c,Hg:air} \approx 0.07 P_{c,Hg:air} \quad (8)$$

assuming a $scCO_2$–brine IFT for this work experimental conditions of 27 mN/m [23], θ of the sample S3 (35°), Hg–air θ of 130°, and surface tension of 485 mN/m (laboratory conditions). Equation (8) indicates that the $scCO_2$–brine capillary pressure (Pc) at subsurface conditions for the confining zone S3 is about ~14 times smaller than laboratory Hg–air Pc.

During the short-term experiments, the predominant geochemical reactions were directed toward altering carbonate minerals, leaving the capillary properties of the quartz-rich sandstone largely unaffected within the experimental duration. The sandstone exhibited minimal reactivity, which aligns with findings in similar studies conducted over comparable timeframes and under analogous experimental conditions [91,117–120]. It is conceivable that extended treatment periods (0.5–1.5 years) might reveal some level of reactivity with quartz [69,104], although this is less likely to substantially influence surface wettability, as suggested by the consistently unaltered contact angles reported in this study.

Conversely, the main implication of such reactions for the capillary displacement of the confining zone appears to be an increase in pore size at the surface due to calcite dissolution followed by possible precipitation. The preferential reactivity of confining zones toward carbonate minerals agrees with literature experiments on crushed samples conducted in similar timescales [92,94,95,98,99]. Although these experiments also reported alterations in pore size distributions (PSDs) of crushed samples, we observed that the PSDs of intact samples obtained with three different methods remained unaltered. It is understood that the propagation of the geochemical reactions is controlled not only by mineralogy but experimental conditions (dynamic vs static exposure, water/rock brine ratio) as well as surface area and matrix transport properties. Therefore, short-term geochemical reactions do not seem to significantly impact the capillary displacement in quartz-rich storage reservoirs or primary capillary trapping mechanisms in either siltstone or mudstone confining zones in the condition studied.

5. Conclusions

In this experimental study, sandstone (storage zone), clay-rich siltstone, and carbonate-rich mudstone (confining zones) thin disk samples were treated under a $scCO_2$-enriched brine condition (10:1 brine/rock volume ratio) for 21 days at 20.7 MPa and 65 °C using a 2.5 wt.% KCl brine solution. Before and after treatment, geochemical alterations were assessed at the surface level using scanning electron microscopy (SEM) coupled with energy-dispersive X-ray spectroscopy (EDS) and X-ray fluorescence (XRF). Fluid-equilibrated $scCO_2$–brine–rock contact angle measurements were conducted to monitor wettability changes using the captive-bubble method. Pore size alterations were determined using mercury injection capillary pressure (MICP), nuclear magnetic resonance (NMR), and isothermal N_2 adsorption. The main conclusions of this work are as follows:

- Carbonate minerals preferentially react at the surface level in the form of calcite dissolution, while other potentially reactive minerals such as feldspars, clays, dolomite, and pyrite remain preserved. A reduction in the relative content of calcium oxide (CaO) from 55 wt.% to 14 wt.% is observed at the surface.
- Before treatment, fluid-equilibrated $scCO_2$–brine–rock contact angles are comparable with the literature values for storage and confining zones. Contact angles of $42 \pm 1°$, $37 \pm 1°$, and $35 \pm 2°$ were determined for sandstone, siltstone, and mudstone, respectively. After treatment, negligible alteration in the contact angles is observed, despite surface dissolution in the forms of etching and pitting for the carbonate-rich mudstone (S3). The rocks studied here maintain their water wettability after geochemical reactivity and at elevated pressures (~20 MPa).
- Although reactions are observed at the surface, the pore size distributions (PSDs) of the thin disk samples obtained with three different methods (MICP, NMR, and N_2 adsorption) show

little alteration. Minor changes are consistently observed for the carbonate-rich mudstone (S3). Coupled with SEM/EDS images, these changes could be indicative of surface dissolution of calcite followed by possible precipitation. Overall, the short-term geochemical reactions observed here did not significantly impact the cumulative porosities or PSD of the three samples.

Our research demonstrates that short-term geochemical reactions do not substantially impact the capillary displacement mechanism in the studied quartz-rich sandstone storage reservoirs and siltstone/mudstone confining zones. These findings contribute valuable insights into the behavior of scCO$_2$–brine–rock systems and highlight the preservation of rock properties under the conditions examined.

Author Contributions: Conceptualization, F.C.; Methodology, F.C.; Validation, F.C. and S.D.; Formal analysis, F.C.; Investigation, F.C.; Resources, S.D. and M.C.; Data curation, F.C.; Writing—original draft, F.C.; Writing—review & editing, S.D., M.C. and C.R.; Visualization, F.C.; Supervision, S.D.; Project administration, C.R.; Funding acquisition, F.C. All authors have read and agreed to the published version of the manuscript.

Funding: Financial support for publication was provided by the University of Oklahoma Libraries' Open Access Fund.

Data Availability Statement: The data will be made available on specific user request.

Acknowledgments: We would like to thank the experimental efforts of Micaela Langevin, Gary Stowe, Sidi Mamoudou, Abdelali Guezei, Quan Nguyen, and Abdullah Al Raisi. Integrated Core Characterization Center (IC3) contribution (#198).

Conflicts of Interest: The authors declare no conflict of interest.

References

1. Davis, S.J.; Lewis, N.S.; Shaner, M.; Aggarwal, S.; Arent, D.; Azevedo, I.L.; Benson, S.M.; Bradley, T.; Brouwer, J.; Chiang, Y.-M.; et al. Net-zero emissions energy systems. *Science* **2018**, *360*, eaas9793. [CrossRef] [PubMed]
2. Zhao, H. *Phase Equilibria in CO$_2$-Brine System for CO$_2$ Storage*; ProQuest Dissertations Publishing: Ann Arbor, MI, USA, 2014.
3. Heinemann, N.; Wilkinson, M.; Pickup, G.E.; Haszeldine, R.S.; Cutler, N.A. CO$_2$ storage in the offshore UK Bunter Sandstone Formation. *Int. J. Greenh. Gas Control* **2012**, *6*, 210–219. [CrossRef]
4. Medici, G.; West, L.J. Reply to discussion on 'Review of groundwater flow and contaminant transport modelling approaches for the Sherwood Sandstone aquifer, UK insights from analogous successions worldwide' by Medici and West (QJEGH, 55, qjegh2021-176). *Q. J. Eng. Geol. Hydrogeol.* **2023**, *56*, qjegh2022-097. [CrossRef]
5. Zhang, C.P.; Ranjith, P.G.; Perera, M.S.A.; Zhao, J.; Zhang, D.; Wanniarachchi, W.A.M. A novel approach to precise evaluation of carbon dioxide flow behaviour in siltstone under tri-axial drained conditions. *J. Nat. Gas Sci. Eng.* **2016**, *34*, 331–340. [CrossRef]
6. Jayasekara, D.W.; Ranjith, P.G.; Wanniarachchi, W.A.M.; Rathnaweera, T.D.; Chaudhuri, A. Effect of salinity on supercritical CO$_2$ permeability of caprock in deep saline aquifers: An experimental study. *Energy* **2020**, *191*, 116486. [CrossRef]
7. Lima, V.d.; Einloft, S.; Ketzer, J.M.; Jullien, M.; Bildstein, O.; Petronin, J.-C. CO$_2$ Geological storage in saline aquifers: Paraná Basin caprock and reservoir chemical reactivity. *Energy Procedia* **2011**, *4*, 5377–5384. [CrossRef]
8. Shi, J.-Q.; Sinayuc, C.; Durucan, S.; Korre, A. Assessment of carbon dioxide plume behaviour within the storage reservoir and the lower caprock around the KB-502 injection well at In Salah. *Int. J. Greenh. Gas Control* **2012**, *7*, 115–126. [CrossRef]
9. Zhang, K.; Sang, S.; Zhou, X.; Liu, C.; Ma, M.; Niu, Q. Influence of supercritical CO$_2$-H$_2$O-caprock interactions on the sealing capability of deep coal seam caprocks related to CO$_2$ geological storage: A case study of the silty mudstone caprock of coal seam no. 3 in the Qinshui Basin, China. *Int. J. Greenh. Gas Control* **2021**, *106*, 103282. [CrossRef]
10. Lu, J.; Wilkinson, M.; Haszeldine, R.S.; Fallick, A.E. Long-term performance of a mudrock seal in natural CO$_2$ storage. *Geology* **2009**, *37*, 35–38. [CrossRef]
11. Lake, L.W. *Enhanced Oil Recovery*; Prentice Hall: Saddle River, NJ, USA, 1989.
12. Bachu, S.; Bonijoly, D.; Bradshaw, J.; Burruss, R.; Holloway, S.; Christensen, N.P.; Mathiassen, O.M. CO$_2$ storage capacity estimation: Methodology and gaps. *Int. J. Greenh. Gas Control* **2007**, *1*, 430–443. [CrossRef]
13. Bachu, S. Review of CO$_2$ storage efficiency in deep saline aquifers. *Int. J. Greenh. Gas Control* **2015**, *40*, 188–202. [CrossRef]
14. Al Hameli, F.; Belhaj, H.; Al Dhuhoori, M. CO$_2$ Sequestration Overview in Geological Formations: Trapping Mechanisms Matrix Assessment. *Energies* **2022**, *15*, 7805. [CrossRef]
15. Ajayi, T.; Gomes, J.S.; Bera, A. A review of CO$_2$ storage in geological formations emphasizing modeling, monitoring and capacity estimation approaches. *Pet. Sci.* **2019**, *16*, 1028–1063. [CrossRef]
16. Kaszuba, J.; Yardley, B.; Andreani, M. Experimental perspectives of mineral dissolution and precipitation due to carbon dioxide-water-rock interactions. *Rev. Mineral. Geochem.* **2013**, *77*, 153–188. [CrossRef]
17. Song, J.; Zhang, D. Comprehensive Review of Caprock-Sealing Mechanisms for Geologic Carbon Sequestration. *Environ. Sci. Technol.* **2013**, *47*, 9–22. [CrossRef] [PubMed]

18. Abidoye, L.K.; Khudaida, K.J.; Das, D.B. Geological Carbon Sequestration in the Context of Two-Phase Flow in Porous Media: A Review. *Crit. Rev. Environ. Sci. Technol.* **2015**, *45*, 1105–1147. [CrossRef]
19. Cai, J.; Jin, T.; Kou, J.; Zou, S.; Xiao, J.; Meng, Q. Lucas–Washburn Equation-Based Modeling of Capillary-Driven Flow in Porous Systems. *Langmuir* **2021**, *37*, 1623–1636. [CrossRef]
20. Behroozi, F. A Fresh Look at the Young-Laplace Equation and Its Many Applications in Hydrostatics. *Phys. Teach.* **2022**, *60*, 358–361. [CrossRef]
21. Chiquet, P.; Daridon, J.-L.; Broseta, D.; Thibeau, S. CO_2/water interfacial tensions under pressure and temperature conditions of CO_2 geological storage. *Energy Convers. Manag.* **2007**, *48*, 736–744. [CrossRef]
22. Van der Meer, L.G.H.; Hofstee, C.; Orlic, B. The fluid flow consequences of CO_2 migration from 1000 to 600 metres upon passing the critical conditions of CO_2. *Energy Procedia* **2009**, *1*, 3213–3220. [CrossRef]
23. Chalbaud, C.; Robin, M.; Lombard, J.M.; Martin, F.; Egermann, P.; Bertin, H. Interfacial tension measurements and wettability evaluation for geological CO_2 storage. *Adv. Water Resour.* **2009**, *32*, 98–109. [CrossRef]
24. Bikkina, P.K.; Shoham, O.; Uppaluri, R. Equilibrated Interfacial Tension Data of the CO_2–Water System at High Pressures and Moderate Temperatures. *J. Chem. Eng. Data* **2011**, *56*, 3725–3733. [CrossRef]
25. Li, X.; Boek, E.; Maitland, G.C.; Trusler, J.P.M. Interfacial Tension of (Brines + CO_2): (0.864 NaCl + 0.136 KCl) at Temperatures between (298 and 448) K, Pressures between (2 and 50) MPa, and Total Molalities of (1 to 5) mol·kg^{-1}. *J. Chem. Eng. Data* **2012**, *57*, 1078–1088. [CrossRef]
26. Pereira, L.M.C.; Chapoy, A.; Burgass, R.; Oliveira, M.B.; Coutinho, J.A.P.; Tohidi, B. Study of the impact of high temperatures and pressures on the equilibrium densities and interfacial tension of the carbon dioxide/water system. *J. Chem. Thermodyn.* **2016**, *93*, 404–415. [CrossRef]
27. Hebach, A.; Oberhof, A.; Dahmen, N.; Kögel, A.; Ederer, H.; Dinjus, E. Interfacial Tension at Elevated PressuresMeasurements and Correlations in the Water + Carbon Dioxide System. *J. Chem. Eng. Data* **2002**, *47*, 1540–1546. [CrossRef]
28. Aggelopoulos, C.A.; Robin, M.; Perfetti, E.; Vizika, O. CO_2/$CaCl_2$ solution interfacial tensions under CO_2 geological storage conditions: Influence of cation valence on interfacial tension. *Adv. Water Resour.* **2010**, *33*, 691–697. [CrossRef]
29. Aggelopoulos, C.A.; Robin, M.; Vizika, O. Interfacial tension between CO_2 and brine (NaCl + $CaCl_2$) at elevated pressures and temperatures: The additive effect of different salts. *Adv. Water Resour.* **2011**, *34*, 505–511. [CrossRef]
30. Bennion, B.; Bachu, S. Relative Permeability Characteristics for Supercritical CO_2 Displacing Water in a Variety of Potential Sequestration Zones in the Western Canada Sedimentary Basin. In Proceedings of the SPE Annual Technical Conference and Exhibition, Dallas, TX, USA, 9–12 October 2005.
31. Chiquet, P.; Broseta, D.; Thibeau, S. Wettability alteration of caprock minerals by carbon dioxide. *Geofluids* **2007**, *7*, 112–122. [CrossRef]
32. Yang, D.; Gu, Y.; Tontiwachwuthikul, P. Wettability Determination of the Reservoir Brine—Reservoir Rock System with Dissolution of CO_2 at High Pressures and Elevated Temperatures. *Energy Fuels* **2008**, *22*, 504–509. [CrossRef]
33. Espinoza, D.N.; Santamarina, J.C. Water-CO_2-mineral systems: Interfacial tension, contact angle, and diffusion-Implications to CO_2 geological storage. *Water Resour. Res.* **2010**, *46*. [CrossRef]
34. Bikkina, P. Contact angle measurements of CO_2-water-quartz/calcite systems in the perspective of carbon sequestration. *Int. J. Greenh. Gas Control* **2011**, *5*, 1259–1271. [CrossRef]
35. Mills, J.; Riazi, M.; Sohrabi, M. *Wettability of Common Rock-Forming Minerals in a CO_2-Brine System at Reservoir Conditions*; Society of Core Analysts Fredericton, Canada: Fredericton, NJ, Canada, 2011.
36. Tonnet, N.; Mouronval, G.; Chiquet, P.; Broseta, D. Petrophysical assessment of a carbonate-rich caprock for CO_2 geological storage purposes. *Energy Procedia* **2011**, *4*, 5422–5429. [CrossRef]
37. Broseta, D.; Tonnet, N.; Shah, V. Are rocks still water-wet in the presence of dense CO_2 or H_2S? *Geofluids* **2012**, *12*, 280–294. [CrossRef]
38. Jung, J.-W.; Jiamin, W.A.N. Supercritical CO_2 and Ionic Strength Effects on Wettability of Silica Surfaces: Equilibrium Contact Angle Measurements. *Energy Fuels* **2012**, *26*, 6053–6059. [CrossRef]
39. Kim, Y.; Wan, J.; Kneafsey, T.J.; Tokunaga, T.K. Dewetting of Silica Surfaces upon Reactions with Supercritical CO_2 and Brine: Pore-Scale Studies in Micromodels. *Environ. Sci. Technol.* **2012**, *46*, 4228–4235. [CrossRef]
40. Farokhpoor, R.; Bjørkvik, B.J.A.; Lindeberg, E.; Torsæter, O. CO_2 Wettability Behavior During CO_2 Sequestration in Saline Aquife—An Experimental Study on Minerals Representing Sandstone and Carbonate. *Energy Procedia* **2013**, *37*, 5339–5351. [CrossRef]
41. Saraji, S.; Goual, L.; Piri, M.; Plancher, H. Wettability of Supercritical Carbon Dioxide/Water/Quartz Systems: Simultaneous Measurement of Contact Angle and Interfacial Tension at Reservoir Conditions. *Langmuir* **2013**, *29*, 6856–6866. [CrossRef] [PubMed]
42. Wang, S.; Edwards, I.M.; Clarens, A.F. Wettability Phenomena at the CO_2–Brine–Mineral Interface: Implications for Geologic Carbon Sequestration. *Environ. Sci. Technol.* **2013**, *47*, 234–241. [CrossRef]
43. Andrew, M.; Bijeljic, B.; Blunt, M.J. Pore-scale contact angle measurements at reservoir conditions using X-ray microtomography. *Adv. Water Resour.* **2014**, *68*, 24–31. [CrossRef]
44. Iglauer, S.; Salamah, A.; Sarmadivaleh, M.; Liu, K.; Phan, C. Contamination of silica surfaces: Impact on water-CO_2-quartz and glass contact angle measurements. *Int. J. Greenh. Gas Control* **2014**, *22*, 325–328. [CrossRef]

45. Kaveh, N.S.; Rudolph, E.S.J.; van Hemert, P.; Rossen, W.R.; Wolf, K.H. Wettability Evaluation of a CO_2/Water/Bentheimer Sandstone System: Contact Angle, Dissolution, and Bubble Size. *Energy Fuels* **2014**, *28*, 4002–4020. [CrossRef]

46. Saraji, S.; Piri, M.; Goual, L. The effects of SO_2 contamination, brine salinity, pressure, and temperature on dynamic contact angles and interfacial tension of supercritical CO_2/brine/quartz systems. *Int. J. Greenh. Gas Control* **2014**, *28*, 147–155. [CrossRef]

47. Al-Yaseri, A.; Sarmadivaleh, M.; Saeedi, A.; Lebedev, M.; Barifcani, A.; Iglauer, S. N_2 + CO_2 + NaCl brine interfacial tensions and contact angles on quartz at CO_2 storage site conditions in the Gippsland basin, Victoria/Australia. *J. Pet. Sci. Eng.* **2015**, *129*, 58–62. [CrossRef]

48. Chaudhary, K.; Guiltinan, E.J.; Cardenas, M.B.; Maisano, J.A.; Ketcham, R.A.; Bennett, P.C. Wettability measurement under high P-T conditions using X-ray imaging with application to the brine-supercritical CO_2 system. *Geochem. Geophys. Geosystems G3* **2015**, *16*, 2858–2864. [CrossRef]

49. Iglauer, S.; Al-Yaseri, A.Z.; Rezaee, R.; Lebedev, M. CO_2 wettability of caprocks: Implications for structural storage capacity and containment security. *Geophys. Res. Lett.* **2015**, *42*, 9279–9284. [CrossRef]

50. Liu, Y.; Mutailipu, M.; Jiang, L.; Zhao, J.; Song, Y.; Chen, L. Interfacial tension and contact angle measurements for the evaluation of CO_2-brine two-phase flow characteristics in porous media. *Environ. Prog. Sustain. Energy* **2015**, *34*, 1756–1762. [CrossRef]

51. Al-Yaseri, A.Z.; Lebedev, M.; Barifcani, A.; Iglauer, S. Receding and advancing (CO_2 + brine + quartz) contact angles as a function of pressure, temperature, surface roughness, salt type and salinity. *J. Chem. Thermodyn.* **2016**, *93*, 416–423. [CrossRef]

52. Lv, P.; Liu, Y.; Jiang, L.; Song, Y.; Wu, B.; Zhao, J.; Zhang, Y. Experimental determination of wettability and heterogeneity effect on CO_2 distribution in porous media. *Greenh. Gases Sci. Technol.* **2016**, *6*, 401–415. [CrossRef]

53. Roshan, H.; Al-Yaseri, A.Z.; Sarmadivaleh, M.; Iglauer, S. On wettability of shale rocks. *J. Colloid. Interface Sci.* **2016**, *475*, 104–111. [CrossRef]

54. Shojai Kaveh, N.; Barnhoorn, A.; Wolf, K.H. Wettability evaluation of silty shale caprocks for CO_2 storage. *Int. J. Greenh. Gas Control* **2016**, *49*, 425–435. [CrossRef]

55. Al-Yaseri, A.Z.; Roshan, H.; Zhang, Y.; Rahman, T.; Lebedev, M.; Barifcani, A.; Iglauer, S. Effect of the Temperature on CO_2/Brine/Dolomite Wettability: Hydrophilic versus Hydrophobic Surfaces. *Energy Fuels* **2017**, *31*, 6329–6333. [CrossRef]

56. Arif, M.; Lebedev, M.; Barifcani, A.; Iglauer, S. CO_2 storage in carbonates: Wettability of calcite. *Int. J. Greenh. Gas Control* **2017**, *62*, 113–121. [CrossRef]

57. Botto, J.; Fuchs, S.J.; Fouke, B.W.; Clarens, A.F.; Freiburg, J.T.; Berger, P.M.; Werth, C.J. Effects of Mineral Surface Properties on Supercritical CO_2 Wettability in a Siliciclastic Reservoir. *Energy Fuels* **2017**, *31*, 5275–5285. [CrossRef]

58. Guiltinan, E.J.; Cardenas, M.B.; Bennett, P.C.; Zhang, T.; Espinoza, D.N. The effect of organic matter and thermal maturity on the wettability of supercritical CO_2 on organic shales. *Int. J. Greenh. Gas Control* **2017**, *65*, 15–22. [CrossRef]

59. Lv, P.; Liu, Y.; Wang, Z.; Liu, S.; Jiang, L.; Chen, J.; Song, Y. In Situ Local Contact Angle Measurement in a CO_2–Brine–Sand System Using Microfocused X-ray CT. *Langmuir* **2017**, *33*, 3358–3366. [CrossRef]

60. Qin, C.; Jiang, Y.; Luo, Y.; Xian, X.; Liu, H.; Li, Y. Effect of Supercritical Carbon Dioxide Treatment Time, Pressure, and Temperature on Shale Water Wettability. *Energy Fuels* **2017**, *31*, 493–503. [CrossRef]

61. Tudek, J.; Crandall, D.; Fuchs, S.; Werth, C.J.; Valocchi, A.J.; Chen, Y.; Goodman, A. In situ contact angle measurements of liquid CO_2, brine, and Mount Simon sandstone core using micro X-ray CT imaging, sessile drop, and Lattice Boltzmann modeling. *J. Pet. Sci. Eng.* **2017**, *155*. [CrossRef]

62. Alnili, F.; Al-Yaseri, A.; Roshan, H.; Rahman, T.; Verall, M.; Lebedev, M.; Sarmadivaleh, M.; Iglauer, S.; Barifcani, A. Carbon dioxide/brine wettability of porous sandstone versus solid quartz: An experimental and theoretical investigation. *J. Colloid. Interface Sci.* **2018**, *524*, 188–194. [CrossRef]

63. Pan, B.; Li, Y.; Wang, H.; Jones, F.; Iglauer, S. CO_2 and CH_4 Wettabilities of Organic-Rich Shale. *Energy Fuels* **2018**, *32*, 1914–1922. [CrossRef]

64. Prem, B.; Imran, S. Interfacial Tension and Contact Angle Data Relevant to Carbon Sequestration. In *Carbon Capture, Utilization and Sequestration*; Ramesh, K.A., Ed.; IntechOpen: Rijeka, Croatia, 2018; p. 10.

65. Arif, M.; Abu-Khamsin, S.A.; Iglauer, S. Wettability of rock/CO_2/brine and rock/oil/CO_2-enriched-brine systems:Critical parametric analysis and future outlook. *Adv. Colloid. Interface Sci.* **2019**, *268*, 91–113. [CrossRef]

66. Mutailipu, M.; Liu, Y.; Jiang, L.; Zhang, Y. Measurement and estimation of CO_2–brine interfacial tension and rock wettability under CO_2 sub- and super-critical conditions. *J. Colloid. Interface Sci.* **2019**, *534*, 605–617. [CrossRef] [PubMed]

67. Yekeen, N.; Padmanabhan, E.; Sevoo, T.A.L.; Kanesen, K.A.L.; Okunade, O.A. Wettability of rock/CO_2/brine systems: A critical review of influencing parameters and recent advances. *J. Ind. Eng. Chem.* **2020**, *88*, 1–28. [CrossRef]

68. Baban, A.; Al-Yaseri, A.; Keshavarz, A.; Amin, R.; Iglauer, S. CO_2—brine—sandstone wettability evaluation at reservoir conditions via Nuclear Magnetic Resonance measurements. *Int. J. Greenh. Gas Control* **2021**, *111*, 103435. [CrossRef]

69. Gholami, R.; Raza, A.; Andersen, P.; Escalona, A.; Cardozo, N.; Marín, D.; Sarmadivaleh, M. Long-term integrity of shaly seals in CO_2 geo-sequestration sites: An experimental study. *Int. J. Greenh. Gas Control* **2021**, *109*, 103370. [CrossRef]

70. Hashemi, L.; Glerum, W.; Farajzadeh, R.; Hajibeygi, H. Contact angle measurement for hydrogen/brine/sandstone system using captive-bubble method relevant for underground hydrogen storage. *Adv. Water Resour.* **2021**, *154*, 103964. [CrossRef]

71. Al-Yaseri, A.; Abbasi, G.R.; Yekeen, N.; Al-Shajalee, F.; Giwelli, A.; Xie, Q. Effects of cleaning process using toluene and acetone on water-wet-quartz/CO_2 and oil-wet-quartz/CO_2 wettability. *J. Pet. Sci. Eng.* **2022**, *208*, 109555. [CrossRef]

72. Baban, A.; Keshavarz, A.; Amin, R.; Iglauer, S. Impact of Wettability Alteration on CO_2 Residual Trapping in Oil-Wet Sandstone at Reservoir Conditions Using Nuclear Magnetic Resonance. *Energy Fuels* **2022**, *36*, 13722–13731. [CrossRef]

73. Qin, C.; Jiang, Y.; Zhou, J.; Zuo, S.; Chen, S.; Liu, Z.; Yin, H.; Li, Y. Influence of supercritical CO_2 exposure on water wettability of shale: Implications for CO_2 sequestration and shale gas recovery. *Energy* **2022**, *242*, 122551. [CrossRef]

74. Song, J.Y.; Jeong, Y.J.; Yun, T.S. Effects of Anisotropy and CO_2 Wettability on CO_2 Storage Capacity in Sandstone. *Geofluids* **2022**, *2022*, 5900255. [CrossRef]

75. Al-Khdheeawi, E.A.; Vialle, S.; Barifcani, A.; Sarmadivaleh, M.; Iglauer, S. Effect of wettability heterogeneity and reservoir temperature on CO_2 storage efficiency in deep saline aquifers. *Int. J. Greenh. Gas Control* **2018**, *68*, 216–229. [CrossRef]

76. Iglauer, S.; Pentland, C.H.; Busch, A. CO_2 wettability of seal and reservoir rocks and the implications for carbon geo-sequestration. *Water Resour. Res.* **2015**, *51*, 729–774. [CrossRef]

77. Rahman, T.; Lebedev, M.; Barifcani, A.; Iglauer, S. Residual trapping of supercritical CO_2 in oil-wet sandstone. *J. Colloid. Interface Sci.* **2016**, *469*, 63. [CrossRef]

78. Espinoza, D.N.; Santamarina, J.C. CO_2 breakthrough—Caprock sealing efficiency and integrity for carbon geological storage. *Int. J. Greenh. Gas Control* **2017**, *66*, 218–229. [CrossRef]

79. Stavropoulou, E.; Laloui, L. Evaluating CO_2 breakthrough in a shaly caprock material: A multi-scale experimental approach. *Sci. Rep.* **2022**, *12*, 10706. [CrossRef]

80. Gaus, I.; Azaroual, M.; Czernichowski-Lauriol, I. Reactive transport modelling of the impact of CO_2 injection on the clayey cap rock at Sleipner (North Sea). *Chem. Geol.* **2005**, *217*, 319–337. [CrossRef]

81. Gaus, I. Role and impact of CO_2–rock interactions during CO_2 storage in sedimentary rocks. *Int. J. Greenh. Gas Control* **2010**, *4*, 73–89. [CrossRef]

82. Rosenbauer, R.J.; Koksalan, T.; Palandri, J.L. Experimental investigation of CO_2-brine-rock interactions at elevated temperature and pressure: Implications for CO_2 sequestration in deep-saline aquifers. *Fuel Process. Technol.* **2005**, *86*, 1581–1597. [CrossRef]

83. Espinoza, D.N.; Kim, S.H.; Santamarina, J.C. CO_2 geological storage—Geotechnical implications. *KSCE J. Civ. Eng.* **2011**, *15*, 707–719. [CrossRef]

84. Palandri, J.L.; Kharaka, Y.K. *A Compilation of Rate Parameters of Water—Mineral Interaction Kinetics for Application to Geochemical Modeling*; U.S. Geological Survey: Reston, VA, USA, 2004.

85. Pokrovsky, O.S.; Golubev, S.V.; Schott, J.; Castillo, A. Calcite, dolomite and magnesite dissolution kinetics in aqueous solutions at acid to circumneutral pH, 25 to 150 °C and 1 to 55 atm pCO_2: New constraints on CO_2 sequestration in sedimentary basins. *Chem. Geol.* **2009**, *260*, 317. [CrossRef]

86. Golubev, S.V.; Bénézeth, P.; Schott, J.; Dandurand, J.L.; Castillo, A. Siderite dissolution kinetics in acidic aqueous solutions from 25 to 100 °C and 0 to 50 atm pCO_2. *Chem. Geol.* **2009**, *265*, 13–19. [CrossRef]

87. Dresel, P.E. The Dissolution Kinetics of Siderite and Its Effect on Acid Mine Drainage. Ph.D. Thesis, The Pennsylvania State University, State College, PA, USA, 1989.

88. Gunter, W.D.; Wiwehar, B.; Perkins, E.H. Aquifer disposal of CO_2-rich greenhouse gases: Extension of the time scale of experiment for CO_2-sequestering reactions by geochemical modelling. *Mineral. Petrol.* **1997**, *59*, 121–140. [CrossRef]

89. Seyyedi, M.; Mahmud, H.K.B.; Verrall, M.; Giwelli, A.; Esteban, L.; Ghasemiziarani, M.; Clennell, B. Pore Structure Changes Occur During CO_2 Injection into Carbonate Reservoirs. *Sci. Rep.* **2020**, *10*, 3624. [CrossRef]

90. Wang, H.; Alvarado, V.; Smith, E.R.; Kaszuba, J.P.; Bagdonas, D.A.; McLaughlin, J.F.; Quillinan, S.A. Link Between CO_2-Induced Wettability and Pore Architecture Alteration. *Geophys. Res. Lett.* **2020**, *47*. [CrossRef]

91. Kim, K.; Kundzicz, P.M.; Makhnenko, R.Y. Effect of CO_2 Injection on the Multiphase Flow Properties of Reservoir Rock. *Transp. Porous Media* **2023**, *147*, 429–461. [CrossRef]

92. Mouzakis, K.M.; Navarre-Sitchler, A.K.; Rother, G.; Bañuelos, J.L.; Wang, X.; Kaszuba, J.P.; Heath, J.E.; Miller, Q.R.S.; Alvarado, V.; McCray, J.E. Experimental Study of Porosity Changes in Shale Caprocks Exposed to CO_2-Saturated Brines I: Evolution of Mineralogy, Pore Connectivity, Pore Size Distribution, and Surface Area. *Environ. Eng. Sci.* **2016**, *33*, 725–735. [CrossRef]

93. Pan, Y.; Hui, D.; Luo, P.; Zhang, Y.; Sun, L.; Wang, K. Experimental Investigation of the Geochemical Interactions between Supercritical CO_2 and Shale: Implications for CO_2 Storage in Gas-Bearing Shale Formations. *Energy Fuels* **2018**, *32*, 1963–1978. [CrossRef]

94. Sanguinito, S.; Goodman, A.; Tkach, M.; Kutchko, B.; Culp, J.; Natesakhawat, S.; Fazio, J.; Fukai, I.; Crandall, D. Quantifying dry supercritical CO_2-induced changes of the Utica Shale. *Fuel* **2018**, *226*, 54–64. [CrossRef]

95. Goodman, A.; Sanguinito, S.; Tkach, M.; Natesakhawat, S.; Kutchko, B.; Fazio, J.; Cvetic, P. Investigating the role of water on CO_2-Utica Shale interactions for carbon storage and shale gas extraction activities—Evidence for pore scale alterations. *Fuel* **2019**, *242*, 744–755. [CrossRef]

96. Fatah, A.; Mahmud, H.B.; Bennour, Z.; Hossain, M.; Gholami, R. Effect of supercritical CO_2 treatment on physical properties and functional groups of shales. *Fuel* **2021**, *303*, 121310. [CrossRef]

97. Medina, B.X.; Kohli, A.; Kovscek, A.R.; Alvarado, V. Effects of Supercritical CO_2 Injection on the Shale Pore Structures and Mass Transport Rates. *Energy Fuels* **2023**, *37*, 1151–1168. [CrossRef]

98. Wang, S.; Zhou, S.; Pan, Z.; Elsworth, D.; Yan, D.; Wang, H.; Liu, D.; Hu, Z. Response of pore network fractal dimensions and gas adsorption capacities of shales exposed to supercritical CO_2: Implications for CH_4 recovery and carbon sequestration. *Energy Rep.* **2023**, *9*, 6461–6485. [CrossRef]

99. Miller, Q.R.S.; Wang, X.; Kaszuba, J.P.; Mouzakis, K.M.; Navarre-Sitchler, A.K.; Alvarado, V.; McCray, J.E.; Rother, G.; Bañuelos, J.L.; Heath, J.E. Experimental Study of Porosity Changes in Shale Caprocks Exposed to Carbon Dioxide-Saturated Brine II: Insights from Aqueous Geochemistry. *Environ. Eng. Sci.* **2016**, *33*, 736–744. [CrossRef]

100. Cruz, F.; Tinni, A.; Brackeen, S.; Sondergeld, C.; Rai, C. Impact of Capillary Pressure on Total Porosity in Unconventional Shales. In Proceedings of the SPE/AAPG/SEG Unconventional Resources Technology Conference, Online, 20–22 July 2020.

101. Sondergeld, C.H.; Rai, C.S. A New Exploration Tool: Quantitative Core Characterization. In *Experimental Techniques in Mineral and Rock Physics: The Schreiber Volume*; Liebermann, R.C., Sondergeld, C.H., Eds.; Birkhäuser Basel: Basel, Switzerland, 1994; pp. 249–268.

102. Ballard, B. Quantitative Mineralogy of Reservoir Rocks Using Fourier Transform Infrared Spectroscopy. In Proceedings of the SPE Annual Technical Conference and Exhibition, Anaheim, CA, USA, 1 January 2007; p. 8.

103. Curtis, M.; Mamoudou, S.; Cruz, F.; Dang, S.; Rai, C.; Devegowda, D. Effects of CO_2 Exposure on Unconventional Reservoir Rock Microstructure. In Proceedings of the SPE/AAPG/SEG Unconventional Resources Technology Conference, Denver, CO, USA, 13–15 June 2023.

104. Rathnaweera, T.D.; Ranjith, P.G.; Perera, M.S.A. Experimental investigation of geochemical and mineralogical effects of CO_2 sequestration on flow characteristics of reservoir rock in deep saline aquifers. *Sci. Rep.* **2016**, *6*, 19362. [CrossRef] [PubMed]

105. Sokama-Neuyam, Y.A.; Boakye, P.; Aggrey, W.N.; Obeng, N.O.; Adu-Boahene, F.; Woo, S.H.; Ursin, J.R. Theoretical Modeling of the Impact of Salt Precipitation on CO_2 Storage Potential in Fractured Saline Reservoirs. *ACS Omega* **2020**, *5*, 14776–14785. [CrossRef]

106. Vilarrasa, V.; Rutqvist, J. Thermal effects on geologic carbon storage. *Earth-Sci. Rev.* **2017**, *165*, 245–256. [CrossRef]

107. Comisky, J.T.; Santiago, M.; McCollom, B.; Buddhala, A.; Newsham, K.E. Sample Size Effects on the Application of Mercury Injection Capillary Pressure for Determining the Storage Capacity of Tight Gas and Oil Shales. In Proceedings of the Canadian Unconventional Resources Conference, Calgary, AB, Canada, 1 January 2011; p. 23.

108. Wang, R.; Sang, S.; Zhu, D.; Liu, S.; Yu, K. Pore characteristics and controlling factors of the Lower Cambrian Hetang Formation shale in Northeast Jiangxi, China. *Energy Explor. Exploit.* **2017**, *36*, 43–65. [CrossRef]

109. Tian, Y.; Chen, Q.; Yan, C.; Deng, H.; He, Y. Classification of Adsorption Isotherm Curves for Shale Based on Pore Structure. *Petrophysics* **2020**, *61*, 417–433. [CrossRef]

110. Fuhua, S.; Ke, M.; Yanming, Z.H.U.; Meng, W.; Xin, T.; Yang, W.; Haitao, G.A.O.; Guangjun, F.; Wentian, M.I. Pore structure, adsorption capacity and their controlling factors of shale in complex structural area. *Méitàn Kēxué Jìshù* **2023**, *51*, 269–282. [CrossRef]

111. Thommes, M.; Kaneko, K.; Neimark, A.V.; Olivier, J.P.; Rodriguez-Reinoso, F.; Rouquerol, J.; Sing, K.S.W. Physisorption of gases, with special reference to the evaluation of surface area and pore size distribution (IUPAC Technical Report). *Pure Appl. Chem.* **2015**, *87*, 1051–1069. [CrossRef]

112. Sing, K.S.W. Reporting physisorption data for gas/solid systems with special reference to the determination of surface area and porosity (Recommendations 1984). *Pure Appl. Chem.* **1985**, *57*, 603–619. [CrossRef]

113. Dang, S.; Mukherjee, S.; Sondergeld, C.; Rai, C. Measurement of Effective Tortuosity in Unconventional Tight Rock using Nuclear Magnetic Resonance. In Proceedings of the SPE/AAPG/SEG Unconventional Resources Technology Conference, Online, 26–28 July 2021.

114. Odiachi, J.; Cruz, F.; Tinni, A. Diffusional and Electrical Tortuosity in Unconventional Shale Reservoirs. In Proceedings of the SPE Annual Technical Conference and Exhibition, Houston, TX, USA, 3–5 October 2022.

115. Tinni, A.; Odusina, E.; Sulucarnain, I.; Sondergeld, C.; Rai, C.S. Nuclear-Magnetic-Resonance Response of Brine, Oil, and Methane in Organic-Rich Shales. *SPE Reserv. Eval. Eng.* **2015**, *18*, 400–406. [CrossRef]

116. Mukherjee, S.; Dang, S.T.; Rai, C.; Sondergeld, C. Revisiting the Concept of Wettability for Organic-Rich Tight Rocks: Application in Formation Damage-Water Blockage. *Petrophysics* **2020**, *61*, 473–481. [CrossRef]

117. Wigand, M.; Carey, J.W.; SchÜTt, H.; Spangenberg, E.; Erzinger, J. Geochemical effects of CO_2 sequestration in sandstones under simulated in situ conditions of deep saline aquifers. *Appl. Geochem.* **2008**, *23*, 2735–2745. [CrossRef]

118. Lu, J.; Kharaka, Y.K.; Thordsen, J.J.; Horita, J.; Karamalidis, A.; Griffith, C.; Hakala, J.A.; Ambats, G.; Cole, D.R.; Phelps, T.J.; et al. CO_2–rock–brine interactions in Lower Tuscaloosa Formation at Cranfield CO_2 sequestration site, Mississippi, USA. *Chem. Geol.* **2012**, *291*, 269–277. [CrossRef]

119. Shi, Z.; Sun, L.; Haljasmaa, I.; Harbert, W.; Sanguinito, S.; Tkach, M.; Goodman, A.; Tsotsis, T.T.; Jessen, K. Impact of Brine/CO_2 exposure on the transport and mechanical properties of the Mt Simon sandstone. *J. Pet. Sci. Eng.* **2019**, *177*, 295–305. [CrossRef]

120. Kim, K.; Makhnenko, R.Y. Short- and Long-Term Responses of Reservoir Rock Induced by CO_2 Injection. *Rock. Mech. Rock. Eng.* **2022**, *55*, 6605–6625. [CrossRef]

Article

Carbon Emission Prediction and the Reduction Pathway in Industrial Parks: A Scenario Analysis Based on the Integration of the LEAP Model with LMDI Decomposition

Dawei Feng [1], Wenchao Xu [1], Xinyu Gao [2], Yun Yang [1], Shirui Feng [1], Xiaohu Yang [2,*] and Hailong Li [3,*]

[1] China Energy Engineering Group Jiangsu Power Design Institute Co., Ltd., Nanjing 210012, China; fengdawei@jspdi.com.cn (D.F.); xuwenchao@jspdi.com.cn (W.X.); yangyun@jspdi.com.cn (Y.Y.); fengshirui@jspdi.com.cn (S.F.)

[2] Institute of the Building Environment & Sustainability Technology, School of Human Settlements and Civil Engineering, Xi'an Jiaotong University, Xi'an 710049, China; gaoxinyu@stu.xjtu.edu.cn

[3] School of Business Society and Engineering, Mälardalen University, 721 23 Västerås, Sweden

* Correspondence: xiaohuyang@xjtu.edu.cn (X.Y.); hailong.li@mdh.se (H.L.)

Abstract: Global climate change imposes significant challenges on the ecological environment and human sustainability. Industrial parks, in line with the national climate change mitigation strategy, are key targets for low-carbon revolution within the industrial sector. To predict the carbon emission of industrial parks and formulate the strategic path of emission reduction, this paper amalgamates the benefits of the "top-down" and "bottom-up" prediction methodologies, incorporating the logarithmic mean divisia index (LMDI) decomposition method and long-range energy alternatives planning (LEAP) model, and integrates the Tapio decoupling theory to predict the carbon emissions of an industrial park cluster of an economic development zone in Yancheng from 2020 to 2035 under baseline (BAS) and low-carbon scenarios (LC1, LC2, and LC3). The findings suggest that, in comparison to the BAS scenario, the carbon emissions in the LC1, LC2, and LC3 scenarios decreased by 30.4%, 38.4%, and 46.2%, respectively, with LC3 being the most suitable pathway for the park's development. Finally, the paper explores carbon emission sources, and analyzes emission reduction potential and optimization measures of the energy structure, thus providing a reference for the formulation of emission reduction strategies for industrial parks.

Keywords: carbon emissions; industrial parks; scenario analysis; LEAP model; LMDI model; Tapio decoupling theory

Citation: Feng, D.; Xu, W.; Gao, X.; Yang, Y.; Feng, S.; Yang, X.; Li, H. Carbon Emission Prediction and the Reduction Pathway in Industrial Parks: A Scenario Analysis Based on the Integration of the LEAP Model with LMDI Decomposition. *Energies* **2023**, *16*, 7356. https://doi.org/10.3390/en16217356

Academic Editor: Kun Mo Lee

Received: 1 August 2023
Revised: 19 September 2023
Accepted: 20 September 2023
Published: 31 October 2023

1. Introduction

Since the Industrial Revolution, the high-carbon industrial modus operandi, characterized by significant natural resource consumption and vast greenhouse gas emissions, has escalated numerous environmental issues [1–4]. Climate change, environmental pollution, and energy security have emerged as global challenges, posing severe threats to human survival and development [5,6]. In such a milieu, the globe has reached the consensus to develop a low-carbon economy; it is also one of China's critical measures to address climate change issues [7,8]. Presently, China is amidst rapid urbanization and industrialization, making the industry the chief consumer of energy and producer of carbon emissions. Among them, energy consumption in the industrial sector accounts for 65% of the country's energy consumption, and carbon emissions are as high as 70%, which means the country has an enormous potential for energy conservation and emission reduction [9,10].

Industrial parks are both carriers of industrialization and crucial units for implementing low-carbon emission reduction goals [11]. Corresponding to the national climate change mitigation strategy, industrial parks are being earmarked as key targets for green transformation in the industrial sector. Therefore, macro strategy adjustment and micro technology optimization are getting more and more attention in industrial park emission reduction. Various scholars have evaluated emission reduction strategies for industrial parks, exploring avenues such as improvements in industrial development [12–15] and emission reduction technologies [16–18]. Chen et al. [19] proposed a composite evaluation method to understand the impact of emissions on energy consumption. Wei et al. [20] deliberated the feasibility and implementation of an electric-thermal carbon-neutral industrial park and pursued economic, energy, and environmental analyses across different scenarios.

Undeniably, predicting carbon emissions provides the foundational basis for formulating emission reduction strategies within a set timeline, chiefly within the economy–energy–environment system of a park [21]. Existing prediction models primarily utilize either the "top-down" or "bottom-up" models, typically applicable at country, province, city, or industry scales [22–25]. Implementing these models at a park level is a future challenge and necessity. The "top-down" macro forecasting approach analyzes the impact of changes in population, economy, and other social aspects on regional carbon emissions, though offering limited detail on individual energy consumption sectors and energy varieties. Representative models include the computable general equilibrium (CGE) [26,27], stochastic impacts by regression on population affluence and technology (STIRPAT) [28,29], and the logarithmic mean divisia index (LMDI) models [30–32]. Usman et al. [33] applied the STIRPAT model to explore the correlation between nuclear energy and the ecological footprint. In 1989, Kaya introduced the Kaya identity, which segmented carbon emissions into four macro-influencing factors. Subsequently, numerous scholars have analyzed the influence of these factors on carbon emissions within countries [34], cities [35], heating systems [36], and more using LMDI decomposition.

On the contrary, the "bottom-up" method emulates carbon emissions from different processes such as energy transportation and consumption, yet struggles to encapsulate the impact of macro conditions on carbon emissions. Noteworthy models include the long-range energy alternatives planning system (LEAP) [37] and MARKAL [38]. Countries [39], provinces, cities [40], buildings [41], and energy generation sectors [42] often use the LEAP model to calculate carbon emissions. Zou et al. [43] and Duan et al. [44] employed the LEAP model to investigate carbon emission pathways and formulate climate change mitigation strategies, respectively. Given the preceding research and model characteristics, LMDI decomposition can accommodate multi-scale carbon emissions, while the LEAP model's long-term energy system planning capability makes it suited for formulating emission reduction strategies. Consequently, this study adopts the LEAP and LMDI models to devise a hybrid framework for predicting and calculating park carbon emissions. The LEAP model and LMDI decomposition hybrid analysis framework combines the advantages of the two models, which can reflect the contribution of various types of commercial developments and the application of energy technology to carbon emissions at the micro level, reflect the impact of policy variation on carbon emissions at the macro level, and expand the calculation of carbon emissions to the industrial park scale, consequently providing a reference for the development of emission reduction strategies in parks.

In conclusion, research on carbon emission prediction and calculation at an industrial park level is sparse, and existing forecasting models fall short in providing both macro- and micro-perspective analyses. Bridging this gap, this study combines the strengths of "top-down" and "bottom-up" prediction methods to establish a LEAP–LMDI hybrid analysis framework, incorporating multiple data analysis approaches including the Tapio decoupling theory. It also takes into account diverse low-carbon scenarios such as macro-control and micro-technology, to simulate carbon emissions and emission structures in various scenarios within an industrial park cluster of an economic development zone in Yancheng from 2020 to 2035. This provides a reference point for the industry in their evaluation of park emission reduction strategies.

The paper is organized as follows: Section 1 summarizes the research on and literature review of carbon emission prediction and calculation. The methodology used in this research is introduced in Section 2. Section 3 describes the scenario setting and the analyses of LMDI decomposition and decoupling theory. Section 4 presents the results and detailed discussion of the carbon emissions forecast and pathways for low-carbon industrial park development. Conclusions are shown in Section 4.

2. Methodology

2.1. Analysis Framework

Taking an industrial park cluster of an economic development zone in Yancheng as the subject of study, this research assesses and forecasts carbon dioxide emissions associated with macroeconomic strategies and micro-energy activities from 2020 to 2035. We construct a hybrid analysis framework to investigate the park's path towards reducing emissions, as represented in Figure 1. This framework is principally integrated with logarithmic mean divisia index (LMDI) decomposition and the long-range energy alternatives planning (LEAP) system, leveraging various data processing methods and Tapio decoupling theory. This integration enables the functionality of predicting regional carbon emission trends, assessing energy-saving capabilities, and forming emission reduction strategies. The framework comprises four primary modules: (1) the Parameter Acquisition Module, (2) the Driving Factor Analysis Module, (3) the Carbon Emission Simulation Module, and (4) the Assessment Module.

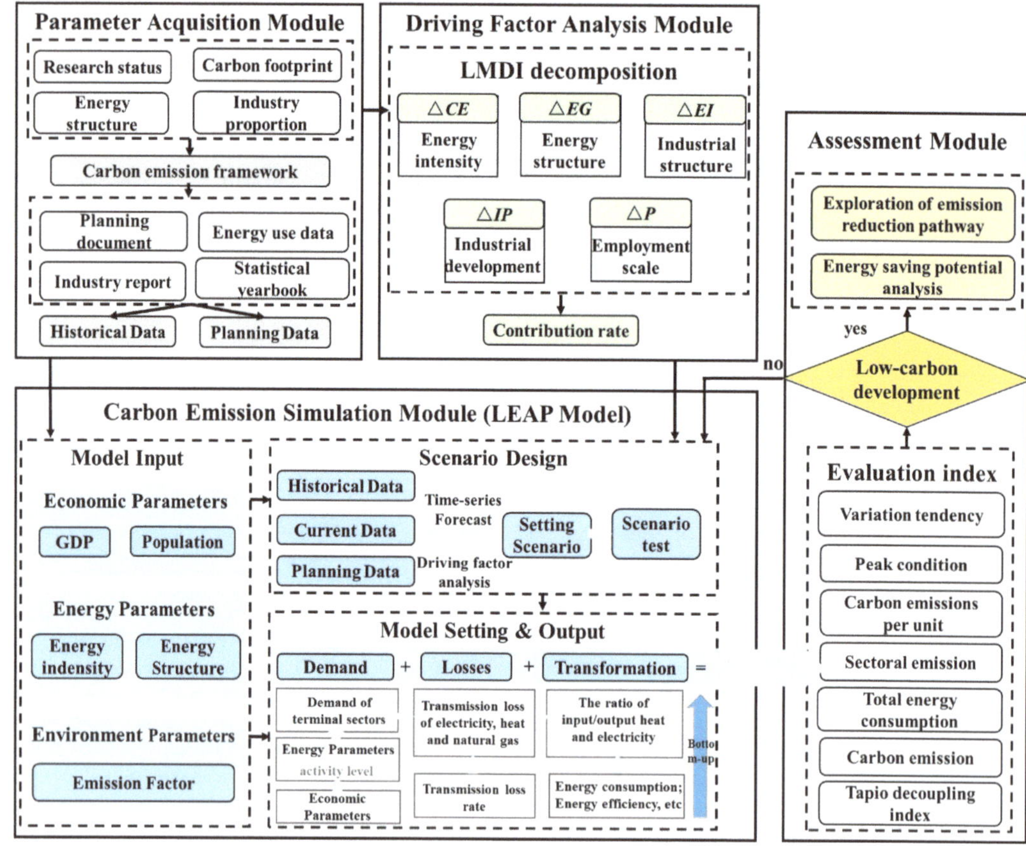

Figure 1. Analysis framework of the research.

First, the industrial park's accounting boundaries are determined, and an emission calculation framework is created. Subsequently, we obtain input parameters from historical records and planning documents, improve the Kaya identity equation, and utilize the LMDI model to analyze the driving factors contributing to the park's carbon emissions. Furthermore, we quantify the contribution rate of various factors, feeding these statistics into the subsequent LEAP model's scenario settings. The Carbon Emission Simulation Module, integral to our hybrid optimization framework, combines data from the two previous modules to set macro, energy, and environmental parameters within the LEAP model. Scenarios considering activity level, energy demand, transformation and losses are designed, and resultant carbon emission trends are simulated. In conclusion, using evaluation indicators such as the carbon emission output of the LEAP model and the decoupling theory, we ascertain whether the scenario aligns with development needs and low-carbon construction. This process aids in determining future development pathways for the park.

2.2. LEAP Model

The extensively used LEAP model facilitates the assessment of energy policies and climate change mitigation. Using scenario analysis, a critical method within the LEAP model, we measure regional carbon emissions and explore the effects of macro policies and technical tools on carbon dioxide mitigation actions. The LEAP model simulates energy demand, transformation, and corresponding carbon emissions using the given setting scenario, taking the form of [45]:

$$C_{total} = CD + CT = \sum_i \sum_j A_{j,i} \times E_{j,i} \times EF_j + \sum_n \sum_j \frac{TE_{n,j}}{r} \times EF_j \tag{1}$$

where C_{total} presents total carbon emissions; CD and CT denote the carbon emissions of energy demand and transformation, respectively; A stands for the activity level; E denotes the energy consumption per unit activity level; EF represents the emission factor, which is displayed in Table 1; TE indicates the secondary energy production from the transformation output; r accounts for the conversion efficiency; i is different energy consumption sectors, j denotes different fuel types; and m means the type of energy produced.

Table 1. CO_2 emission factors of different fuel types.

Fuel	CO_2 Emission Factor
Coal bituminous	2.78 kgCO$_2$/kg
Coal sub bituminous	3.09 kgCO$_2$/kg
Petroleum coke	3.45 kgCO$_2$/kg
Crude oil	3.32 kgCO$_2$/L
Gasoline	3.04 kgCO$_2$/L
Diesel	3.41 kgCO$_2$/L
LPG	3.41 kgCO$_2$/L
Natural gas	2.38 kgCO$_2$/m^3

2.3. Decomposition Model

The decomposition model used in this research is based on the Kaya identity equation [46]. It aims to identify the driving factors of carbon emissions in industrial parks and decompose these emissions using the logarithmic mean divisia index (LMDI) method. The Kaya identity equation is commonly used in large-scale studies of countries and cities, and it considers various factors such as the industrial structure of energy consumption, the gross product of parks, and the employed population. In this study, we replace the gross domestic product (GDP) with the gross product of parks (*EI*), the population with the employed population (*P*), and introduce the industrial added value (*EG*) to describe the industrial structure. Equation (2) shows the decomposition of carbon emissions from park energy activities into five driving factors: energy intensity (ΔCE), energy structure (ΔEG), industrial structure (ΔEI), industrial economic development (ΔIP), and employment scale (ΔP).

$$C = \sum_i C_i = \sum_i \left(\frac{C_i}{E_i} \times \frac{E_i}{EG_i} \times \frac{EG_i}{IG} \times \frac{IG}{P} \times P \right) = \sum_i (CE_i \times EG_i \times EI_i \times IP \times P) \quad (2)$$

where C represents the carbon emissions of energy consumption in the park, E stands for the consumption of various energy sources, EG is the added value of industries, IG accounts for the gross product of the park, P denotes the employed population of the park, and i represents different industries.

According to the additive decomposition principle of the LMDI model, the carbon emission influence of each driving factor from the base year (year 0) to the target year (year t) is displayed in Equations (3)–(8). The carbon emission contribution rate represents the ratio of the carbon emission influence of each driving factor to the carbon emission increment (ΔC).

$$\Delta C = \sum_i (C_i^t - C_i^0) = \sum_i (\Delta CE_i + \Delta EG_i + \Delta EI_i + \Delta IP + \Delta P) \quad (3)$$

$$\Delta CE = \sum_i \left(\frac{C_i^t - C_i^0}{\ln C_i^t - \ln C_i^0} \times \ln \frac{CE_i^t}{CE_i^0} \right) \quad (4)$$

$$\Delta EG = \sum_i \left(\frac{C_i^t - C_i^0}{\ln C_i^t - \ln C_i^0} \times \ln \frac{EG_i^t}{EG_i^0} \right) \quad (5)$$

$$\Delta EI = \sum_i \left(\frac{C_i^t - C_i^0}{\ln C_i^t - \ln C_i^0} \times \ln \frac{EI_i^t}{EI_i^0} \right) \quad (6)$$

$$\Delta IP = \sum_i \left(\frac{C_i^t - C_i^0}{\ln C_i^t - \ln C_i^0} \times \ln \frac{IP_i^t}{IP_i^0} \right) \quad (7)$$

$$\Delta P = \sum_i \left(\frac{C_i^t - C_i^0}{\ln C_i^t - \ln C_i^0} \times \ln \frac{P_i^t}{P_i^0} \right) \quad (8)$$

2.4. Tapio Decoupling Theory

To verify the rationality of the scenario setting and the stability of peak results, the Tapio decoupling index is employed to explore the short-term changes in carbon emissions and economic development in industrial parks [47]. In simple terms, the decoupling index is a relative rate of change that reflects the ratio between the relative change degree of carbon emissions and the relative change degree of economic indicators in a certain period. The specific expression of the decoupling model is shown in Equation (9), and the decoupling types are summarized in Table 2.

$$\varepsilon_{(C,GDP)} = \frac{\Delta C/C}{\Delta IG/IG} \quad (9)$$

2.5. Data Sources

The researched industrial park cluster established a planning system in 2016, which mainly consists of five industries: the chemical industry (CH), steel industry (ST), high-tech equipment manufacturing industry (HI), pharmaceutical industry (PH), and paper industry (PA). The historical data of the park include the gross product, industrial proportion, and energy-use information from 2020 to 2022. The infrastructure construction scale and settlement of enterprises in the park from 2023 to 2025 are complete and detailed. Based on this, a scenario setting and carbon emission prediction for the park from 2023 to 2035 are carried out. The research data mainly include the gross production value, industrial added value, employed population, and different types of energy consumption in the historical, base, and forecast years. These data are obtained from park planning documents, statistical yearbooks, and industry development reports.

Table 2. Tapio decoupling index summary of decoupling types.

Category	ΔC	ΔIG	Decoupling Index
Expansion negative decoupling (EN)	>0	>0	$\varepsilon > 1.2$
Strong negative decoupling (SN)	>0	<0	$\varepsilon < 0$
Weak negative decoupling (WN)	<0	<0	$0 < \varepsilon < 0.8$
Recessionary decoupling (R)	<0	<0	$\varepsilon > 1.2$
Strong decoupling (S)	<0	>0	$\varepsilon < 0$
Weak decoupling (W)	>0	>0	$0 < \varepsilon < 0.8$
Growth linkage (GL)	>0	>0	$0.8 < \varepsilon < 1.2$

3. Model Analysis and Scenario Setting

3.1. Analysis of Driving Factors

The contribution rates of carbon emissions from different driving factors and the industrial effect are presented in Figure 2, which is determined through the decomposition of carbon emissions from 2020 to 2022 using the LMDI model. It can be seen that only energy intensity and energy structure have a negative contribution to CO_2 emissions, and other factors have a significant promoting effect on CO_2 emissions. This is because the construction of the industrial park cluster is in its initial stage, the carbon emissions of the industrial park and economic benefits have not been decoupled, and the contribution rate of the employment population and per capita GDP is positive and occupies a large proportion. Furthermore, with the construction of high energy consumption industries, the contribution of industry ratio is also positive. However, because the chemical industry, steel and other industries should strive for an energy intensity and energy consumption structure at the industry benchmark level, some process optimization has been carried out, so the contribution rate of these two items is negative. The contribution rates of the employment population and industrial economic development are significantly higher than those of other factors, indicating that the development of the park is closely linked to the economy. Additionally, the positive contribution rate of the industrial structure can be attributed to the growth of the steel industry and chemical industry. On the other hand, the contribution rates of energy consumption intensity and energy consumption structure are negative, aligning with the park's plan to improve energy utilization efficiency and optimize energy structure. Although the contribution rate is small, there is potential to further increase technology investment to achieve energy conservation and emission reduction targets. Furthermore, when considering the contribution rates of carbon emissions from different industries, the steel industry has the largest effect at 63%, followed by the chemical industry at approximately 22%. In contrast, the contribution rates of other high-tech industries in the park are relatively small. Consequently, the park should focus on adjusting the steel industry, utilizing technical methods to improve the quality of enterprises with high energy consumptions, optimizing process flows, and promoting industrial chain circulation.

3.2. Scenario Setting

By quantifying the influencing factors of the park's gross product, energy structure, energy intensity, and industrial structure based on the driving factor analysis results and the development plan of the industrial park, six scenarios are devised and illustrated in the upper part of Figure 3. These scenarios involve optimizing the industrial chain, adjusting the energy structure, and incorporating energy technology, and the stars are used to mark development measures adopted. The scenarios are as follows: the baseline scenario (BAS) serves as a reference scenario based on short-term planning without any macro-control or energy-saving technical means, while the low-carbon pathway scenarios 1, 2, 3, 4, and 5 (LP1, LP2, LP3, LP4, and LP5) explore various strategies. LP1 involves macro-control on the BAS scenario, reducing the economic growth rate of the park, enhancing high-quality green development, limiting the growth rate of the steel and chemical industries, focusing on adjusting the industrial chain of the steel industry, and optimizing the industrial ratio of the park planning. LP4 is similar to LP1, but with the economic growth rate of the park aligning with the BAS scenario. LP2 adjusts the industrial structure of the park, improves the technology used in high-energy consumption industries, and gradually introduces energy storage technology in 2025 to decrease energy intensity. Likewise, LP5 is similar to LP2, but maintains the industrial structure of the BAS scenario and does not introduce energy storage technology. In the LP3 scenario, the gross product growth rate increases in the preliminary stage, and the renewable energy structure is adjusted. However, the gross product growth rate decreases in the later stage to account for economic development. Moreover, the LP3 scenario introduces renewable energy utilization devices such as solar and geothermal energy, as well as carbon capture devices in 2030, which is a necessary technological means to achieve carbon neutrality [48–52].

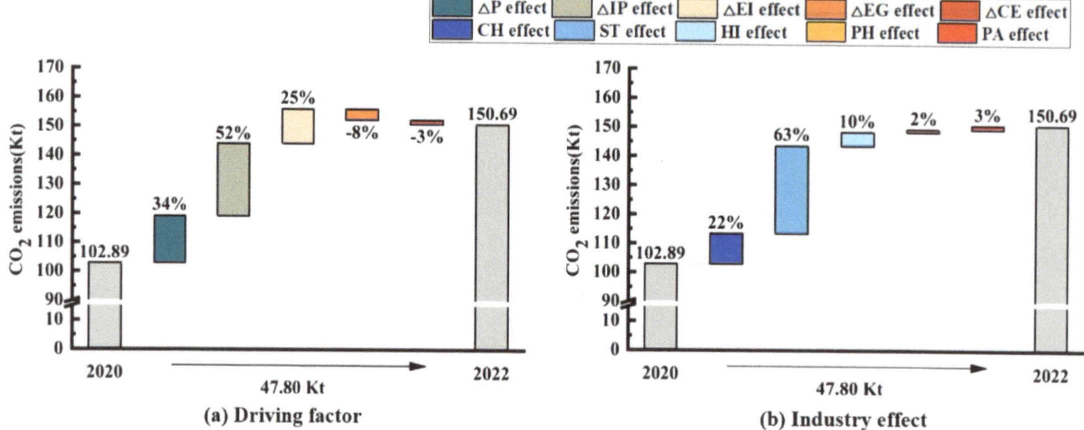

Figure 2. Carbon emissions during the period of 2020–2022: (**a**) Driving factor (**b**) Industry effect.

3.3. Decoupling Analysis

The decoupling index, displayed in Figure 4 for the decoupling categories of the six scenarios, depicts the relationship between annual carbon emissions and economic growth, which is determined by the Tapio decoupling theory in Section 2.4. During the initial stage of industrial park development from 2020 to 2025, the main trend is growth linkage. Both the BAS scenario and LP5 scenario continue to exhibit growth linkage throughout the planning period, which clearly contradicts the requirements of low-carbon development. The LP4 scenario is expected to transition to weak decoupling in 2032, while the LP1, LP2, and LP3 scenarios are projected to stabilize in a relatively ideal state of weak decoupling and strong decoupling after 2025. Therefore, these three scenarios are selected for further output calculations and comparison with the BAS scenario, as indicated on the right side of Figure 3.

Scenario setting		Main index	Scenario					
			BAS	LC1	LC2	LC3	LC4	LC5
Benchmark setting		Current macro policy and technology level	★	★	★	★	★	★
Macro-control	Economic growth rate	The gross domestic product of the park increases first and then decreases, and the gross domestic product in 2035 is consistent with that in the BAS scenario.				★		
		The gross domestic product of the park has slowed down, and green and coordinated development has been achieved.		★		★		
	Industrial structure	Optimize the industrial structure and limit the growth of industries with high energy consumption, such as steel and chemical industries.		★	★		★	
Micro-technology	Optimization	Optimize industrial processes reduce energy consumption intensity, and introduce energy storage devices.			★			★
	Strengthening	Optimize the industrial process, reduce the intensity of energy consumption, and strengthen the utilization of renewable energy devices to introduce efficient terminals.				★		

Tapio decoupling theory

Scenario	Description	Year	GDP (Hundred million yuan)	Industry ratio CH:ST:HI:PH:PA (%)	Total energy consumption (KtCe)
BAS	Reference scenario, the industrial structure and energy structure remain unchanged, and energy consumption increases naturally	2020	2.9	25:10:60:3:2	39.57
		2025	8.1	15:40:40:3:2	86.20
		2030	13.1	15:40:40:3:2	123.60
		2035	17.5	15:40:40:3:2	137.18
LC1	Optimize the industrial structure and reduce the gross domestic product of the park	2020	2.9	25:10:60:3:2	39.57
		2025	8.1	15:40:40:3:2	86.20
		2030	12.0	12:35:45:5:3	91.86
		2035	15.7	10:30:50:5:5	94.08
LC2	Optimize the industrial structure and reduce the energy consumption intensity of the park	2020	2.9	25:10:60:3:2	39.57
		2025	8.1	15:40:40:3:2	86.20
		2030	13.1	12:35:45:5:3	87.32
		2035	17.5	10:30:50:5:5	83.06
LC3	Improve the industrial structure, adjust gross domestic product growth and the energy structure, and reduce energy intensity	2020	2.9	25:10:60:3:2	39.57
		2025	8.1	15:40:40:3:2	86.20
		2030	14.1	12:35:45:5:3	93.04
		2035	17.5	10:30:50:5:5	97.42

Figure 3. Scenario setting, selection, and details.

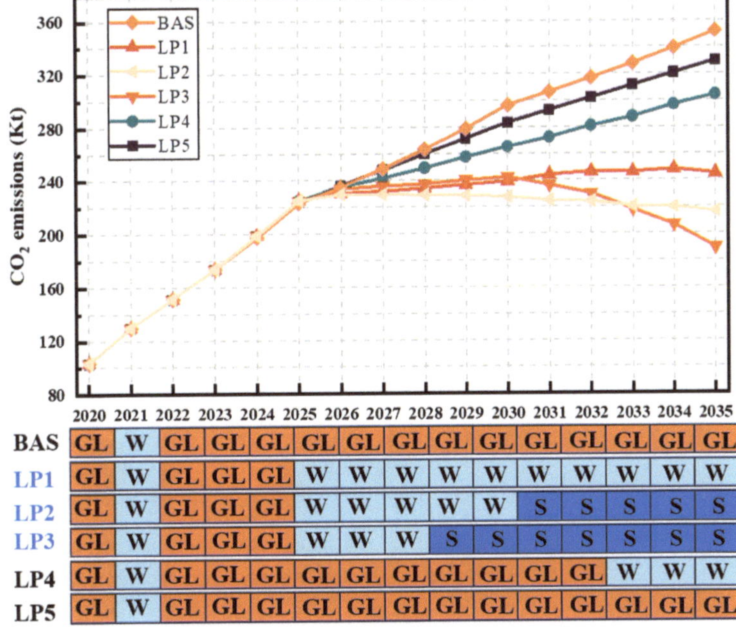

Figure 4. Decoupling categories for different scenarios.

4. Discussion

4.1. Total Carbon Emissions under Different Scenarios

The comparison between carbon emissions under the low-carbon scenarios of the park and the BAS scenario can be obtained from the simulation results of the LEAP model, as demonstrated in Figure 5. In the BAS scenario, carbon emissions steadily increase each year, reaching 351.4 $KtCO_2e$ by 2035. The other three emission reduction scenarios, LC1, LC2, and LC3, exhibit significantly lower carbon emissions than the baseline scenario, emitting 244.6 $KtCO_2e$, 216.2 $KtCO_2e$, and 189.0 $KtCO_2e$, respectively, in 2035. This translates to emission reduction rates of 30.4%, 38.5%, and 46.2% in 2035, respectively. Although emissions in the LC1 scenario are slightly lower than those in the LC3 scenario by 2030 due to slower gross product growth, the advantages of optimized energy technology under the LC2 and LC3 scenarios become more evident as the park economy continues to develop. Specifically, LC2 will reach its peak in 2028 due to the gradual introduction of energy storage technology in 2025. Furthermore, the LC3 scenario demonstrates a downward trend after an inflection point in 2030, thanks to the introduction of carbon capture technology for power generation units in the chemical industry, resulting in significantly higher emission reduction compared to other scenarios.

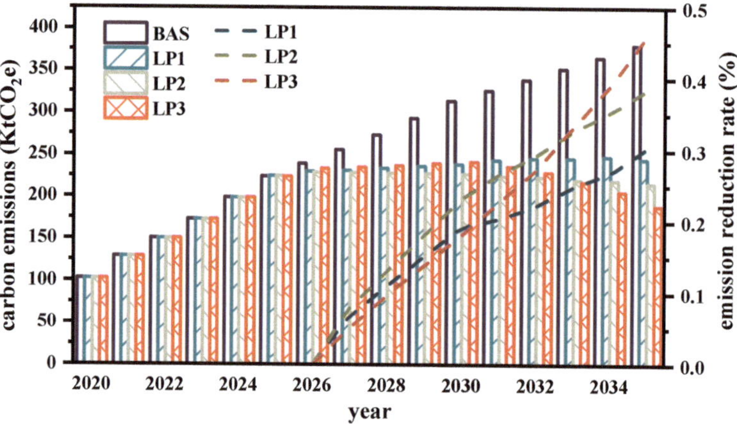

Figure 5. Total carbon emissions of the industrial park under different scenarios.

Figure 6 illustrates the carbon intensity of different scenarios in 2020, 2025, 2030, and 2035. In the initial stage of the BAS scenario, high-energy industries such as the steel and chemical industries in the park continue to develop, leading to a high carbon emission intensity. From 2020 to 2025, the park is required to vigorously realize industrial planning in the initial stage of construction, and the introduction of emission reduction technologies and macro policy adjustments are not considered in four scenarios. As a result, the carbon intensities in 2020 and 2025 are the same. From 2025 to 2035, the carbon emission intensity slowly decreases due to the establishment of high-energy consumption industries and the continuous development of high-tech industries, reaching 2.2 tCO_2/ten thousand yuan by 2035. Meanwhile, the LC1, LC2, and LC3 scenarios exhibit carbon emission intensities of 1.6 tCO_2/ten thousand yuan, 1.2 tCO_2/ten thousand yuan, and 1.1 tCO_2/ten thousand yuan, respectively. It can be concluded that adjusting the industrial ratio effectively reduces carbon emission intensity, but the effect gradually weakens as the industrial park is established and economic development occurs. At this stage, introducing energy storage technology, efficient terminals, and carbon capture devices becomes necessary to achieve further emissions reduction.

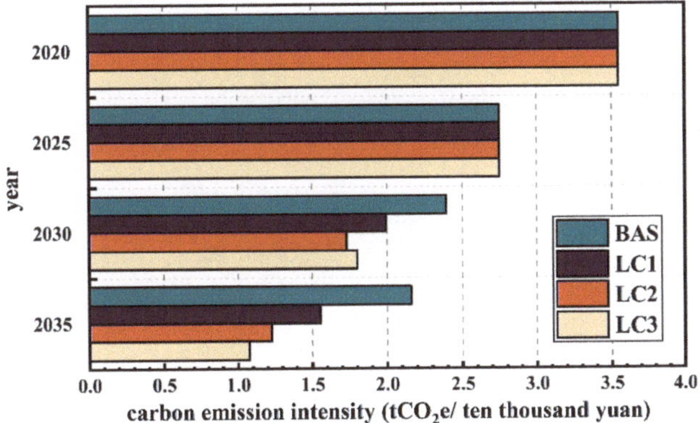

Figure 6. Carbon emissions per unit gross product under different scenarios.

4.2. Carbon Emission Source Analysis

To further analyze the emission reduction potential of different scenarios, the sources of carbon dioxide emissions in the park are explored in detail. As shown in Figure 7, energy consumption emissions (E) greatly surpass non-energy consumption emissions (N) in all four scenarios, accounting for over 95% of the total emissions. Non-energy consumption emissions, depicted in the top-right thumbnail, primarily originate from steel production, aluminum production, magnesium production, and fluoride production. With the rapid development of industries such as iron and steel smelting, non-ferrous metals, and ferrous metals, non-energy consumption emissions show an upward trend, emphasizing the need to optimize the process flow of the steel industry as a key strategy for reducing emissions. From the perspective of energy consumption emissions, the BAS scenario exhibits significantly higher energy consumption than other scenarios, strongly indicating that the optimization of the industrial structure has a prominent effect on reducing carbon emissions. A more detailed analysis of carbon emissions from different industries is provided below.

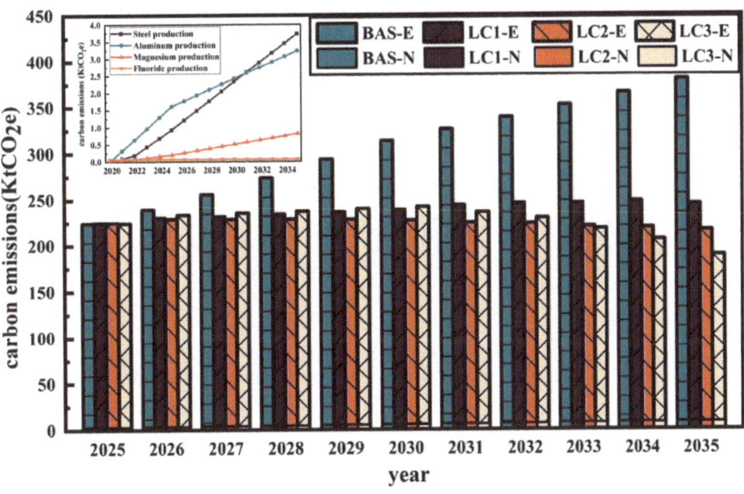

Figure 7. Carbon emissions of energy and non-energy sections under different scenarios.

As illustrated in Figure 8, carbon emissions from energy consumption originate from the chemical industry, steel industry, high-tech equipment manufacturing industry, pharmaceutical industry, and paper industry, with the largest proportion coming from the CH, ST, and HJ industries. In the BAS scenario, the steel industry accounts for the highest emissions, reaching 236.98 KtCO$_2$e by 2035, representing 67.54% of the total energy emissions. Therefore, optimizing the steel industry becomes the focal point of the energy-saving pathway for the industrial park. In the other three scenarios, carbon emissions are significantly reduced compared to the BAS scenario due to a reduction in the steel industry's growth rate, improvements in the industrial chain, and the introduction of emission reduction technology. Specifically, emissions from the steel industry are reduced by 51%, 52%, and 61% in the LC1, LC2, and LC3 scenarios, respectively.

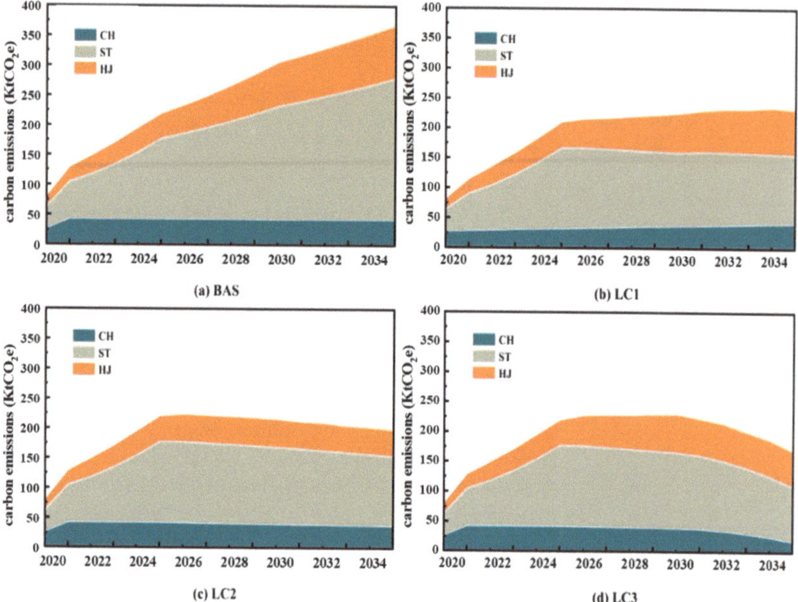

Figure 8. Carbon emissions from various industries under different scenarios.

Figure 9 displays the proportion of carbon emissions from coal, coke, natural gas, oil, LPG, diesel, gasoline, and purchased electricity consumption under different scenarios. Overall, coal emissions account for the largest proportion, reaching 88% in 2020. As the industrial park develops, the proportion of coal emissions gradually decreases, while the proportion of natural gas and purchased power emissions increases. Comparing the four scenarios, LC1 adjusts the economic growth rate and industrial structure of the park compared to the BAS scenario, resulting in reduced carbon emissions from coal. However, the proportion of emissions varies little compared to the BAS scenario, accounting for 57% of total energy emissions by 2035. In the LC2 and LC3 scenarios, due to improved terminal energy utilization efficiency and the adjustment of the industrial energy consumption structure, the proportion of coal emissions continues to decline to 51% and 44% by 2035, respectively. In summary, reducing carbon emissions from coal is crucial in formulating energy conservation and emission reduction strategies for the park, and the introduction of energy-saving technologies in the LC2 and LC3 scenarios is the key to achieving this objective.

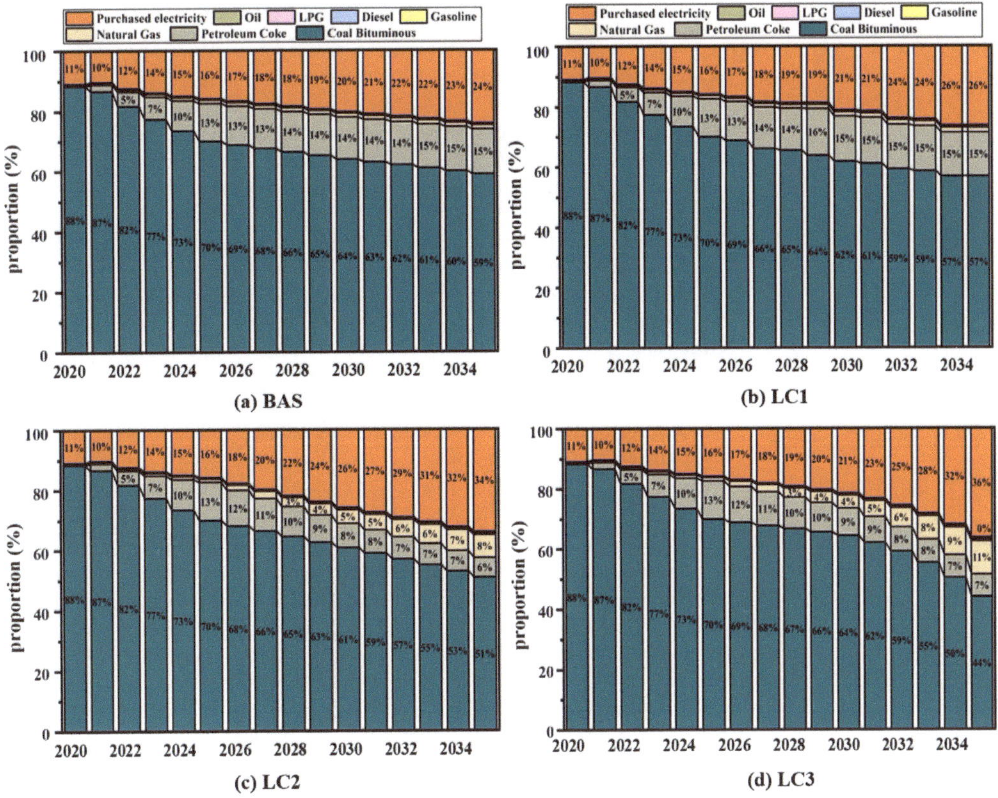

Figure 9. Carbon emissions from different fuel types under different scenarios.

4.3. Pathways for Low-Carbon Industrial Park Development

In accordance with the output results of the model, the carbon emission intensity steadily increases in the BAS scenario, contradicting the low-carbon development strategy of the park. The other three emission reduction scenarios demonstrate significant effects due to the optimization of the industrial chain and adjustment of the industrial structure. However, the LC1 scenario meets the development demand from 2025 to 2035 without adopting energy-saving technologies. As a consequence, carbon emissions progressively increase in the LC1 and LC2 scenarios with economic development, contradicting the concept of promoting regional economic development of the industrial park and proving unsuitable for the park. In contrast, the introduction of technology in the LC2 and LC3 scenarios results in a substantial reduction in coal emissions, enabling low-carbon development by incorporating technological means while safeguarding the park's economy and realizing high-quality development of the green economy. Theoretically, the LC3 scenario represents the most suitable development pathway for the park thanks to the carbon capture devices. However, it requires the introduction of numerous emission reduction processes and low-carbon technologies. These two emission reduction scenarios should be selected and adjusted based on the actual economic development in the initial stage.

5. Conclusions

This paper constructed a hybrid analysis framework that integrates LMDI decomposition with the LEAP model. The framework is used to decompose the driving factors and calculate and predict the carbon emissions of the industrial park cluster under various low-carbon scenarios from 2020 to 2035. The study yields the following conclusions and suggestions for reduction pathways:

1. In accordance with the characteristics of the industrial park, the driving factors can be categorized as follows: energy intensity, energy structure, industrial structure, industrial economic development, and employment scale. Among these factors, industrial economic development accounts for 52%, indicating that the park has not achieved decoupling in the initial stage of development and remains highly dependent on the economy.

2. Under the BAS scenario, carbon emissions will reach 351.4 $KtCO_2e$ by 2035. However, the carbon emissions of the LC1, LC2, and LC3 scenarios decrease by 30.4%, 38.4%, and 46.2%, respectively, compared to the BAS scenario. Among these scenarios, the LC3 scenario emerges as the most suitable pathway for reducing emissions in the park.

3. Additionally, investment in emission reduction technology; an increased proportion of clean energy; measures aimed at reducing carbon emissions from coal, such as improving the efficiency of terminal energy devices; optimizing process flows; and introducing carbon capture devices, are crucial for controlling total emissions and achieving sustainable emission reduction in the park.

To sum up: the development of an emission reduction pathway for industrial park clusters is based on a comprehensive study of economic development and energy supply. In order to achieve low-carbon development of the park, on the one hand, the economic development and industrial development potential of the park should be taken into account to develop a green economy and optimize the industrial chain. On the other hand, we should consider the technical conditions and application costs, develop renewable energies, improve energy efficiency, and reduce coal emissions. Meanwhile, the efficient utilization of carbon capture technology and renewable energy is the key to achieving green and low-carbon development of industrial parks, as they greatly reduce energy consumption and carbon emissions, which is crucial in terms of both economic and environmental benefits.

This paper only puts forward the mixed analysis model of industrial park cluster carbon emission measurement and low-carbon pathway formulation. With the continuous development of the carbon trading market, the model needs to be further improved in the future. At the same time, the impact of carbon capture technology on the total cost is required to be explored in future studies. The model framework can also be employed for large-scale regional energy planning and carbon emission measurement. In addition, social development and technological input can be incorporated into the model to provide data support for the expansion and application of the model.

Author Contributions: Conceptualization, X.Y. and H.L.; Methodology, D.F., W.X., X.G., S.F. and H.L.; Software, W.X., X.G. and S.F.; Validation, D.F. and W.X.; Formal analysis, D.F. and W.X.; Investigation, D.F.; Resources, X.Y.; Data curation, W.X., Y.Y., S.F. and X.Y.; Writing—original draft, D.F. and W.X.; Writing—review & editing, D.F. and X.Y.; Visualization, Y.Y.; Supervision, X.Y. and H.L. All authors have read and agreed to the published version of the manuscript.

Funding: This work was supported by Research on hybrid optimization modeling and application of regional carbon emission estimation (China Energy Engineering Group Jiangsu Power Design Institute Co., Ltd., 32-JK-2022-009).

Data Availability Statement: No data were used in this study.

Conflicts of Interest: The authors declare no conflict of interest.

References

1. Giama, E.; Kyriaki, E.; Papaevaggelou, A.; Papadopoulos, A. Energy and Environmental Analysis of Renewable Energy Systems Focused on Biomass Technologies for Residential Applications: The Life Cycle Energy Analysis Approach. *Energies* **2023**, *16*, 4433. [CrossRef]
2. Gao, X.; Wei, P.; Yu, J.; Huang, X.; Yang, X.; Sundén, B. Design and assessments on graded metal foam in heat storage tank: An experimental and numerical study. *Int. Commun. Heat Mass Transf.* **2023**, *146*, 106902. [CrossRef]

3. Li, Y.; Huang, X.; Huang, X.; Gao, X.; Hu, R.; Yang, X.; He, Y.L. Machine learning and multilayer perceptron enhanced CFD approach for improving design on latent heat storage tank. *Appl. Energy* **2023**, *347*, 121458. [CrossRef]

4. Li, Y.; Niu, Z.; Gao, X.; Guo, J.; Yang, X.; He, Y.-L. Effect of filling height of metal foam on improving energy storage for a thermal storage tank. *Appl. Therm. Energy* **2023**, *229*, 120584. [CrossRef]

5. Du, Z.; Liu, G.; Huang, X.; Xiao, T.; Yang, X.; He, Y.-L. Numerical studies on a fin-foam composite structure towards improving melting phase change. *Int. J. Heat Mass Transf.* **2023**, *208*, 124076. [CrossRef]

6. Pedersen, J.T.S.; van Vuuren, D.; Gupta, J.; Santos, F.D.; Edmonds, J.; Swart, R. IPCC emission scenarios: How did critiques affect their quality and relevance 1990–2022? *Glob. Environ. Change* **2022**, *75*, 102538. [CrossRef]

7. Gao, X.; Niu, Z.; Huang, X.; Yang, X.; Yan, J. Thermo-economic assessments on building heating by a thermal energy storage system with metal foam. *Case Stud. Therm. Eng.* **2023**, *49*, 103307. [CrossRef]

8. Ekonomou, G.; Halkos, G. Exploring the Impact of Economic Growth on the Environment: An Overview of Trends and Developments. *Energies* **2023**, *16*, 4497. [CrossRef]

9. Misrol, M.A.; Wan Alwi, S.R.; Lim, J.S.; Manan, Z.A. Optimising renewable energy at the eco-industrial park: A mathematical modelling approach. *Energy* **2022**, *261*, 125345. [CrossRef]

10. Feng, J.-C.; Yan, J.; Yu, Z.; Zeng, X.; Xu, W. Case study of an industrial park toward zero carbon emission. *Appl. Energy* **2018**, *209*, 65–78. [CrossRef]

11. Yu, X.; Zheng, H.; Sun, L.; Shan, Y. An emissions accounting framework for industrial parks in China. *J. Clean. Prod.* **2020**, *244*, 118712. [CrossRef]

12. Geng, Y.; Zhao, H. Industrial park management in the Chinese environment. *J. Clean. Prod.* **2009**, *17*, 1289–1294. [CrossRef]

13. Wang, N.; Guo, J.; Zhang, X.; Zhang, J.; Li, Z.; Meng, F.; Zhao, B.; Ren, X. The circular economy transformation in industrial parks: Theoretical reframing of the resource and environment matrix. *Resour. Conserv. Recycl.* **2021**, *167*, 105251. [CrossRef]

14. Wang, S.; Lu, C.; Gao, Y.; Wang, K.; Zhang, R. Life cycle assessment of reduction of environmental impacts via industrial symbiosis in an energy-intensive industrial park in China. *J. Clean. Prod.* **2019**, *241*, 118358. [CrossRef]

15. Zhu, D.; Yang, B.; Liu, Y.; Wang, Z.; Ma, K.; Guan, X. Energy management based on multi-agent deep reinforcement learning for a multi-energy industrial park. *Appl. Energy* **2022**, *311*, 118636. [CrossRef]

16. Li, F.; Huang, X.; Li, Y.; Lu, L.; Meng, X.; Yang, X.; Sundén, B. Application and analysis of flip mechanism in the melting process of a triplex-tube latent heat energy storage unit. *Energy Rep.* **2023**, *9*, 3989–4004. [CrossRef]

17. Liu, G.; Xiao, T.; Wei, P.; Meng, X.; Yang, X.; Yan, J.; He, Y. Experimental and numerical studies on melting/solidification of PCM in a horizontal tank filled with graded metal foam. *Sol. Energy Mater. Sol. Cells* **2023**, *250*, 112092. [CrossRef]

18. Shu, G.; Xiao, T.; Guo, J.; Wei, P.; Yang, X.; He, Y.-L. Effect of charging/discharging temperatures upon melting and solidification of PCM-metal foam composite in a heat storage tube. *Int. J. Heat Mass Transf.* **2023**, *201*, 123555. [CrossRef]

19. Chen, X.; Dong, M.; Zhang, L.; Luan, X.; Cui, X.; Cui, Z. Comprehensive evaluation of environmental and economic benefits of industrial symbiosis in industrial parks. *J. Clean. Prod.* **2022**, *354*, 131635. [CrossRef]

20. Wei, X.; Qiu, R.; Liang, Y.; Liao, Q.; Klemeš, J.J.; Xue, J.; Zhang, H. Roadmap to carbon emissions neutral industrial parks: Energy, economic and environmental analysis. *Energy* **2022**, *238*, 121732. [CrossRef]

21. Cai, L.; Luo, J.; Wang, M.; Guo, J.; Duan, J.; Li, J.; Li, S.; Liu, L.; Ren, D. Pathways for municipalities to achieve carbon emission peak and carbon neutrality: A study based on the LEAP model. *Energy* **2023**, *262*, 125435. [CrossRef]

22. Fan, W.; Zhang, J.; Zhou, J.; Li, C.; Hu, J.; Hu, F.; Nie, Z. LCA and Scenario Analysis of Building Carbon Emission Reduction: The Influencing Factors of the Carbon Emission of a Photovoltaic Curtain Wall. *Energies* **2023**, *16*, 4501. [CrossRef]

23. Kolahchian Tabrizi, M.; Famiglietti, J.; Bonalumi, D.; Campanari, S. The Carbon Footprint of Hydrogen Produced with State-of-the-Art Photovoltaic Electricity Using Life-Cycle Assessment Methodology. *Energies* **2023**, *16*, 5190. [CrossRef]

24. Liao, X.; Qian, B.; Jiang, Z.; Fu, B.; He, H. Integrated Energy Station Optimal Dispatching Using a Novel Many-Objective Optimization Algorithm Based on Multiple Update Strategies. *Energies* **2023**, *16*, 5216. [CrossRef]

25. Xu, C.; Zhang, Y.; Yang, Y.; Gao, H. Carbon Peak Scenario Simulation of Manufacturing Carbon Emissions in Northeast China: Perspective of Structure Optimization. *Energies* **2023**, *16*, 5527. [CrossRef]

26. Cai, Y.; Woollacott, J.; Beach, R.H.; Rafelski, L.E.; Ramig, C.; Shelby, M. Insights from adding transportation sector detail into an economy-wide model: The case of the ADAGE CGE model. *Energy Econ.* **2023**, *123*, 106710. [CrossRef] [PubMed]

27. O'Ryan, R.; Nasirov, S.; Osorio, H. Assessment of the potential impacts of a carbon tax in Chile using dynamic CGE model. *J. Clean. Prod.* **2023**, *403*, 136694. [CrossRef]

28. Wei, L.; Feng, X.; Liu, P.; Wang, N. Impact of intelligence on the carbon emissions of energy consumption in the mining industry based on the expanded STIRPAT model. *Ore Geol. Rev.* **2023**, *159*, 105504. [CrossRef]

29. Xing, L.; Khan, Y.A.; Arshed, N.; Iqbal, M. Investigating the impact of economic growth on environment degradation in developing economies through STIRPAT model approach. *Renew. Sustain. Energy Rev.* **2023**, *182*, 113365. [CrossRef]

30. Wang, J.; Dong, K.; Hochman, G.; Timilsina, G.R. Factors driving aggregate service sector energy intensities in Asia and Eastern Europe: A LMDI analysis. *Energy Policy* **2023**, *172*, 113315. [CrossRef]

31. Xin, M.; Guo, H.; Li, S.; Chen, L. Can China achieve ecological sustainability? An LMDI analysis of ecological footprint and economic development decoupling. *Ecol. Indic.* **2023**, *151*, 110313. [CrossRef]

32. Ratomski, P.; Hawrot-Paw, M.; Koniuszy, A.; Golimowski, W.; Kwaśnica, A.; Marcinkowski, D. Indicators of Engine Performance Powered by a Biofuel Blend Produced from Microalgal Biomass: A Step towards the Decarbonization of Transport. *Energies* **2023**, *16*, 5376. [CrossRef]

33. Usman, A.; Ozturk, I.; Naqvi, S.M.M.A.; Ullah, S.; Javed, M.I. Revealing the nexus between nuclear energy and ecological footprint in STIRPAT model of advanced economies: Fresh evidence from novel CS-ARDL model. *Prog. Nucl. Energy* **2022**, *148*, 104220. [CrossRef]

34. Huang, Y.; Wang, Y.; Peng, J.; Li, F.; Zhu, L.; Zhao, H.; Shi, R. Can China achieve its 2030 and 2060 CO_2 commitments? Scenario analysis based on the integration of LEAP model with LMDI decomposition. *Sci. Total Environ.* **2023**, *888*, 164151. [CrossRef] [PubMed]

35. Luo, X.; Liu, C.; Zhao, H. Driving factors and emission reduction scenarios analysis of CO_2 emissions in Guangdong-Hong Kong-Macao Greater Bay Area and surrounding cities based on LMDI and system dynamics. *Sci. Total Environ.* **2023**, *870*, 161966. [CrossRef] [PubMed]

36. You, K.; Yu, Y.; Li, Y.; Cai, W.; Shi, Q. Spatiotemporal decomposition analysis of carbon emissions on Chinese residential central heating. *Energy Build.* **2021**, *253*, 111485. [CrossRef]

37. Yang, D.; Liu, D.; Huang, A.; Lin, J.; Xu, L. Critical transformation pathways and socio-environmental benefits of energy substitution using a LEAP scenario modeling. *Renew. Sustain. Energy Rev.* **2021**, *135*, 110116. [CrossRef]

38. Tsai, M.-S.; Chang, S.-L. Taiwan's 2050 low carbon development roadmap: An evaluation with the MARKAL model. *Renew. Sustain. Energy Rev.* **2015**, *49*, 178–191. [CrossRef]

39. Emodi, N.V.; Emodi, C.C.; Murthy, G.P.; Emodi, A.S.A. Energy policy for low carbon development in Nigeria: A LEAP model application. *Renew. Sustain. Energy Rev.* **2017**, *68*, 247–261. [CrossRef]

40. Amo-Aidoo, A.; Kumi, E.N.; Hensel, O.; Korese, J.K.; Sturm, B. Solar energy policy implementation in Ghana: A LEAP model analysis. *Sci. Afr.* **2022**, *16*, e01162. [CrossRef]

41. Zhang, C.; Luo, H. Research on carbon emission peak prediction and path of China's public buildings: Scenario analysis based on LEAP model. *Energy Build.* **2023**, *289*, 113053. [CrossRef]

42. Shin, H.-C.; Park, J.-W.; Kim, H.-S.; Shin, E.-S. Environmental and economic assessment of landfill gas electricity generation in Korea using LEAP model. *Energy Policy* **2005**, *33*, 1261–1270. [CrossRef]

43. Zou, X.; Wang, R.; Hu, G.; Rong, Z.; Li, J. CO_2 Emissions Forecast and Emissions Peak Analysis in Shanxi Province, China: An Application of the LEAP Model. *Sustainability* **2022**, *14*, 637. [CrossRef]

44. Duan, H.; Hou, C.; Yang, W.; Song, J. Towards lower CO_2 emissions in iron and steel production: Life cycle energy demand-LEAP based multi-stage and multi-technique simulation. *Sustain. Prod. Consum.* **2022**, *32*, 270–281. [CrossRef]

45. Feng, Y.; Zhang, L. Scenario analysis of urban energy saving and carbon abatement policies: A case study of Beijing city, China. *Procedia Environ. Sci.* **2012**, *13*, 632–644. [CrossRef]

46. Kaya, Y. *Impact of Carbon Dioxide Emission Control on GNP Growth: Interpretation of Proposed Scenarios*; Intergovernmental Panel on Climate Change/Response Strategies Working Group: Geneva, Switzerland, 1989.

47. Tapio, P. Towards a theory of decoupling: Degrees of decoupling in the EU and the case of road traffic in Finland between 1970 and 2001. *Transp. Policy* **2005**, *12*, 137–151. [CrossRef]

48. Hou, Z.; Luo, J.; Xie, Y.; Wu, L.; Huang, L.; Xiong, Y. Carbon Circular Utilization and Partially Geological Sequestration: Potentialities, Challenges, and Trends. *Energies* **2023**, *16*, 324. [CrossRef]

49. Huang, L.; Hou, Z.; Fang, Y.; Liu, J.; Shi, T. Evolution of CCUS Technologies Using LDA Topic Model and Derwent Patent Data. *Energies* **2023**, *16*, 2556. [CrossRef]

50. Sitinjak, C.; Ebennezer, S.; Ober, J. Exploring Public Attitudes and Acceptance of CCUS Technologies in JABODETABEK: A Cross-Sectional Study. *Energies* **2023**, *16*, 4026. [CrossRef]

51. Wetzel, M.; Otto, C.; Chen, M.; Masum, S.; Thomas, H.; Urych, T.; Kempka, T. Hydromechanical Impacts of CO_2 Storage in Coal Seams of the Upper Silesian Coal Basin (Poland). *Energies* **2023**, *16*, 3279. [CrossRef]

52. Xie, Y.; Qi, J.; Zhang, R.; Jiao, X.; Shirkey, G.; Ren, S. Toward a Carbon-Neutral State: A Carbon–Energy–Water Nexus Perspective of China's Coal Power Industry. *Energies* **2022**, *15*, 4466. [CrossRef]

 *energies*

Article

Techno-Economic Assessment for the Best Flexible Operation of the CO_2 Removal Section by Potassium Taurate Solvent in a Coal-Fired Power Plant

Stefania Moioli *, Elvira Spatolisano and Laura A. Pellegrini

GASP—Group on Advanced Separation Processes & GAS Processing, Dipartimento di Chimica, Materiali e Ingegneria Chimica "Giulio Natta", Politecnico di Milano, Piazza Leonardo da Vinci 32, I-20133 Milan, Italy
* Correspondence: stefania.moioli@polimi.it; Tel.: +39-02-2399-4721

Abstract: Alternative solvents based on aqueous solutions of amino acids have been recently developed as possible substitutes for Mono Ethanol Amine (MEA) for CO_2 removal from flue gas streams. The potassium taurate solvent has the advantages of degradation resistance, low toxicity and low energy requirements for its regeneration. With any type of solvent, CO_2 removal applied to a power production plant decreases the revenues obtained from selling electricity because of the energy requirements. Operating the CO_2 removal section in flexible mode avoids significant effects on the profits of the power plant, while accomplishing environmental regulations. This work is the first journal paper focusing on the application in flexible mode of the potassium taurate system for treating a flue gas stream from a 500 MW coal-fired power plant. Techno-economic evaluations are performed to determine the best operating conditions considering the variation in the electricity demand and its price, and different values of carbon tax. In the summer period, with high electricity prices and demands, carbon tax values between 45 EUR/t_{CO2} and 60 EUR/t_{CO2} favor CO_2 absorption in the flexible mode, without periods of full CO_2 emissions during the day.

Keywords: CO_2 removal; potassium taurate solvent; process design; energy requirement

Citation: Moioli, S.; Spatolisano, E.; Pellegrini, L.A. Techno-Economic Assessment for the Best Flexible Operation of the CO_2 Removal Section by Potassium Taurate Solvent in a Coal-Fired Power Plant. *Energies* **2024**, *17*, 1736. https://doi.org/10.3390/en17071736

Academic Editors: Michael Zhengmeng Hou, Hejuan Liu, Cheng Cao, Liehui Zhang and Yulong Zhao

Received: 14 February 2024
Revised: 21 March 2024
Accepted: 26 March 2024
Published: 4 April 2024

1. Introduction

With the aim of reducing the amounts of greenhouse gases (GHGs), mainly CO_2, emitted into the atmosphere due to power and industrial production [1] in order to limit the global temperature increase to 1.5 °C, different technologies can be employed as absorption, mainly with chemical solvents, adsorption, membranes or low-temperature separations [2–5]. The aqueous solution of Mono Ethanol Amine (MEA), generally 30 wt. %, is the benchmark solvent considered in chemical absorption for removing CO_2 from flue gases of power plants, and its thermal requirement can significantly reduce the production of the power plant. The energy consumption is a key parameter for determining the advantages of a process, also taking into account the global energy demand. The latest Monthly Statistics report by the International Energy Agency (IEA) states that in the countries of the Organization for Economic Cooperation and Development (OECD), the production of natural gas increased by 3.9% compared to December 2022, and the imports of natural gas have been 6.4% lower on a year-on-year basis, while total OECD exports have decreased by 2.5% in the same period. The total OECD production of crude oil, NGL and refinery feedstocks increased by 8.7% in December 2023 compared to December 2022.

In addition to the high energy requirement, MEA is characterized by toxicity and degradation, so it cannot be considered a sustainable solvent given its effects on health and the environment, relevant topics for the construction of new industrial plants in general. For this reason, innovative solvents are being taken into account [6], and research on new species and their influence on the CO_2 removal process [7–11] is being carried out. Among the innovative solvents proposed in the last few years, amino acids can be considered environmentally friendly [12], in addition to being advantageous also for their energy requirements. Their salts in aqueous solution have been studied, as in Lerche [13], Majchrowicz [14], Majchrowicz, et al. [15], Sanchez-Fernandez and Goetheer [16], Sanchez Fernandez, et al. [17] and Sanchez Fernandez, et al. [18], who found that amino acids are less corrosive, more stable, and have lower enthalpy in their reactions than the traditional MEA solvent [19]. The potassium taurate (KTau) process, developed by Sanchez-Fernandez et al. [18], can be used for CO_2 removal, with lower energy requirements than MEA. In any case, after an overall study of more than 1000 patents focusing on different technologies by Li et al. [20], it was concluded that no processes have been developed for this purpose that could favor a low added Cost of Electricity (CoE), so the CO_2 removal section of a power plant is always a consuming section of the power plant, representing a cost for the company and, therefore, reducing the profits related to selling the produced electricity.

The flexible operation of the CO_2 removal section makes the plant's profitability improve [21,22], because, during peak demands or when the price of electricity is high, the consumption of the CO_2 removal plant is reduced and the production of electricity is increased. Several studies in the literature, with tests on pilot plants [23], focused on the flexible operation of Carbon Capture and Storage (CCS) power stations, considering MEA as the chemical absorption solvent, and studying the bypass and solvent storage techniques [24–28], the operation at variable capture level [29,30] and the variable solvent regeneration mode [31,32]. Bui et al. [33,34] demonstrated that flexible operation is technically feasible in a large-scale CO_2 capture process via simulations and experimental tests using MEA solvent.

This work was carried out at the Process Design laboratory (PD lab) of GASP of Politecnico di Milano, which is provided with the Process Design and Process Thermodynamics laboratory (PD&PT lab). This paper is the first in the literature aiming at studying the possible operation in flexible mode of the CO_2 removal section of a 500 MW coal-fired power plant with an aqueous solution of potassium taurate (4 M potassium 4 M taurine) to determine the best operating conditions for each hour of key days. It follows previous works in the literature focusing on the same topic applied only to the use of the MEA solvent for CO_2 removal in power plants. Techno-economic evaluations have been carried out, taking into account the price of and demand for electricity in Italy, and the influence of the carbon tax, determining its value of breakeven point.

2. Materials and Methods

2.1. The Electricity Market in Italy

This work was carried out by taking as a basis the price of and the demand for electricity in Italy, for which official detailed values are available.

Terna [35] reports that the electricity market demand in Italy in 2015 was 316.9 TWh and the consumption was equal to 297.2 TWh. In total, 85% of the electricity demand was supplied by Italy and 15% was taken from other countries. The gross national production was about 283.0 TWh. 51% (192 TWh) of the national production was supplied by thermoelectric power plants, 15% (47 TWh) by hydroelectric and 20% (44 TWh) by renewable sources (wind, solar, geothermal and bioenergy). The gross efficient power was 120 GW.

In 2015, the gross thermoelectric production was 198.24 TWh, with geothermal sources accounting for 6.185 TWh. The conventional power production was 192.053 GWh. Natural gas covered provided 110.86 TWh and solid fuel 43.2 TWh, the remaining 37.993 TWh was produced by other fuels (gas derivates, petrochemicals, etc.). Solid fuel accounted for 43.25 TWh of the yearly production of electricity in 2015 [35].

The typical trends of electricity consumption have been detailed in a previous work focusing on chemical absorption using an MEA solvent [36], and the analysis of the electricity price for the year taken as a reference in this work (2015) has been reported by Moioli et al. [28].

Throughout the whole year, the requested power output of fossil fueled power plants depends on variations in the demand, with peaks occurring hourly and daily. The energy profiles vary every month, with the highest difference between months in Italy being seen during summer, when the price of electricity is higher than 110 EUR/MWh. Lowering the electricity sold to the market because of the steam consumption due to the operation of a CO_2 removal plant during the peak hours in summer would be less advantageous compared to paying a carbon tax.

2.2. Details of the Considered Plant

The gaseous stream to be treated is a flue gas stream from a 500 MW coal-fired power plant, with a flowrate of 19.60 kmol/s and a composition (mole fraction %) of 7% water, 13% CO_2, 75% N_2 and 5% O_2 [37].

The scheme (Figure 1) is similar to the one for traditional amine solvents, with an absorption section wherein the solvent is concurrently contacted with the flue gas and with a regeneration section, where CO_2 is removed from the solvent and is recycled, after cooling, to the absorption columns. Moreover, differently from only liquid amine aqueous solutions, a dissolution heat exchanger is added before the rich solution is fed to the lean–rich heat exchanger for heat recovery. The aim of the additional unit is to dissolve all the precipitated taurine by heating, using condensed water as the service fluid, before entering the regeneration section.

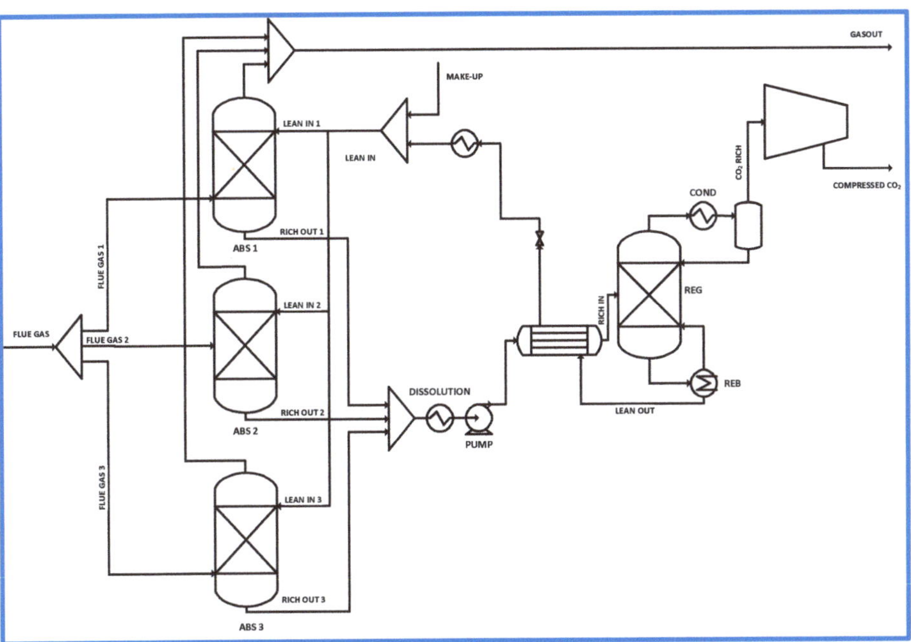

Figure 1. Scheme for the process considered in this work (*"ABS"* refers to the absorption column, *"REG"* refers to the regeneration column, *"COND"* refers to the condenser and *"REB"* refers to the reboiler).

In this work three parallel absorption columns for treating the large flowrate of flue gas are considered [38]. The number of columns has been selected on the basis of the maximum diameter considered at the industrial level [39–41], reported with values up to 15 m, and considering that this choice has already been reported to work in previous literature [42,43] related to the treatment of high flue gas flowrates with chemical absorption for CO_2 removal. The sizes of the absorber and of the regeneration column were selected with the aim of minimizing the total costs [38].

The lean loading has been chosen as done in previous works carried out on this topic [18,37]. The rich loading is a result obtained from simulation, with a value agreeing with common values usually employed for chemical solvents based on amine.

This work does not consider the option of adding a solid–liquid separator, which would allow the bottom products from the absorber to be split into a slurry and a liquid stream (recycled to the absorber) for further lowering the energy requirement.

Table 1 reports the main characteristics of the considered process.

Table 1. Main process parameters for the considered scheme.

Characteristic	Absorber		Regeneration Column
Number of columns	3		1
Diameter (m)	12		12.7
Packed height (m)	30		17.6
Packing type	Mellapak 250X		Mellapak 250X
Lean loading		0.27	
Rich loading		0.44	

The CO_2 obtained from the top of the regeneration column is compressed to 150 bar [44].

2.3. Theory

This paper is the first focusing on the topic of the flexible operation of a CO_2 removal plant applied to a power plant that considers the use of a potassium taurate solvent (no previous analyses of this type applied to Natural Gas Combined Cycle (NGCC) or coal-fired power plants have been found). In this work, both operation for fixed CO_2 removal and operation in flexible mode have been considered. The Capture Level Reduction (CLR) mode is a mode of flexible operation considered to be the best performing mode for coal-fired power plants in previous literature focusing on the MEA solvent. In this work, it has been analyzed for a plant located in Italy. The fixed CO_2 removal operating mode and the no-capture operation have also been considered, with a comparison among all the modes. The degree of removal of CO_2 can be higher or lower in the flexible operation mode than in the fixed configuration, and varies during the day [45] in order to minimize the losses of energy and of revenues. The value of the carbon tax to be paid can influence the process' operation in flexible mode.

This paper analyzes the CLR mode, which involves operating a bypass by splitting part of the rich solvent to be recirculated to the absorption column without being regenerated, and then regenerating the other smaller part of the solvent. The amount of CO_2 that is absorbed is reduced, so a reduction in the thermal power needed at the reboiler is achieved. The CLR mode of operation does not necessitate modifications to the equipment, thus avoiding additional investment costs (for instance, additional tanks and a higher solvent inventory are needed in the case of solvent storage). In particular, in this work, we have considered and compared to the case of no capture the following cases:

- Fixed operation at 90% CO_2 removal;
- Fixed operation at a given % ratio;
- Flexible operation.

In the fixed operation mode, a fixed ratio of 90%, 80%, 70% or 60% of CO_2 removal from the base case has been considered (100% ratio refers to 90% CO_2 removal).

In the CLR mode, CO_2 is vented to the atmosphere over specified time intervals. In this case, the energy required for the regeneration of the solvent and the CO_2 compression decreases as the rich solvent flowrate to be regenerated is lower, with a higher production of electric power. In this period of time, the overall emission of CO_2 is higher than the one at 90% fixed CO_2 removal, and a higher carbon tax must be paid. A minimum value equal to 30% has been set for the % operation ratio, as reported in Cohen et al. [46], to avoid issues in the operation of columns, such as drying out in the regeneration unit.

2.3.1. Profit Minimization

An in-house tool developed by GASP of Politecnico di Milano for economic optimization, on the basis of the assumptions and the methodology detailed in Moioli et al. [36], has been employed. The inputs to the tool are the results of the simulations in ASPEN Plus$^{®}$ V11 that had been previously integrated with new ionic species and has been modified for the thermodynamic model by introducing reactions and values of parameters to represent the vapor–liquid–solid equilibrium of the potassium taurate system. The details of the thermodynamic model, with the comparison to the experimental data, are reported in Moioli et al. [47].

It is assumed that no external sources are employed. Therefore, for each time of the day, the available energy has been calculated on the basis of the electricity needed by the final users.

If needed, a lower amount of electricity is sold to the market to run the CO_2 capture section. In this case, the remaining power needed for the demand of the market should be bought by the final user from another source.

For flexible operation, the objective function is the profit, which must be maximized by operating the capture plant with a varying % ratio. The value of the % ratio that minimizes the profit associated to a power station with a CO_2 capture system is selected on an on hourly basis. The profit is calculated as

$$P = W_{out}C_{energy} - F_{CO_2}C_{CO_2Tax} - F_{fuel}C_{Fuel} - C_{b,O\&M} \tag{1}$$

where the following pertains:

- W_{out}, the net production of electrical power exiting from the power station (MW = MWh/h);
- C_{energy}, the price of electrical energy (EUR/MWh);
- C_{CO2Tax}, the carbon tax (EUR/t$_{CO2}$);
- F_{CO2}, the amount of CO_2 vented in one hour (tCO_2/h);
- F_{fuel}, the fuel consumption (kg/h);
- C_{fuel}, the fuel cost (EUR/kg);
- $C_{b,O\&M}$, the cost of base plant operation and maintenance (EUR/h), excluding the cost of W_{out}.

W_{out} takes into account the reduction in power sold to the users because of the extraction of steam from the turbine to operate the reboiler and because of the electricity needed to operate the CO_2 removal section.

The method employed in this work is suitable for this application, also for other CO_2 removal processes with different solvents.

2.3.2. Assumptions

The assumptions at the basis of this work are:

1. The electricity price is based on Italian historical electricity prices referring to 2015, set out by Gestore dei Mercati Energetici [48] (the methodology could be extended to other electricity markets, and also applied when considering the electricity prices of the following years, when available);
2. The electricity demand was set with reference to the historical data by Terna [35], scaling for the considered power plant production;
3. Carbon tax varies in the range of 5 EUR/t$_{CO2}$ to 200 EUR/t$_{CO2}$, in steps of 5 EUR/t$CO2$;
4. Cold start-up and shut-down costs are neglected;
5. Losses in efficiency due to variations in % ratio are considered with a ramp of 5%/min, as also done by Cohen et al. [49];
6. In cases when additional power is produced in each hour (because of the production set defined on a provision based on the expected trend of electricity consumption) that is not required by the market, it is assumed to be lost;

7. January (generally the coldest month in Italy) and July (when the highest variations in electricity demand and prices occur [50]) have been selected;
8. The time step for the analysis is 1 h, corresponding to the time step at which data on the prices of electricity have been found.

The study considers different modes of operation:

- Base case, i.e., 90% CO_2 removal at fixed operation;
- Fixed capture at 90%, 80%, 70% and 60% of the base case;
- CLR flexible operation;
- No capture.

The characteristics of energy consumption have been considered by taking the whole country as uniform. If detailed data related to each region of Italy become available, an analysis of different regions of Italy may be of interest in order to better understand the influence on the results.

3. Results

In Figures 2–4, the % ratio of operation in CLR mode during the whole day is reported as obtained by the optimization. Figure 5 reports the electric power sold to the market in CLR mode while also considering different carbon taxes, and Figure 6 details the carbon tax paid for emitting CO_2 in the CLR mode, as a function of the carbon tax.

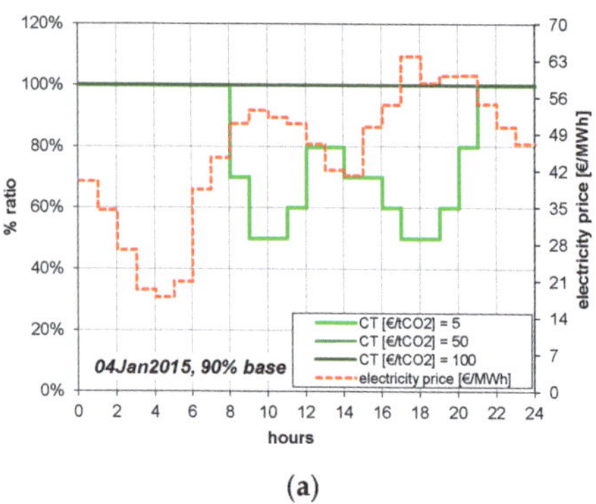

(a)

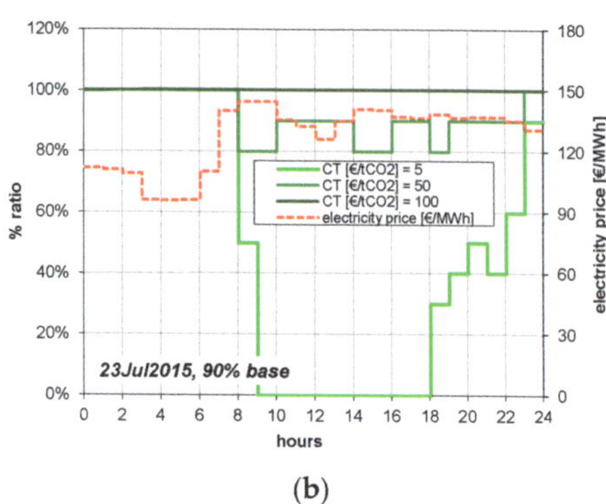

(b)

Figure 2. Optimal operation with CLR on (**a**) 4 January 2015 and (**b**) 23 July 2015 for values of carbon tax (CT) equal to 5, 50 and 100 EUR/t_{CO2}.

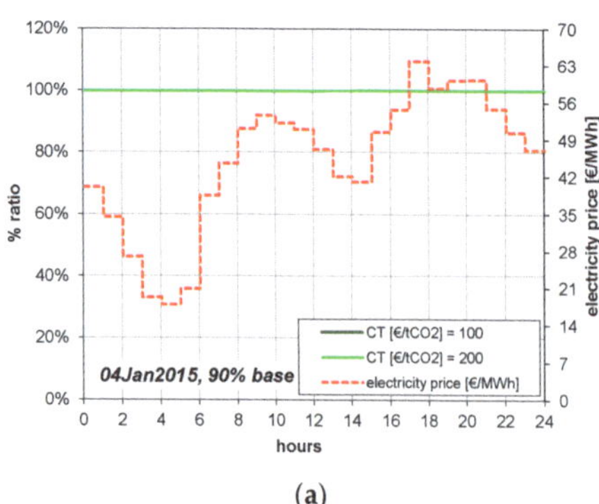

(a)

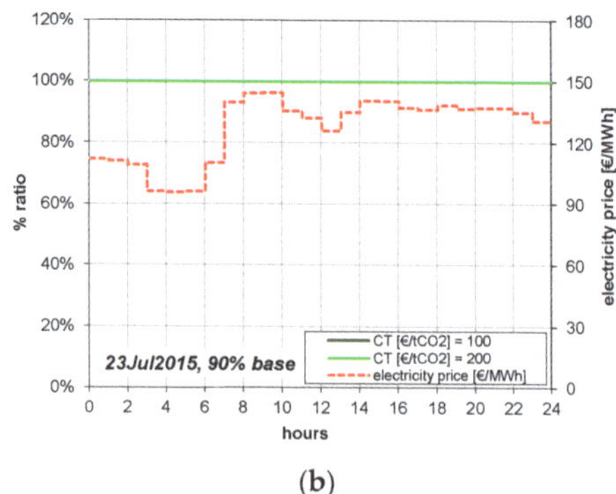

(b)

Figure 3. Optimal operation with CLR on (**a**) 4 January 2015 and (**b**) 23 July 2015 for values of carbon tax (CT) equal to 100 and 200 EUR/t_{CO2}.

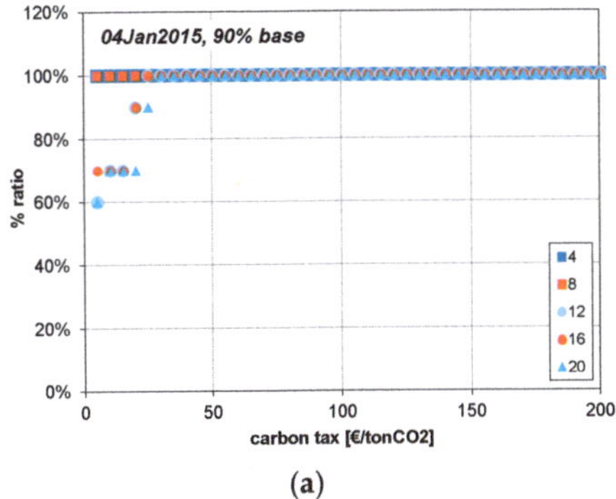
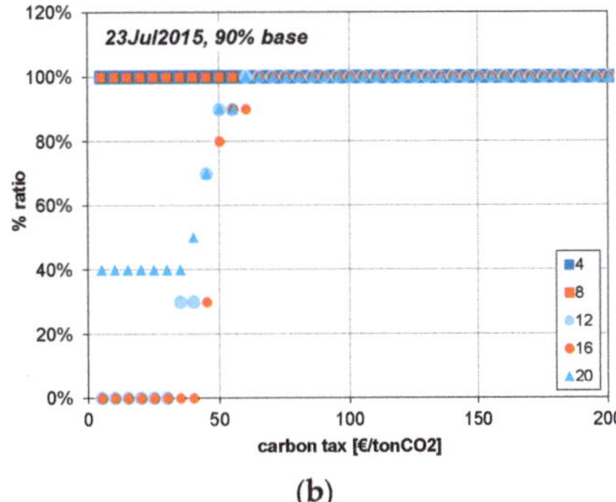

(a)

(b)

Figure 4. Optimum % ratio as affecting the carbon tax in CLR mode at different hours of the day (8, 12, 16 and 20) on (**a**) January 4th 2015 and (**b**) 23 July 2015.

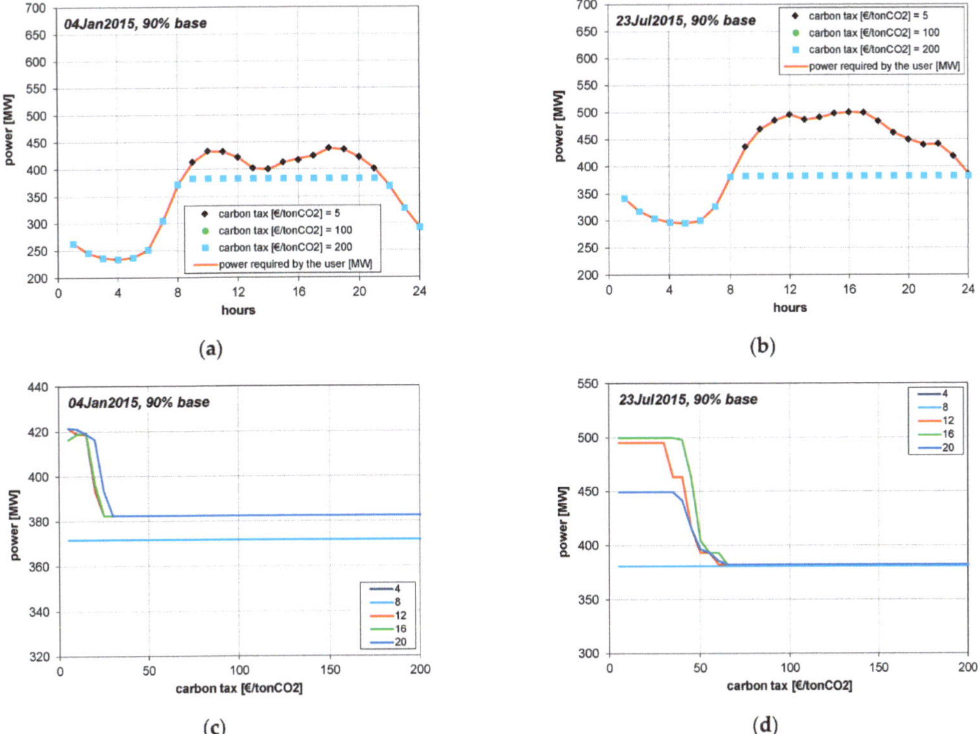

(a)

(b)

(c)

(d)

Figure 5. Power (electric) sold to the market in CLR mode for (**a**) 4 January 2015 and (**b**) 23 July 2015, and for different carbon taxes on (**c**) 4 January 2015 and (**d**) 23 July 2015.

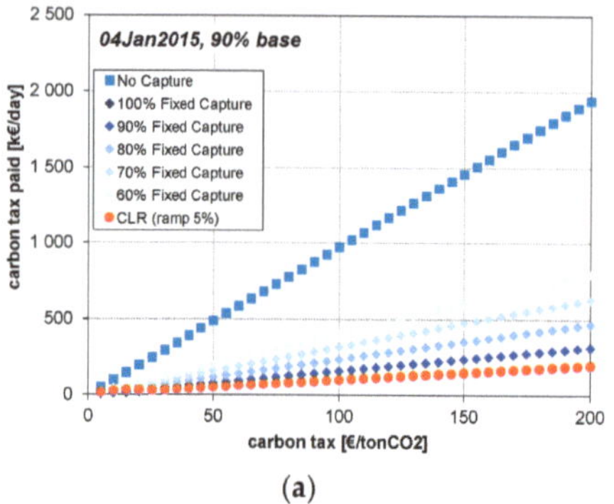
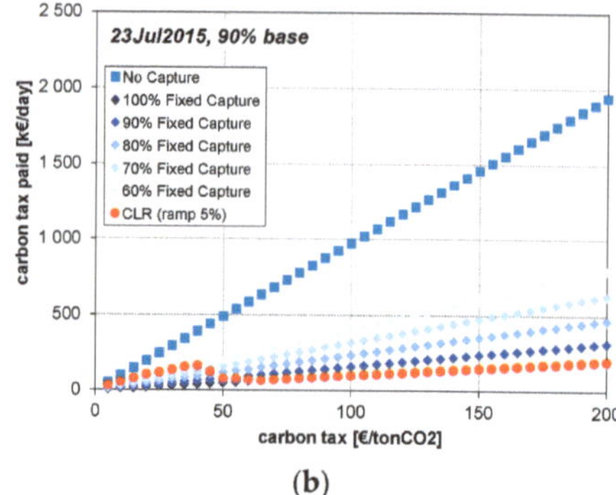

(**a**)   (**b**)

Figure 6. Carbon tax paid for emitting CO_2 in CLR mode as a function of the carbon tax for (**a**) 4 January 2015 and (**b**) 23 July 2015.

4. Discussion

4.1. CLR Mode

The optimal values depend on the electricity's price and demand during the daily period. The % ratio is lower at times of high electricity price and demand, if paying the carbon tax is advantageous. In July, the peaks in the power demand and the high price of electrical energy favor high revenues, meaning that the operation in flexible mode is advantageous (Figures 2b and 3b). In January (Figures 2a and 3a), the level of electricity sold is lower than that produced, and the revenues are lower. Therefore, the expenses related to the payment of the carbon tax due to CO_2 emissions can be avoided, and 90% CO_2 removal is applied (a % ratio equal to 100%). For a carbon tax of 5 EUR/t_{CO2}, emitting into the atmosphere was shown to be advantageous from 9:00 to 21:00 on 4 January 2015 and from 9:00 to the end of the day on 23 July 2015. In summer, CO_2 removal is advantageous for hours with low electricity demand and prices, occurring only in the early mornings and during nights.

The carbon tax influences the optimum % ratio (Figure 4). The threshold value of carbon tax for 4 January 2015 has been shown to be 25 EUR/t_{CO2}, and the one for 23 July 2015 is 65 EUR/t_{CO2}. For higher values, the operation at 90% CO_2 removal is carried out, because the amount of CO_2 emitted would be too expensive in terms of carbon tax to be paid. In winter, the no capture mode is not advantageous, while in summer, with low values of carbon tax, emitting all the CO_2 present in the flue gas stream would seem to be more advantageous than removing it in terms of avoiding the emission of CO_2 into the atmosphere.

For January, the obtained profit decreases with increases in the carbon tax and with reductions in the % ratio in fixed mode. A different trend is obtained for July, due to the high amounts, and high prices, of the electricity sold to the market.

For a carbon tax equal to 5 EUR/t_{CO2}, the electric power remaining after the consumption of steam for CO_2 removal corresponds to the amount of power requested by the market, because CO_2 removal is not in operation (Figure 5). When there are higher carbon taxes, the power sold to the market is decreased, and greater lowering occurs in summer, when much electricity is needed by the user. The power consumed by the CO_2 removal and compression section is provided by significantly reducing the power sold to the market, which case is different from that in winter.

Figure 6 shows that the trend obtained for the CLR mode is different to a monotonic profile, in particular for 23 July 2015, because the carbon tax paid depends on the % ratio at each hour, which is selected on the basis of techno-economic analyses taking into account the variation in the market. In the fixed operation mode, the amount of carbon tax to be paid is constantly increasing with the unitary value of the carbon tax.

This work is the first to focus on the potassium taurate process, and therefore no comparison with other works employing the same solvent can be performed. The obtained trends are in agreement with the trends obtained in previous papers published in the literature referring to the flexible operation mode, although taking into account a different solvent for CO_2 removal, and confirm the validity of the methodology.

4.2. Case of No CO_2 Removal

The no capture case has also been considered. The amount paid in this case is up to one order of magnitude higher (if compared to the CLR mode of operation) than any case of operation for removal (also partial) of carbon dioxide (Figure 6). Treating flue gas is profitable when the cost of the emitted CO_2 is higher than the price of electricity. This work confirms the relevance of policy decisions related to the application of carbon tax, which leads to CO_2 removal being more advantageous than CO_2 emission in economic terms.

The results obtained for the potassium taurate system, though differing in terms of the % ratio and values of carbon tax at which operating the CO_2 removal plant at a 100% ratio is advantageous, are in line with the results obtained in the literature for the MEA solvent used for chemical absorption to treat flue gases produced by coal-fired power plants [24,25,51–54].

4.3. Cases of Fixed Operation

When operating in a fixed operation mode at certain % ratios, if more CO_2 is removed, a lower amount of electric power is available to the final user. In summer, the profit obtained by paying the carbon tax for emitting carbon dioxide could prove to be more advantageous than attempting the absorption of high amounts of CO_2. When the demand for electricity is lower, as in winter, the more CO_2 is absorbed, the better it is for the economics of the plant, because a lower carbon tax is paid. Therefore, the full 90% CO_2 removal must be selected (instead of a lower % ratio of fixed CO_2 removals). It can be concluded that if the operation in fixed mode is preferred, the % ratio could be selected on the basis of the period of the year, and on the required energy. In any case, the CLR operation would ensure lower losses of revenues for the company, because the variation in the % ratio will be optimized for this aim.

5. Conclusions

This paper presents the first techno-economic evaluation in the literature on the application of the potassium taurate process in flexible mode for the treatment of a flue gas stream produced by a 500 MW coal-fired power plant. A process using this type of solvent has been chosen because it is characterized by lower energy requirements than other processes based on the MEA solvent, in addition to employing a solution with reduced toxicity and corrosive effects.

In this paper, the effects of the variation in the electricity demand and price in Italy, and in the value of carbon tax, in the range of 5 EUR/t_{CO2} to 200 EUR/t_{CO2}, have been analyzed in detail for two relevant days in different seasons of the year (winter and summer) for which information was available.

On the basis of the results, it can be concluded, as expected, that the higher the value of the carbon tax, the higher the average % ratio of the CO_2 capture system.

In summer, because of the high energy demand and the high price of electricity, venting CO_2 into the atmosphere and paying a carbon tax as high as 40 EUR/t_{CO2} will be more profitable at given hours because of the high value of electricity, losses in which caused by carbon capture operation heavily affect the economics of the system. With a carbon tax ranging from 45 EUR/t_{CO2} to 60 EUR/t_{CO2}, flexible absorption without any hours of full CO_2 emission is the best option. The obtained results for 4th July 2015 are different, and here, no operation at 0% ratio is needed, thus precluding the emission of the CO_2 present in the flue gas stream at given hours of the day.

Higher carbon tax values lead to the carbon dioxide removal system being operated at higher % ratios, both in winter and in summer.

The demand for and the price of electricity in different periods of the year cause the maximum % ratio (except 0%) to be different in summer and in winter, with the minimum values being 30% in July and 50% in January.

The proposed methodology could be applied to any following year for which detailed data on electricity price and demand are available. A future development of this work could focus on the analysis of the characteristics of energy consumption in different regions of Italy (when detailed data are available) to evaluate the effects on the obtained results.

Author Contributions: Conceptualization, S.M. and L.A.P.; methodology, S.M. and E.S.; software, S.M.; validation, S.M. and E.S.; formal analysis, S.M. and E.S.; investigation, S.M. and E.S.; resources, E.S.; data curation, S.M. and E.S.; writing—original draft preparation, S.M.; writing—review and editing, E.S. and L.A.P.; visualization, S.M., E.S. and L.A.P.; supervision, L.A.P.; project administration, L.A.P. All authors have read and agreed to the published version of the manuscript.

Funding: This research received no external funding.

Data Availability Statement: The original contributions presented in the study are included in the article; further inquiries can be directed to the corresponding author.

Conflicts of Interest: The authors declare no conflicts of interest.

References

1. IEA. *Putting CO$_2$ to Use*; IEA: Paris, France, 2019.
2. De Guido, G.; Pellegrini, L.A. Calculation of solid-vapor equilibria for cryogenic carbon capture. *Comp. Chem. Eng.* **2022**, *156*, 107569. [CrossRef]
3. Pellegrini, L.A.; De Guido, G.; Ingrosso, S. Thermodynamic framework for cryogenic carbon capture. *Comput. Aided Chem. Eng.* **2020**, *48*, 475–480.
4. Ebner, A.D.; Ritter, J.A. State-of-the-art Adsorption and Membrane Separation Processes for Carbon Dioxide Production from Carbon Dioxide Emitting Industries. *Sep. Sci. Technol.* **2009**, *44*, 1273–1421. [CrossRef]
5. Feron, P.H.M.; Jansen, A.E.; Klaassen, R. Membrane technology in carbon dioxide removal. *Energy Convers. Manag.* **1992**, *33*, 421–428. [CrossRef]
6. Yang, C.; Li, T.; Tantikhajorngosol, P.; Sema, T.; Xiao, M.; Tontiwachwuthikul, P. Evaluation of novel aqueous piperazine-based physical-chemical solutions as biphasic solvents for CO$_2$ capture: Initial absorption rate, equilibrium solubility, phase separation and desorption rate. *Chem. Eng. Sci.* **2023**, *277*, 118852. [CrossRef]
7. Pellegrini, L.A.; Gilardi, M.; Giudici, F.; Spatolisano, E. New Solvents for CO$_2$ and H$_2$S Removal from Gaseous Streams. *Energies* **2021**, *14*, 6687. [CrossRef]
8. Lu, Y.; Nielsen, P.; Lu, H. Development and Bench Scale Testing of a Novel Biphasic Solvent Enabled Absorption Process for Post Combustion Carbon Capture DE FE0031600. In Proceedings of the 2022 Carbon Management Project Review Meeting, Pittsburgh, PA, USA, 15–19 August 2022.
9. Wanderley, R.R.; Pinto, D.D.D.; Knuutila, H.K. From hybrid solvents to water-lean solvents—A critical and historical review. *Sep. Purif. Technol.* **2021**, *260*, 118193. [CrossRef]
10. Schuur, B.; Brouwer, T.; Smink, D.; Sprakel, L.M.J. Green solvents for sustainable separation processes. *Curr. Opin. Green Sustain. Chem.* **2019**, *18*, 57–65. [CrossRef]
11. Vanderveen, J.R.; Patiny, L.; Chalifoux, C.B.; Jessop, M.J.; Jessop, P.G. A virtual screening approach to identifying the greenest compound for a task: Application to switchable-hydrophilicity solvents. *Green Chem.* **2015**, *17*, 5182–5188. [CrossRef]
12. Ramezani, R.; Mazinani, S.; Di Felice, R. State-of-the-art of CO$_2$ capture with amino acid salt solutions. *Rev. Chem. Eng.* **2022**, *38*, 273–299. [CrossRef]
13. Lerche, B.M. CO$_2$ Capture from Flue Gas Using Amino Acid Salt Solutions. Ph.D. Thesis, Technical University of Denmark, Lyngby, Denmark, 2012.
14. Majchrowicz, M.E. Amino Acid Salt Solutions for Carbon Dioxide Capture. Ph.D. Thesis, University of Twente, Twente, The Netherlands, 2014.
15. Majchrowicz, M.E.; Brilman, D.W.F.W.; Groeneved, M.J. Precipitation regime for selected amino acid salts for CO$_2$ capture from flue gases. *Energy Procedia* **2009**, *1*, 979–984. [CrossRef]
16. Sanchez-Fernandez, E.; Goetheer, E.L.V. DECAB: Process development of a phase change absorption process. *Energy Procedia* **2011**, *4*, 868–875. [CrossRef]
17. Sanchez Fernandez, E.; Goetheer, E.L.V.; Manzolini, G.; Macchi, E.; Reijvani, S.; Vlugt, T.J.H. Thermodynamic assessment of amine based CO$_2$ capture technologies in power plants based on European Benchmarking Task Force methodology. *Fuel* **2014**, *129*, 318–329. [CrossRef]

18. Sanchez Fernandez, E.; Heffernan, K.; van der Ham, L.V.; Linders, M.J.G.; Eggink, E.; Schrama, F.N.H.; Brilman, D.W.F.; Goetheer, E.L.V.; Vlugt, T.J.H. Conceptual Design of a Novel CO_2 Capture Process Based on Precipitating Amino Acid Solvents. *Ind. Eng. Chem. Res.* **2013**, *52*, 12223–12235. [CrossRef]

19. Brouwer, J.P.; Feron, P.H.M.; ten Asbroek, N.A.M. Amino-Acid Salts for CO_2 Capture from Flue Gases. 2009. Available online: https://www.researchgate.net/publication/228408743_Amino-acid_salts_for_CO2_capture_from_flue_gases (accessed on 15 April 2023).

20. Li, B.; Duan, Y.; Luebke, D.; Morreale, B. Advances in CO_2 capture technology: A patent review. *Appl. Energy* **2013**, *102*, 1439–1447. [CrossRef]

21. Domenichini, R.; Mancuso, L.; Ferrari, N.; Davison, J. Operating Flexibility of Power Plants with Carbon Capture and Storage (CCS). *Enrgy Procedia* **2013**, *37*, 2727–2737. [CrossRef]

22. Bui, M.; Gunawan, I.; Verheyen, V.; Feron, P.; Meuleman, E.; Adeloju, S. Dynamic modelling and optimisation of flexible operation in post-combustion CO_2 capture plants—A review. *Comp. Chem. Eng.* **2014**, *61*, 245–265. [CrossRef]

23. Bui, M.; Tait, P.; Lucquiaud, M.; Mac Dowell, N. Dynamic operation and modelling of amine-based CO_2 capture at pilot scale. *Int. J. Greenh. Gas Control* **2018**, *79*, 134–153. [CrossRef]

24. Chalmers, H.; Gibbins, J. Initial evaluation of the impact of post-combustion capture of carbon dioxide on supercritical pulverised coal power plant part load performance. *Fuel* **2007**, *86*, 2109–2123. [CrossRef]

25. Gibbins, J.R.; Crane, R.I. Scope for reductions in the cost of CO_2 capture using flue gas scrubbing with amine solvents. *Proc. Inst. Mech. Eng. A-J. Power Energy* **2004**, *218*, 231–239. [CrossRef]

26. Delarue, E.; Martens, P.; D'haeseleer, W. Market opportunities for power plants with post-combustion carbon capture. *Int. J. Greenh. Gas Control* **2012**, *6*, 12–20. [CrossRef]

27. Haines, M.R.; Davison, J.E. Designing Carbon Capture power plants to assist in meeting peak power demand. *Greenh. Gas Control Technol.* **2009**, *1*, 1457–1464. [CrossRef]

28. Moioli, S.; Pellegrini, L.A. Operating the CO_2 absorption plant in a post-combustion unit in flexible mode for cost reduction. *Chem. Eng. Res. Des.* **2019**, *147*, 604–614. [CrossRef]

29. Errey, O.; Chalmers, H.; Lucquiaud, M.; Gibbins, J. Valuing Responsive Operation of Post-combustion CCS Power Plants in Low Carbon Electricity Markets. *Energy Procedia* **2014**, *63*, 7471–7484. [CrossRef]

30. Rubin, E.S.; Chen, C.; Rao, A.B. Cost and performance of fossil fuel power plants with CO_2 capture and storage. *Energy Policy* **2007**, *35*, 4444–4454. [CrossRef]

31. Mac Dowell, N.; Shah, N. Optimisation of Post-combustion CO_2 Capture for Flexible Operation. *Energy Procedia* **2014**, *63* (Suppl. C), 1525–1535. [CrossRef]

32. Mechleri, E.; Fennell, P.S.; Dowell, N.M. Flexible Operation Strategies for Coal- and Gas-CCS Power Stations under the UK and USA Markets. *Energy Procedia* **2017**, *114*, 6543–6551. [CrossRef]

33. Bui, M.; Flø, N.E.; de Cazenove, T.; Mac Dowell, N. Demonstrating flexible operation of the Technology Centre Mongstad (TCM) CO_2 capture plant. *Int. J. Greenh. Gas Control* **2020**, *93*, 102879. [CrossRef]

34. Bui, M.; Gunawan, I.; Verheyen, V.; Feron, P.; Meuleman, E. Flexible operation of CSIRO's post-combustion CO_2 capture pilot plant at the AGL Loy Yang power station. *Int. J. Greenh. Gas Control* **2016**, *48*, 188–203. [CrossRef]

35. Terna, S.p.A. Available online: https://www.terna.it/it (accessed on 15 April 2023).

36. Moioli, S.; Pellegrini, L.A. Fixed and Capture Level Reduction operating modes for carbon dioxide removal in a Natural Gas Combined Cycle power plant. *J. Clean. Prod.* **2020**, *254*, 120016. [CrossRef]

37. Moioli, S.; Ho, M.H.; Wiley, D.E.; Pellegrini, L.A. Assessment of carbon dioxide capture by precipitating potassium taurate solvent. *Int. J. Greenh. Gas Control* **2019**, *87*, 159–169. [CrossRef]

38. Baiguini, N. Flexible Operation in Post-Combustion CO_2 Removal Plants by MEA Solvent and Potassium Taurate Solvent. Master's Thesis, Politecnico di Milano, Milano, Italy, 2018.

39. Fluor. Available online: http://www.fluor.com/econamine/pages/default.aspx (accessed on 15 April 2023).

40. Just, P.-E. Advances in the development of CO_2 capture solvents. *Energy Procedia* **2013**, *37*, 314–324. [CrossRef]

41. Shell CANSOLV Carbon Dioxide (CO2) Capture System. Available online: http://www.shell.com/business-customers/global-solutions/gas-processing-licensing/licensed-technologies/shell-cansolv-gas-absorption-solutions/cansolv-co2-capture-system.html (accessed on 15 April 2023).

42. Biliyok, C.; Canepa, R.; Wang, M.; Yeung, H. Techno-Economic Analysis of a Natural Gas Combined Cycle Power Plant with CO_2 Capture. In *Computer Aided Chemical Engineering*; Kraslawski, A., Turunen, I., Eds.; Elsevier: Amsterdam, The Netherlands, 2013; Volume 32, pp. 187–192.

43. Aromada, S.A.; Eldrup, N.H.; Øi, L.E. Cost and Emissions Reduction in CO_2 Capture Plant Dependent on Heat Exchanger Type and Different Process Configurations: Optimum Temperature Approach Analysis. *Energies* **2022**, *15*, 425. [CrossRef]

44. Fout, T.; Zoelle, A.; Keairns, D.; Turner, M.; Woods, M.; Kuehn, N.; Shah, V.; Chou, V.; Pinkerton, L. *Cost and Performance Baseline for Fossil Energy Plants Volume 1a: Bituminous Coal (PC) and Natural Gas to Electricity Revision 3. DOE/NETL-2015/1723*; National Energy Technology Laboratory (NETL): Pittsburgh, PA, USA; Morgantown, WV, USA, 2015.

45. Moioli, S.; Pellegrini, L.A. Comparison of CLR and SS Flexible Systems for Post-Combustion CO_2 Removal in a NGCC Power Plant. *Chem. Eng. Trans.* **2019**, *74*, 829–834.

46. Cohen, S.M.; Rochelle, G.T.; Webber, M.E. Optimal operation of flexible post-combustion CO_2 capture in response to volatile electricity prices. *Energy Procedia* **2011**, *4*, 2604–2611. [CrossRef]

47. Moioli, S.; Ho, M.H.; Wiley, D.E.; Pellegrini, L.A. Thermodynamic Modeling of the System of CO_2 and Potassium Taurate Solution for Simulation of the Process of Carbon Dioxide Removal. *Chem. Eng. Res. Des.* **2018**, *136*, 834–845. [CrossRef]

48. GME Gestore Mercati Energetici. Available online: http://www.mercatoelettrico.org/it/ (accessed on 15 April 2023).

49. Cohen, S.M.; Rochelle, G.T.; Webber, M.E. Optimizing post-combustion CO_2 capture in response to volatile electricity prices. *Int. J. Greenh. Gas Control* **2012**, *8*, 180–195. [CrossRef]

50. Moioli, S.; Pellegrini, L.A. Optimal Operation of a CO_2 Absorption Plant in a Post-Combustion Unit for Cost Reduction. *Chem. Eng. Trans.* **2018**, *69*, 151–156.

51. Chalmers, H.; Leach, M.; Lucquiaud, M.; Gibbins, J. Valuing flexible operation of power plants with CO_2 capture. *Greenh. Gas Control Technol.* **2009**, *1*, 4289–4296. [CrossRef]

52. Chalmers, H.; Lucquiaud, M.; Gibbins, J.; Leach, M. Flexible Operation of Coal Fired Power Plants with Postcombustion Capture of Carbon Dioxide. *J. Environ. Eng.-Asce* **2009**, *135*, 449–458. [CrossRef]

53. Lucquiaud, M.; Chalmers, H.; Gibbins, J. Capture-ready supercritical coal-fired power plants and flexible post-combustion CO_2 capture. *Greenh. Gas Control Technol.* **2009**, *1*, 1411–1418. [CrossRef]

54. Lucquiaud, M.; Fernandez, E.S.; Chalmers, H.; Mac Dowell, N.; Gibbins, J. Enhanced operating flexibility and optimised off-design operation of coal plants with post-combustion capture. *Energy Procedia* **2014**, *63*, 7494–7507. [CrossRef]

MDPI AG
Grosspeteranlage 5
4052 Basel
Switzerland
Tel.: +41 61 683 77 34

Energies Editorial Office
E-mail: energies@mdpi.com
www.mdpi.com/journal/energies